Microwave Systems: Design and Applications

Microwave Systems: Design and Applications

Editor: Mitchel Garth

MURPHY & MOORE
www.murphy-moorepublishing.com

Published by Murphy & Moore Publishing
1 Rockefeller Plaza,
New York City, NY 10020, USA
www.murphy-moorepublishing.com

Microwave Systems: Design and Applications
Edited by Mitchel Garth

International Standard Book Number: 978-1-63987-366-1 (Hardback)

Cataloging-in-Publication Data

Microwave systems : design and applications / edited by Mitchel Garth.
p. cm.
Includes bibliographical references and index.
ISBN 978-1-63987-366-1
1. Microwave communication systems. 2. Microwave communication systems--Design and construction. 3. Telecommunication systems. 4. Microwave devices. 5. Line-of-sight radio links. I. Garth, Mitchel.
TK7876 .M53 2022
621.381 3--dc23

Contents

Preface

Every book is initially just a concept; it takes months of research and hard work to give it the final shape in which the readers receive it. In its early stages, this book also went through rigorous reviewing. The notable contributions made by experts from across the globe were first molded into patterned chapters and then arranged in a sensibly sequential manner to bring out the best results.

Microwave is a form of electromagnetic radiation. Microwaves have various applications in modern technology. It is extensively used in non-broadcast, point-to-point communication. They are more easily focused into narrower beams, have higher frequencies allowing broad bandwidth, and require smaller antennas. Microwaves are also used in satellite and spacecraft communication. Most of the telephone, TV and data communications are transmitted using microwaves. It is also widely used in microwave ovens and radar technology. Other uses of microwave include medical diathermy and cancer treatment, remote sensing, radio astronomy, particle accelerators, spectroscopy, industrial heating and collision avoidance systems. This book unravels the recent studies in the field of microwave systems. There has been rapid progress in this field and its applications are finding their way across multiple industries. Students, researchers, experts and all associated with microwave systems will benefit alike from this book.

It has been my immense pleasure to be a part of this project and to contribute my years of learning in such a meaningful form. I would like to take this opportunity to thank all the people who have been associated with the completion of this book at any step.

Editor

1

Chip-Scale Microwave Photonic Signal Processing

Jian Wang and Yun Long

Abstract

The use of optical technology can provide unprecedented performance to the genera-tion, distribution, and processing of microwave. Recently, on-chip microwave photonics (MWP) has gained significant interests for its numerous advantages, such as robustness, reconfigurability as well as reduction of size, weight, cost, and power consumption. In this chapter, we review our recent progress in ultracompact microwave photonic signal pro-cessing using silicon nanophotonic devices. Using the fabricated silicon waveguide, silicon microring resonators (MRRs) and silicon photonic crystal nanocavities, we demonstrate on-chip analog signal transmission, optically controlled tunable MWP filter, and ultra-high peak rejection notch MWP filter. The performance of analog links and the responses of MWP filters are evaluated in the experiment. In addition, microwave signal multiplica-tion and modulation are also demonstrated based on a silicon Mach-Zehnder modulator in the experiment with favorable operation performance. The demonstrated on-chip analog links, MWP filters, microwave signal multiplication/modulation may help understand on-chip analog signaling and expand novel functionalities of MWP signal processing.

Keywords: optical signal processing, microwave photonics, analog transmission, silicon photonics, microring resonator, photonic crystal nanocavity, microwave photonic filter

1. Introduction

Silicon photonics has become one of the most promising photonic integration platforms owing to its small footprint, low power consumption, and availability of complementary metal-oxide-semiconductor (CMOS) fabrication technology for low-cost mass production [1–4]. Typical silicon nanophotonic devices include silicon waveguides, silicon microring resonators (MRRs), and silicon photonic crystal [5–10], showing unprecedented small size for potential large-scale integration [11–16]. The great success of silicon photonics benefits from the rapid growth of digital optical communications systems such as high capacity optical communications and optical interconnects [4, 17–34].

It has been proved that photonic systems can be also applied to deliver analog signals and process microwave radio frequency (RF) signals in the optical domain, deriving an emerging subject called microwave photonics (MWP) [35, 36]. The initial of MWP was for distributing microwave signal over long distances. However, the applications have evolved dramatically and now include photonic generation of microwave signal [37–40], photonic processing of microwave signal [41–50], frequency measurement of microwave signal [51, 52], and so on. MWP signal processing is of great importance among these applications. As shown in **Figure 1**, MWP signal processing functionalities include microwave signaling [53, 54], filtering [42, 43, 47], differential [48–50], integral [55, 56], pulse shaping [57, 58], Hilbert transformation [59], arbitrary waveform generation [60–63], frequency multiplication [64, 65], beamforming [66, 67], and more.

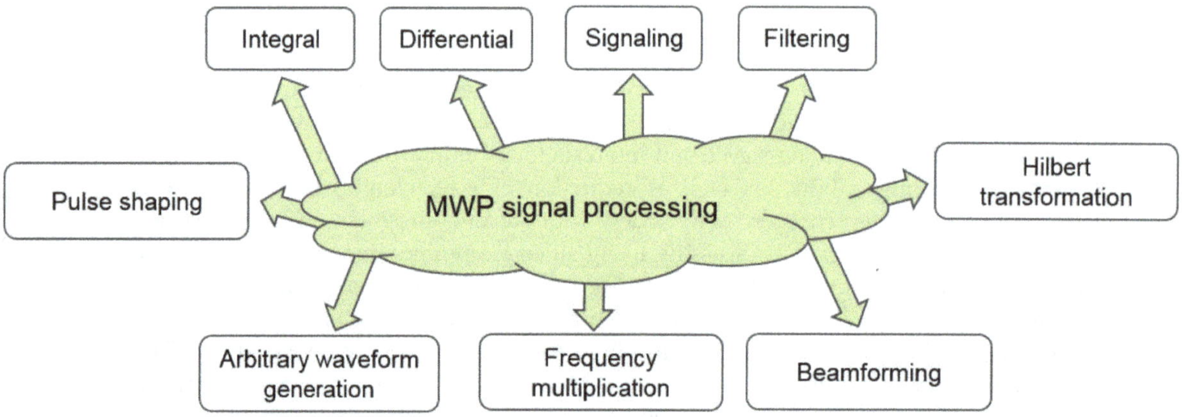

Figure 1. Functionalities classification of MWP signal processing.

Traditional MWP signal processing technologies are mainly based on fibered devices. Recently, there has been an increasing interest on developing integrated components to realize MWP signal processing functionalities. Adopting integrated photonics technologies in MWP signal processing will address the issues such as stability and compactness in traditional fiber-based devices. Very recently, owning to the advance on silicon photonics, some microwave photonic devices such as MWP filters [41, 43, 47, 68], on-chip pulse shapers [57, 58], differentiators [48–50], and Hilbert convertors [59], and ultra wide band (UWB) signal generators [69] have been implemented on silicon on insulator (SOI) platforms. **Figure 2** shows the basic schematic illustration of microwave photonic signal processing using a silicon chip. Usually, a laser source emits a continuous wave (CW) light. The RF signal is generated using a RF generator. After modulation, the RF signal carried by the light carrier is launched to a silicon chip for signal processing. After that, a photodiode is used to convert optical signal to RF signal for detection. The basic silicon photonic devices for MWP signal processing include silicon waveguide, microdisk, microring resonator (MRR), photonic crystal (PhC) nanocavity, Mach-Zehnder modulator (MZM), Bragg grating, and more, as shown in **Figure 3** [70]. Based on the combination of these elementary elements, many complicated MWP processing functionalities can be realized.

In this chapter, we review our recent works in chip-scale microwave photonic signal processing. First, we present the progress of on-chip analog signal transmission. We analyze the performance of on-chip analog signal transmission using the fabricated silicon waveguide and resonators, that is, microring resonators and photonics crystal nanocavities. Second, we report the progress of MWP filters, various types of MWP filters are introduced, and the impact of optical nonlinearities on MWP filters is also discussed. Finally, we show the results of on-chip photonic-assisted microwave signal multiplication and modulation.

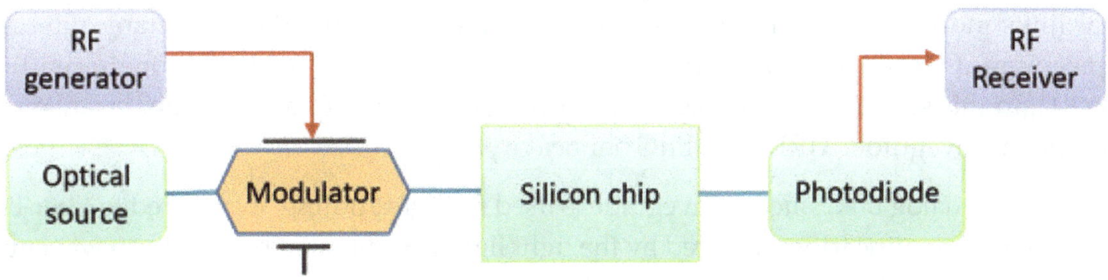

Figure 2. Schematic illustration of chip-scale microwave photonic signal processing.

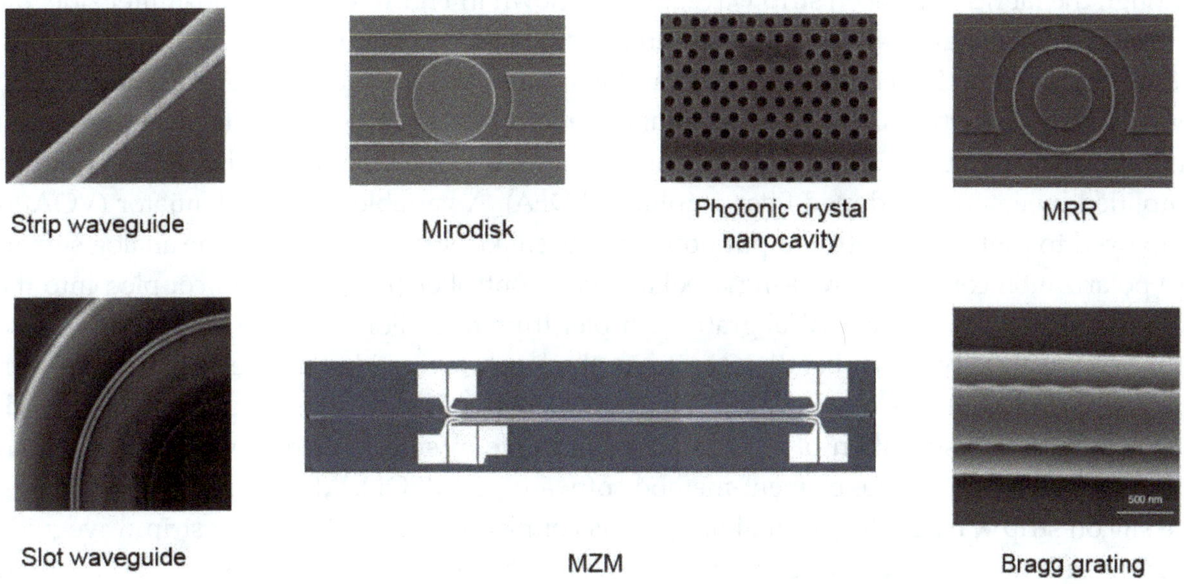

Figure 3. Some photographs of basic silicon photonic devices for MWP signal processing. (The photograph of Bragg grating is reprinted from Ref. [70]).

2. On-chip analog signal transmission

All the MWP systems are essentially analog links. For an analog optical link, linearity is one of the key performance metrics. Traditionally, there are mainly three methods to increase the linearity of analog signal transmission in various systems: (1) improving the driving/bias schemes or optical structures of the modulators to construct linear or quasi-linear modulators [71–74]; (2) using a tunable filter with proper center wavelength and bandwidth to remove the unwanted optical carrier and thus rise the linearity of analog links [75, 76]; (3) using optical waveguides with extremely low nonlinearity, low-loss and high thermal stability for signal transmission [77]. Integrated silicon photonic devices such as silicon waveguides, MRR, microdisk resonators, and PhC cavities have promoted the evolution of on-chip MWP signal processing functions due to the high nonlinear effects of waveguides and diverse filtering properties of resonators. However, these characteristics may also degrade analog signal quality awfully, leading to extra nonlinear signal distortions, which may show different properties compared with traditional signal distortions. To investigate the impacts on analog optical links induced by the nonlinearity of integrated photonic devices, many interesting and meaningful works related to on-chip analog performance have been done recently [53, 54, 78–81]. In this section, we analyze the performance of on-chip analog signal transmission using silicon waveguide, MRR, and PhC nanocavity.

For an on-chip analog link, silicon waveguide is used for light routing. Compared to fiber-based links, the optical nonlinearities induced by the tight light confinement of the silicon waveguides may affect the performance of on-chip analog photonic links. So it is essential to experimentally exploit analog signal transmission distortions induced by the integrated silicon waveguide for future chip-scale analog systems. The experimental setup for analog signal transmission through the fabricated silicon strip waveguide is shown in **Figure 4**. At the transmitter side, the output of an external cavity laser (ECL) is injected to a MZM. Two microwave at frequencies f_1 = 1.98 GHz and f_2 = 2 GHz are combined at the input of MZM. The light source is modulated by the two RFs in the MZM. **Figure 4(a)** and **(b)**describes the typical RF signals at the input and output of MZM. A 1550-nm source emitted from the ECL is modulated by MZM, and then amplified by an erbium-doped fiber amplifier (EDFA). A variable optical attenuator (VOA) is employed to control the optical input power of the silicon strip waveguide. The analog signals are polarization controlled by using a polarization controller (PC), and then coupled into the silicon strip waveguide by vertical grating coupler from fiber. For the dimensions confinement of the SOI wafer, there is a bending region of the long waveguide and the scanning electron microscope (SEM) image of the waveguide bending region is shown in **Figure 4(c)**. **Figure 4(d)** depicts the field distribution of the fundamental mode in silicon strip waveguide calculated by using a full-vector finite-element-method software (COMSOL). After transmission through the silicon strip waveguide, the analog signal is coupled out from the silicon strip waveguide to fiber with the same vertical grating coupler. An electrical spectrum analyzer (ESA) is used to measure the fundamental, second-order harmonic distortion (SHD), third-order harmonic distortion (THD), second-order intermodulation (IM2), and third-order intermodulation (IM3) products.

Figure 5 shows the measured signal-to-noise ratios (SNRs) of IM2, IM3, SHD, and THD as a function of the silicon strip waveguide length. The waveguide length we fabricated is 1, 3.23, 5.25, and 7.38 mm, respectively. One can see that the SNRs of distortions have a negligible change with the increase of the waveguide length.

Figure 6 shows the measured SNRs of IM2, IM3, SHD, and THD as a function of optical input power fed into the silicon strip waveguide. One can see that the SNRs of IM2, IM3, and SHD have a negligible change when increasing the optical input power. The SNR of the THD gradually increases with the input optical power, which might be due to the relatively high sensitivity of THD to the nonlinearity-induced degradation of the silicon strip waveguide.

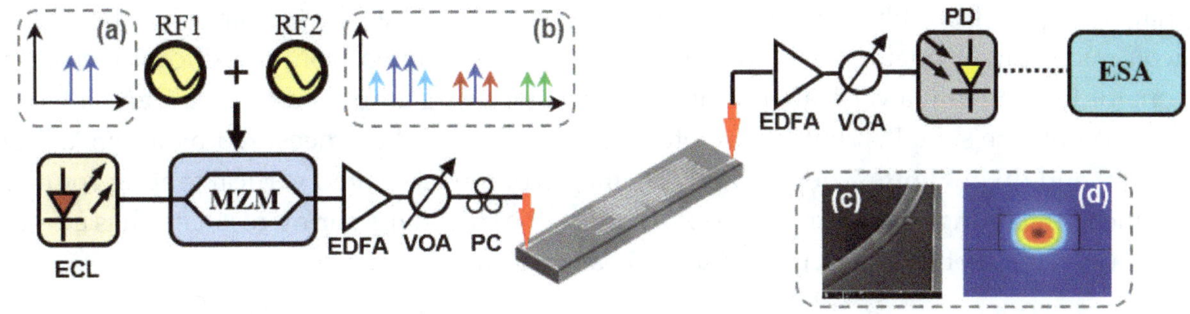

Figure 4. Experimental setup for analog signal transmission through the silicon strip waveguide. Insets: (a) and (b) describe the RF signal at the MZM input and output; (c) is the SEM image and (d) is the simulated mode distribution of silicon strip waveguide.

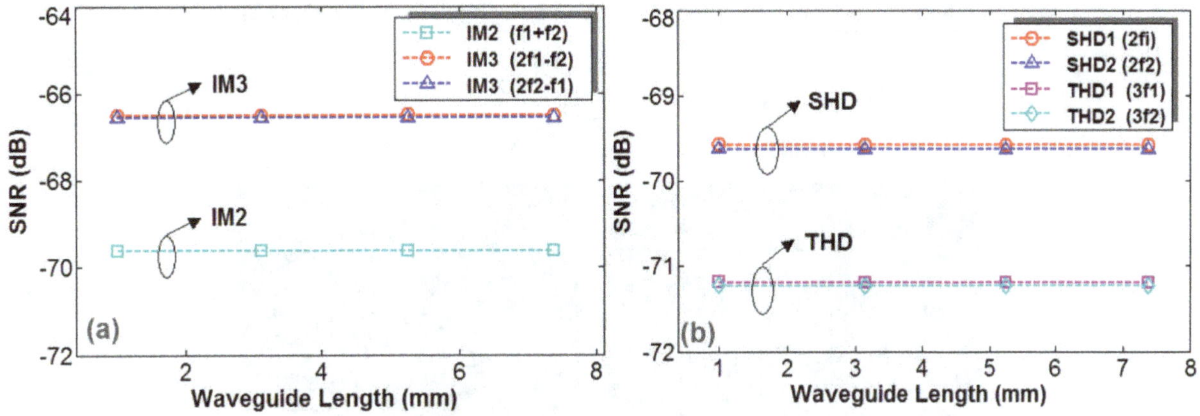

Figure 5. Measured SNRs of (a) IM2, IM3 and (b) SHD, THD versus waveguide length.

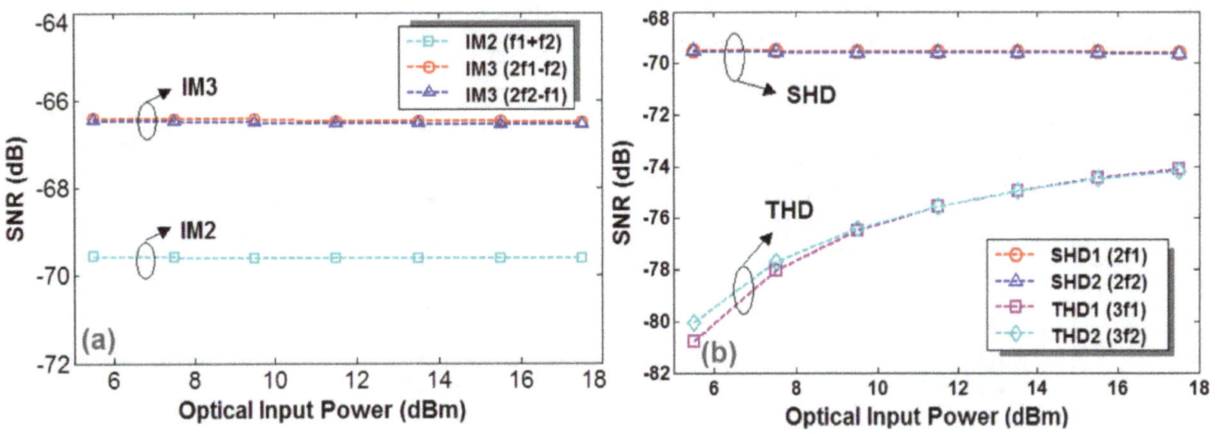

Figure 6. Measured SNRs of (a) IM2, IM3 and (b) SHD, THD versus optical input power.

We then evaluate the performance of on-chip analog signal transmission in a silicon MRR [53]. **Figure 7** shows the experimental setup. At the transmitter, the output of an ECL is sent to a MZM driven by a 6-GHz RF signal. The analog signal is amplified by an EDFA, polarization controlled by a PC, and then adjusted by a VOA. With a vertical grating coupler, the analog signal is coupled from fiber to silicon microring resonator. After the transmission through a silicon microring resonator, at the receiver, the signal is coupled from the silicon microring resonator to fiber with a vertical grating coupler. After being amplified by EDFA and attenuated by VOA, the signal is sent to a photodetector (PD) and then measured by an ESA. The insets (a) and (b) in **Figure 7** depict the SEM images of the fabricated vertical grating coupler and silicon microring resonator. In the experiment, the silicon microring resonator has a resonance wavelength of ~1581.26 nm and a 3-dB bandwidth of ~0.12 nm.

Figure 8(a) and **(b)** shows the acquired output power of the RF carrier and distortions as a function of the RF input power at input signal wavelength of ~1581.07 and ~1581.21 nm, respectively.

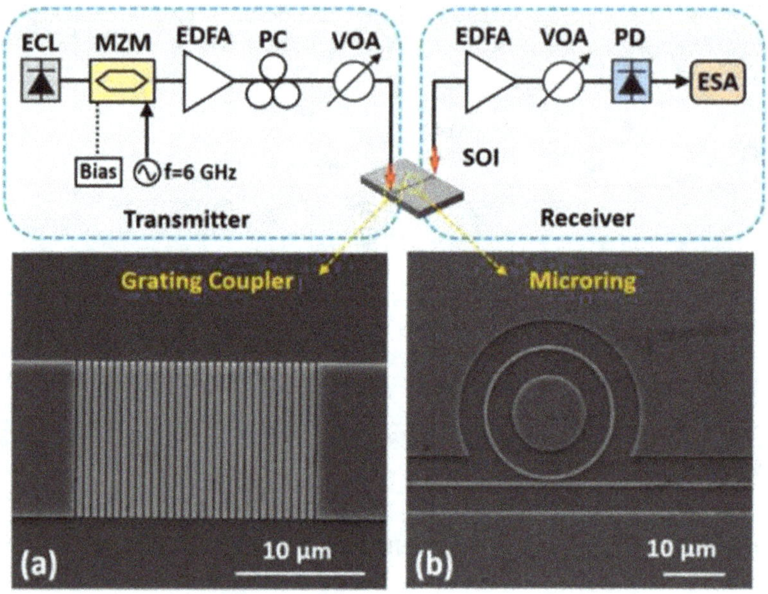

Figure 7. Experimental setup for analog signal transmission in a silicon microring resonator. (a) and (b) SEM images of the grating coupler and silicon microring resonator, respectively.

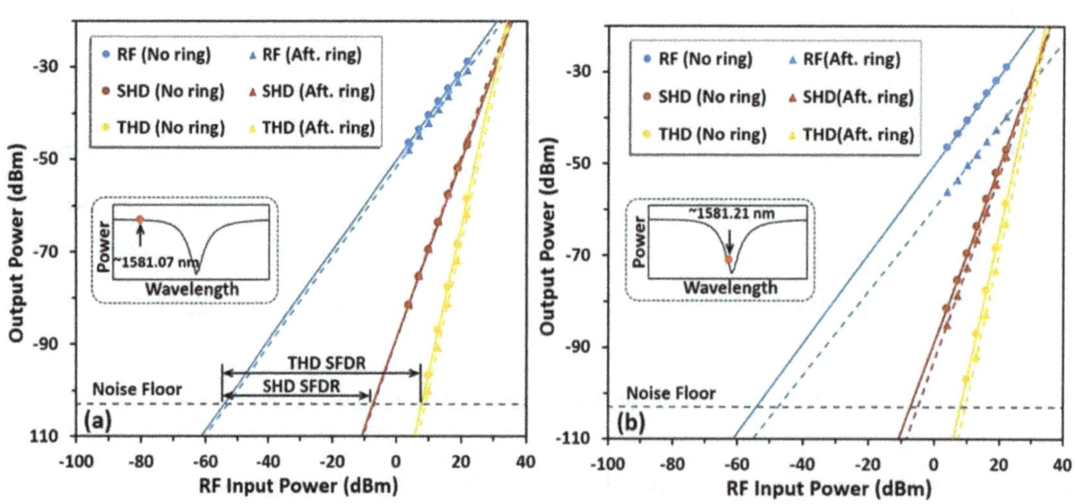

Figure 8. Measured output power of RF carrier and distortions as a function of the RF input power at input signal wavelength of (a) 1581.07 nm and (b) 1581.21 nm, respectively. The optical power of input signal is −17 dBm.

The spurious-free dynamic range (SFDR) of the SHD and THD are used to estimate the analog link performance. A higher SFDR system facilitates a more linear analog signal transmission. As shown in **Figure 8(a)**, for input signal wavelength of 1581.07 nm which is away from the dip resonance wavelength, the output power of RF carrier and distortions after microring change slightly compared with those without microring. The resultant SHD and THD SFDR after microring almost remain the same as those without microring. As shown in **Figure 8(b)**, for input signal wavelength of 1581.21 nm which is close to the dip resonance wavelength, the output power of RF carrier decreases while the distortions vary slightly after microring compared with those without microring. Hence, the SHD and THD SFDR decrease after the transmission, which is due to the notch filtering effect of the silicon microring resonator around the dip resonance wavelength.

To further investigate the analog signal transmission in the 3-dB bandwidth region near the resonance wavelength of 1581.26 nm, we change input signal wavelength from ~1581.07 to ~1581.26 nm with a 0.048-nm spacing. **Figure 9(a)** plots the relationship between SFDR and input signal wavelength at input signal power of 3.5 dBm. One can see two interesting phenomena from **Figure 9(a)**. First, although the minimum optical power transmission wavelength (i.e., notch resonance wavelength) is at ~1581.26nm, the worst SFDRs are obtained at ~1581.21 nm due to the notch filtering effect on the RF carrier sideband. Second, SFDRs at ~1581.12 nm achieve maximum values, which can be explained with the fact that the high-order harmonic sidebands (output SHD and THD) fall into the 3-dB bandwidth region (close to the notch resonance), while the fundamental frequency sideband (output RF carrier) is away from the notch resonance. For low-level input signal power, it is found that the analog signal transmission performance through microring resonator is mainly affected by the notch filtering effect. Input signal power is also taken into consideration to assess its role in the analog signal transmission, as shown in **Figure 9(b)**. For relatively high input signal power, it is interesting to note that the analog signal transmission might be affected by the nonlinearity-induced resonance wavelength shift of the microring with the increase of input signal power.

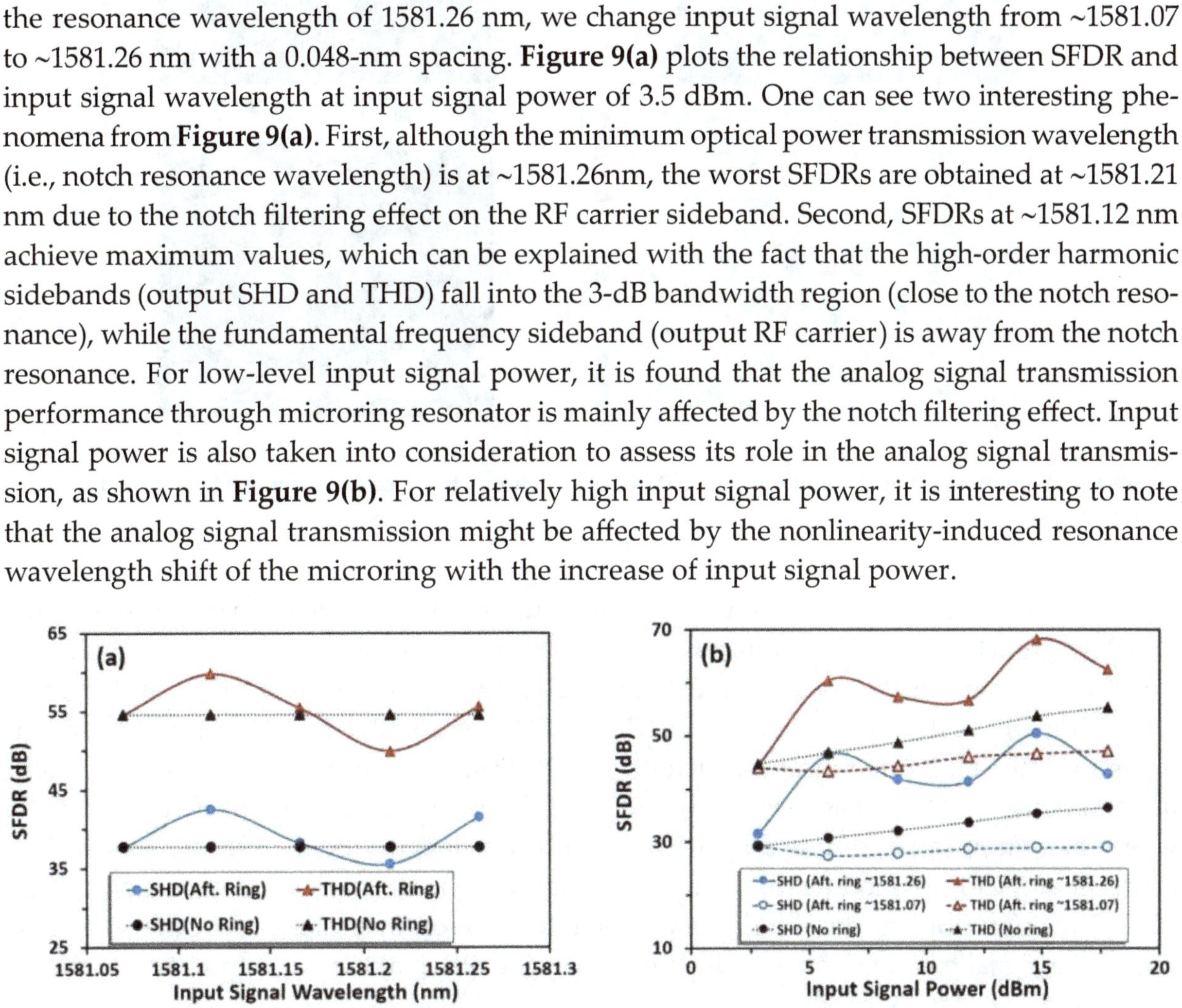

Figure 9. (a) Measured SHD and THD SFDR versus input signal wavelength at input signal power of 3.5 dBm. (b) Measured SHD and THD SFDR versus input signal power.

We further assess the performance of on-chip analog signal transmission using the fabricated silicon photonic crystal nanocavities [54]. The designed and fabricated silicon photonic crystal nanocavity is shown in **Figure 10**. The lattice constant (a) is 420 nm, the radius of air holes (r) is 126 nm, and the three air holes adjacent to the cavity are laterally shifted by 0.175a, 0.025a, and 0.175a, respectively. In the experiment, the silicon photonic crystal nanocavity has a resonance wavelength of ~1552.99 nm, an extinction ratio of ~18.4 dB, and a 3-dB bandwidth of ~0.19 nm.

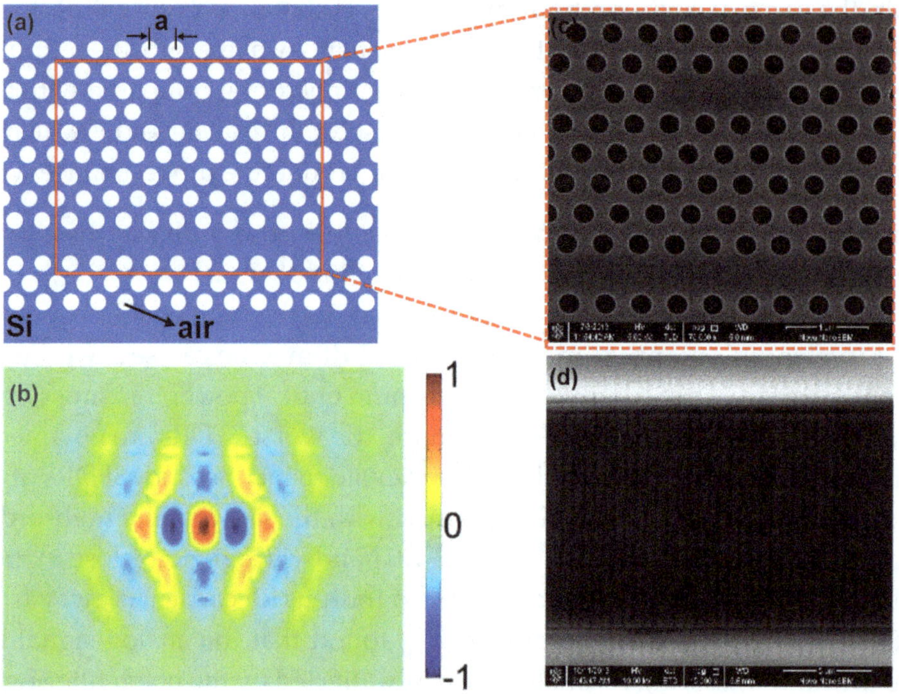

Figure 10. (a) Structure, (b) mode profile and (c) SEM image of silicon photonic crystal nanocavity. (d) SEM image of vertical grating coupler.

Figure 11(a) and **(b)**shows the measured output power of the RF carrier and distortions (SHD, THD) as a function of the RF input power at input optical carrier wavelength of 1552.83 and 1552.99 nm, respectively. As shown in **Figure 11(a)**, for input optical carrier wavelength of 1552.83 nm, the SHD SFDR (~29.8 dB) and THD SFDR (~50.0 dB) degrade slightly. As shown in **Figure 11(b)**, for input optical wavelength of 1552.99 nm (i.e., resonance wavelength), slight degradations of SHD SFDR (~34.6 dB) and THD SFDR (~52.2 dB) are observed. Actually, when the optical carrier is tuned at the resonance wavelength, the modulated optical sidebands after MZM (RF modulation), which are related to the RF carrier and distortions (SHD, THD), are offset from the dip resonance wavelength. As a result, it might not be the worst case for analog signal transmission with the optical carrier sitting at the resonance wavelength of the photonic crystal L3 resonator.

The influences of the optical carrier wavelength and input optical power on the SHD SFDR and THD SFDR are also studied in the experiment. **Figure 12(a)** shows the dependence of SHD SFDR and THD SFDR on the input optical carrier wavelength. **Figure 12(b)** shows the dependence of SHD SFDR and THD SFDR on the optical input power. The obtained results shown in **Figures 11** and **12** indicate favorable analog link performance using the designed and fabricated silicon photonic crystal L3 resonator.

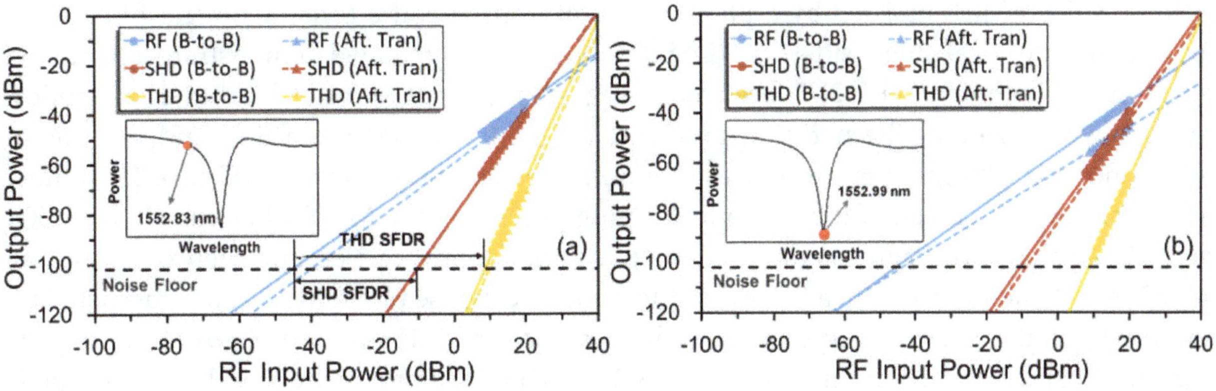

Figure 11. Measured output power of RF carrier and distortions (SHD, THD) as a function of RF input power at different optical carrier wavelengths of (a) 1552.83 and (b) 1552.99 nm.

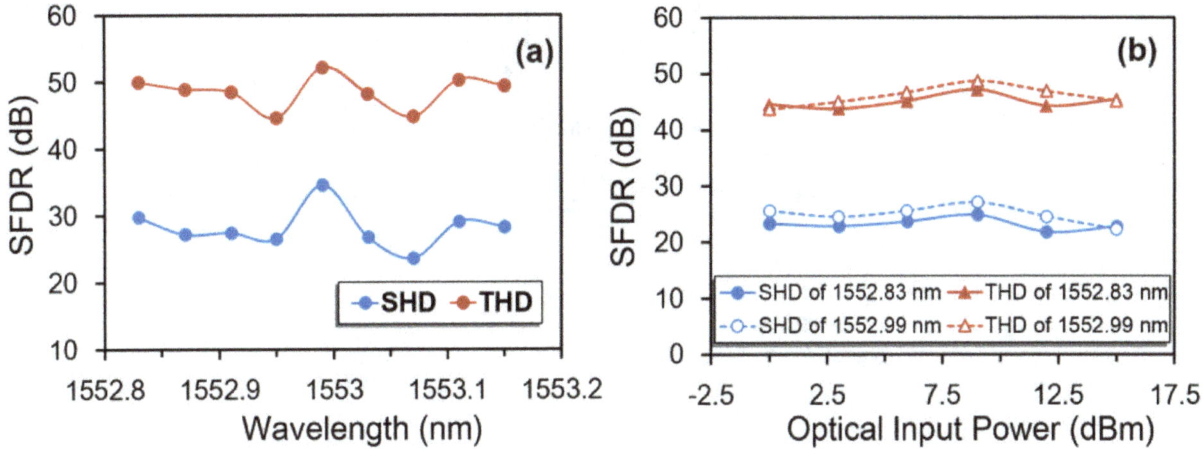

Figure 12. Measured output power of RF carrier and distortions (SHD, THD) as a function of RF input power at different optical carrier wavelengths of (a) 1552.83 and (b) 1552.99 nm.

3. Microwave photonics filters

A MWP filter is a photonic subsystem designed with the aim of carrying equivalent tasks to those of a traditional microwave filter within a RF system [35, 82–85]. Comparing to conventional microwave filters, MWP filters have attracted increasing interest for their advantages, such as huge bandwidth, electromagnetic immunity as well as reconfiguration and tunability. Now, MWP filters are used broadly ranging from radar, satellite to wireless communications. Many approaches to realizing MWP filters have been proposed and demonstrated based on fiber devices in the last decade [86–91]. To enhance the stability and reduce the cost and footprint of the MWP filters, integrated MWP filters have been widely concerned by several groups recently [41, 44, 47, 92–99]. Many of the preliminary approaches have been based mainly on single integrated cavity resonators. For instance, Ref. [96] reports the results for a single passband MWP filters based on a silicon MRR. A few works have also focused on more elaborated designs involving more than one cavity, such as cascaded MRRs [94], MZI-assisted MRRs [41], and more. In this section, we introduce various types of MWP filters.

Using a silicon MRR assisted by a phase modulator (PM) and a tunable optical filter (TBF) to prepare an optical single sideband (OSSB) modulation signal, we demonstrate a bandstop microwave photonic filter [42]. Moreover, using light (pump) to control light (signal optical carrier), we present an alternative approach to tuning the resonance of the bandpass microwave photonic filter, that is, all-optical tuning microwave photonic filter. **Figure 13(a)** illustrates the typical scheme of the all-optical tuning process of the MWP filter based on an SOI MRR. The microwave signal is modulated on a signal optical carrier, and then processed by an SOI MRR and detected by a PD. Since the optical response of the MRR can be adjusted by a pump light, the electrical response of the link can be tuned. The inset of **Figure 13(a)** depicts typical transmission spectrum of an SOI MRR, in which the microwave-modulated signal is located close to one notch resonance frequency and pump light positioned at another notch

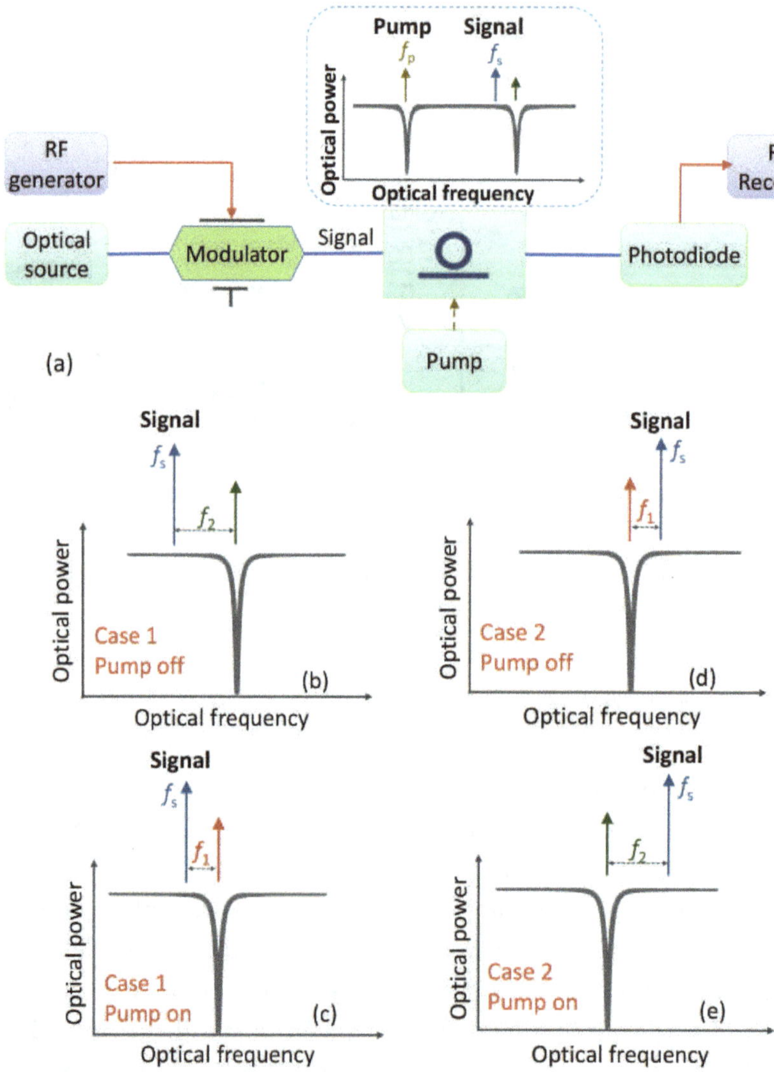

Figure 13. (a) Schematic illustration of the proposed all-optical tuning process of the MWP filter. (b)–(e) Operation principle of the all-optical tuning process of the MWP filter. (b) and (d) Pump off. (c) and (e) Pump on. Case 1: the frequency of the signal optical carrier (fs) is located at the left side of the notch resonance frequency of the MRR. Case 2: the frequency of the signal optical carrier (fs) is located at the right side of the notch resonance frequency of the MRR. The inset of (a) shows typical transmission spectrum of the MRR with relative positions of microwave modulated signal and pump light.

resonance frequency. **Figure 13(b)–(e)** summarizes the operation principle of the proposed all-optical tuning process of the MWP filter. An optical carrier is modulated by an RF signal with OSSB modulation. The output field after modulation is then applied to the MRR for microwave photonic signal processing. The frequency of the signal optical carrier is f_s. There are two cases depending on the relative position between the signal optical carrier and the notch resonance of the MRR. For the Case 1 with the signal optical carrier f_s located at the left side of the notch resonance of the MRR, when the frequency component $(f_s + f_2)$ is just aligned to the notch frequency of the MRR spectrum as shown in **Figure 13(b)**, the output RF response in the absence of the pump vanishes at f_2. Hence, a notch MWP filter with a central frequency of f_2 is obtained. When the pump light is on, one would expect a red shift of the notch peak of the MRR owing to the combined nonlinear effects in the MRR, resulting in a notch MWP filter with a central frequency of f_1, as shown in **Figure 13(c)**. Hence, the central frequency of the notch MWP filter can be tuned from f_2 to f_1 by increasing the pump light power. For the Case 2 with the signal optical carrier f_s located at the right side of the notch resonance of the MRR, as shown in **Figure 13(d)** and **(e)** similar tunable notch MWP filter with its central frequency changed from f_1 to f_2 is also achievable by increasing the pump light power. Both Case 1 and Case 2 shown in **Figure 13(b)–(e)** indicate possible all-optical tunale MWP filter using nonlinear effects of MRR.

We first calculate the RF responses under different pump light power to show the all-optical tuning process of the MWP filter. The parameters used to calculate MRR field transmission are extracted from the measured MRR transmission spectrum. The MWP filter responses under different pump powers in Case 1 and Case 2 are shown in **Figure 14(a)** and **(b)**, respectively. As shown in **Figure 14(a)**, the central frequency of the MWP filter decreases from 15.5 to 9.5 GHz as the pump power increases in Case 1. The central frequency of the MWP filter increases from 9.5 to 15.5 GHz as the pump power increases in Case 2, as shown in **Figure 14(b)**.

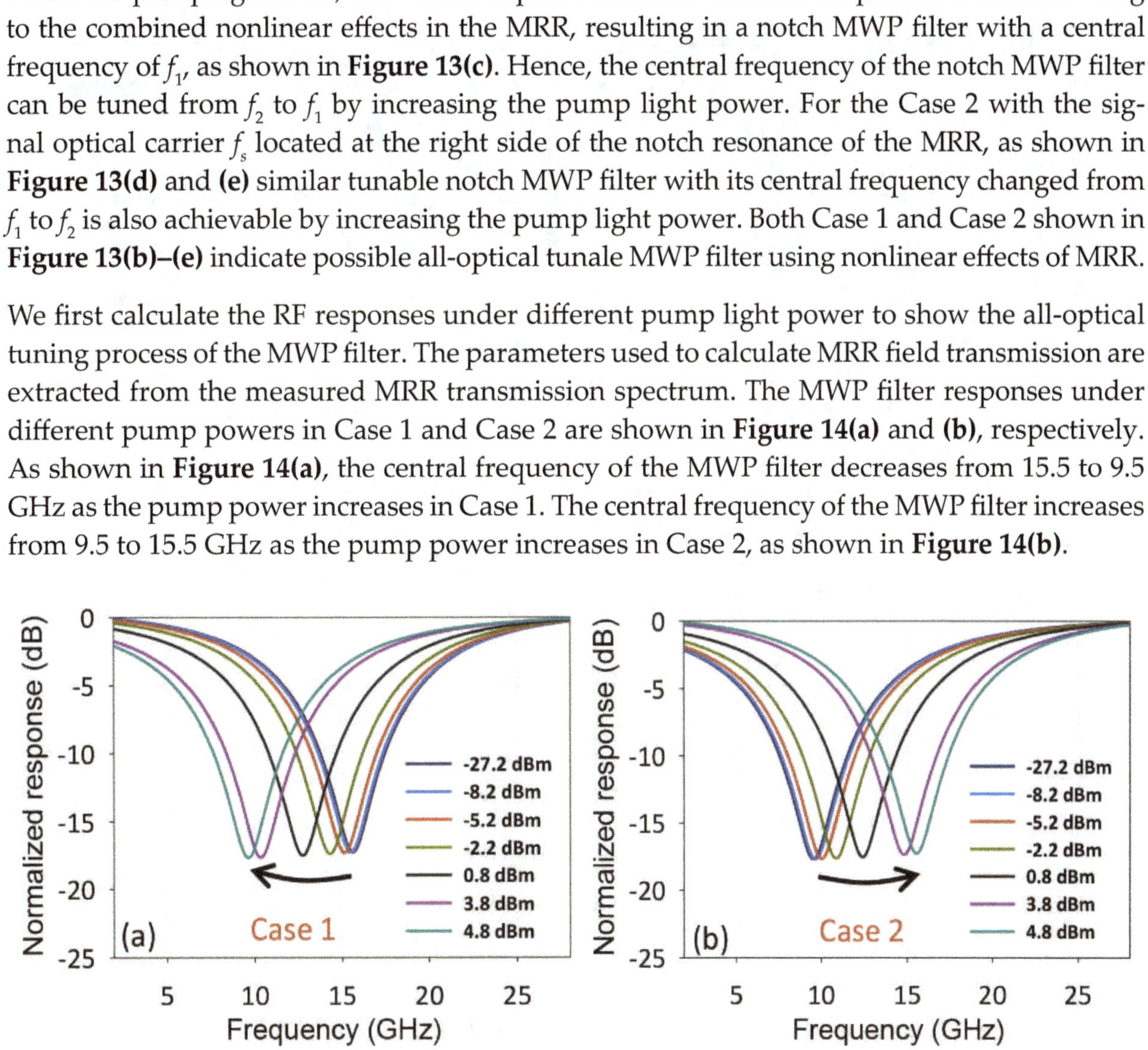

Figure 14. Calculated microwave responses under different pump power levels in (a) Case 1 and (b) Case 2, respectively. Case 1: the central frequency of the MWP filter decreases as the pump power increases. Case 2: the central frequency of the MWP filter increases as the pump power increases.

The measured microwave responses of MWP filter under different pump power levels are shown in **Figure 15**. The pump light wavelength is fixed at 1591.786 nm for all the configurations.

In Case 1, the optical carrier wavelength is 1581.730 nm in both forward and backward pumping configurations. As shown in **Figure 15(a)** and **(b)**, the central frequency of the notch MWP filter is tuned from 15.64 to 8.79 GHz in the forward pumping configuration while from 15.60 to 10.04 GHz in the backward pumping configuration by adjusting the pump light power from −27.2 to 4.8 dBm. In Case 2, the optical carrier wavelength is 1581.564 nm in the forward pumping configuration and 1581.534 nm in the backward pumping configurations. As shown in **Figure 15(c)** and **(d)**, the central frequency of the notch MWP filter is tuned from 5.27 to 12.47 GHz in the forward pumping configuration and from 8.84 to 15.04 GHz in the backward pumping configuration as increasing the pump light power from −27.2 to 4.8 dBm.

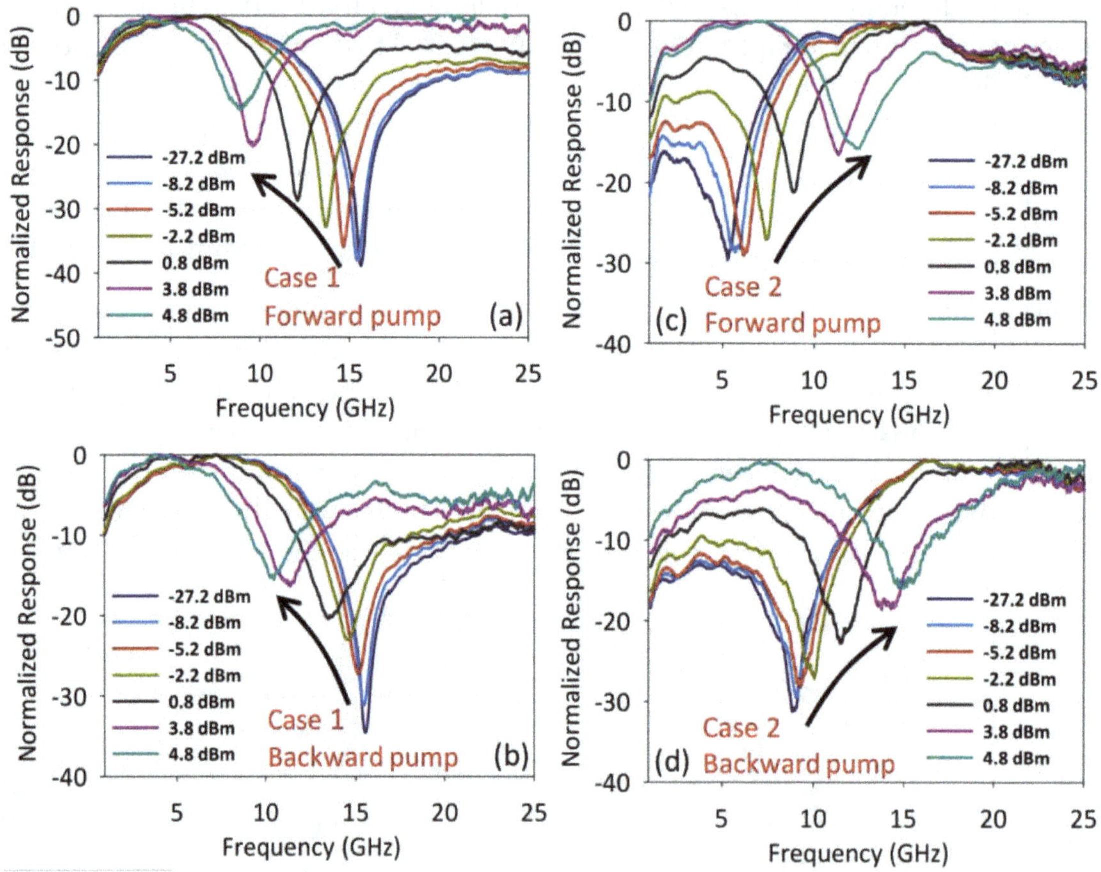

Figure 15. Measured microwave responses of all-optical tunable MWP filter under different pump power levels with (a) Case 1/Forward pump, (b) Case 1/Backward pump, (c) Case 2/Forward pump, and (d) Case 2/Backward pump, respectively.

We also propose a simple yet effective approach to realizing notch MWP filter with an ultra-high peak rejection [43]. We use the combination of a PM, TBF, and an SOI MRR to manipulate the phase and amplitude of optical sidebands for inducing a signal cancellation at the RF notch filter frequency. **Figure 16** shows the concept and operation principle of the proposed approach. A conventional PM driven by an RF signal generates out of phase lower sideband (LSB) and upper sideband (USB). A TBF is used to attenuate the USB signal followed by an SOI MRR. The resonant frequency of the SOI MRR is aligned to the frequency of LSB signal, thus the LSB signal will be filtered by the SOI MRR. Since the SOI MRR has a limited extinction ratio, the amplitude of the remaining LSB signal could be equal to the USB signal. Due to the equal amplitude and π phase difference of the attenuated USB and remained LSB, the signal power at the modulated RF frequency can be cancelled after detection by a photodetector, and a notch microwave photonic filter with infinite rejection in principle can be achieved.

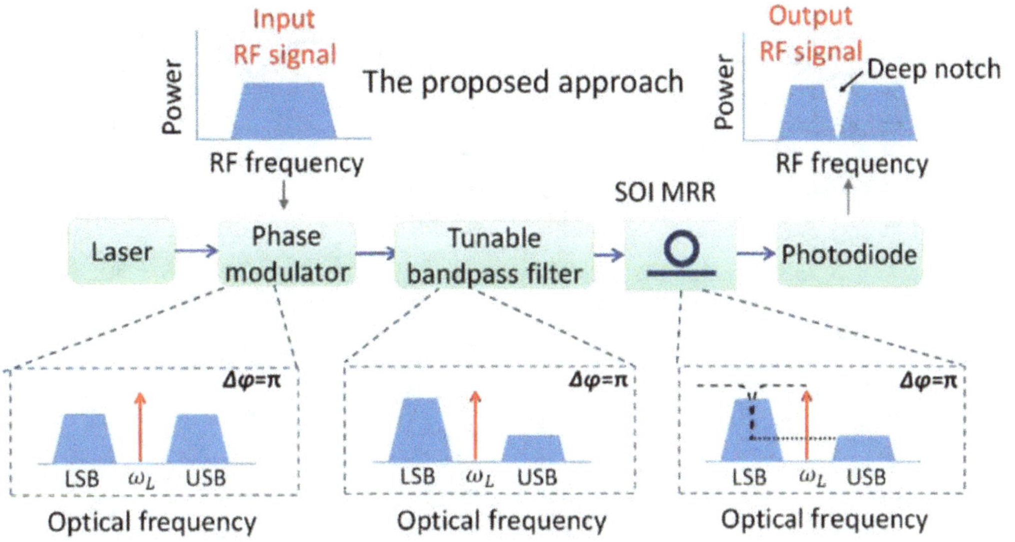

Figure 16. Schematic illustration of the proposed notch MWP filter with ultra-high peak rejection.

The measured optical spectra after the TBF and the corresponding MWP filter responses are shown in **Figure 17**. The optical carrier wavelength is 1581.576 nm. The optical spectra are measured by modulating a microwave signal to the optical carrier. The frequency of the microwave signal is 20 GHz, which is comparable with the central frequency of the MWP filter. **Figure 17(a)–(d)** depicts the optical spectra when the central wavelength of the TBF is 1581.746, 1581.776, 1581.806, and 1581.836 nm, respectively. The transmission spectra of the TBF in these four cases are also plotted in **Figure 17(a)–(d)** for reference. **Figure 17(e)–(h)** shows the corresponding MWP filter responses. As shown in **Figure 17(a)** and **(d)**, when the USB component is slightly attenuated by the TBF, after passing through the MRR, the amplitude difference of the LSB and USB is relatively large, resulting in a shallow RF notch of about 14.5 dB. When we shift the TBF, more power of the USB component is attenuated, and the peak rejection of the MWP filter remarkably increases. The peak rejection of the MWP filter is 29.1 dB when the central wavelength of the TBF is 1581.776 nm as shown in **Figure 17(b)** and **(f)**. A maximum peak rejection of about 61.5 dB is observed with balanced amplitude between LSB and USB components when the central wavelength of the TBF is 1581.806 nm as shown in **Figure 17(c)** and **(g)**. When we further increase the central wavelength of the TBF, the USB signal will be further attenuated, thus the peak rejection decreases again. For example, when the central wavelength of the TBF is 1581.836 nm, the USB almost disappears, and the peak rejection of the MWP filter decreases to 40.9 dB as shown in **Figure 17(d)** and **(h)**.

By changing the carrier light wavelength, the operating frequency of the ultra-high peak rejection MWP filter can be tuned. We measure the operating frequency tunability of the

MWP filter, as shown in **Figure 18**. When the wavelength of the carrier light is changed from 1581.476 to 1581.626 nm, the central frequency of the MWP filter is tuned from 12.4 to 30.6 GHz, maintaining an ultra-high peak rejection. The obtained results shown in **Figure 18** indicate that the proposed ultra-high peak rejection MWP filter can operate over a large tunable frequency range.

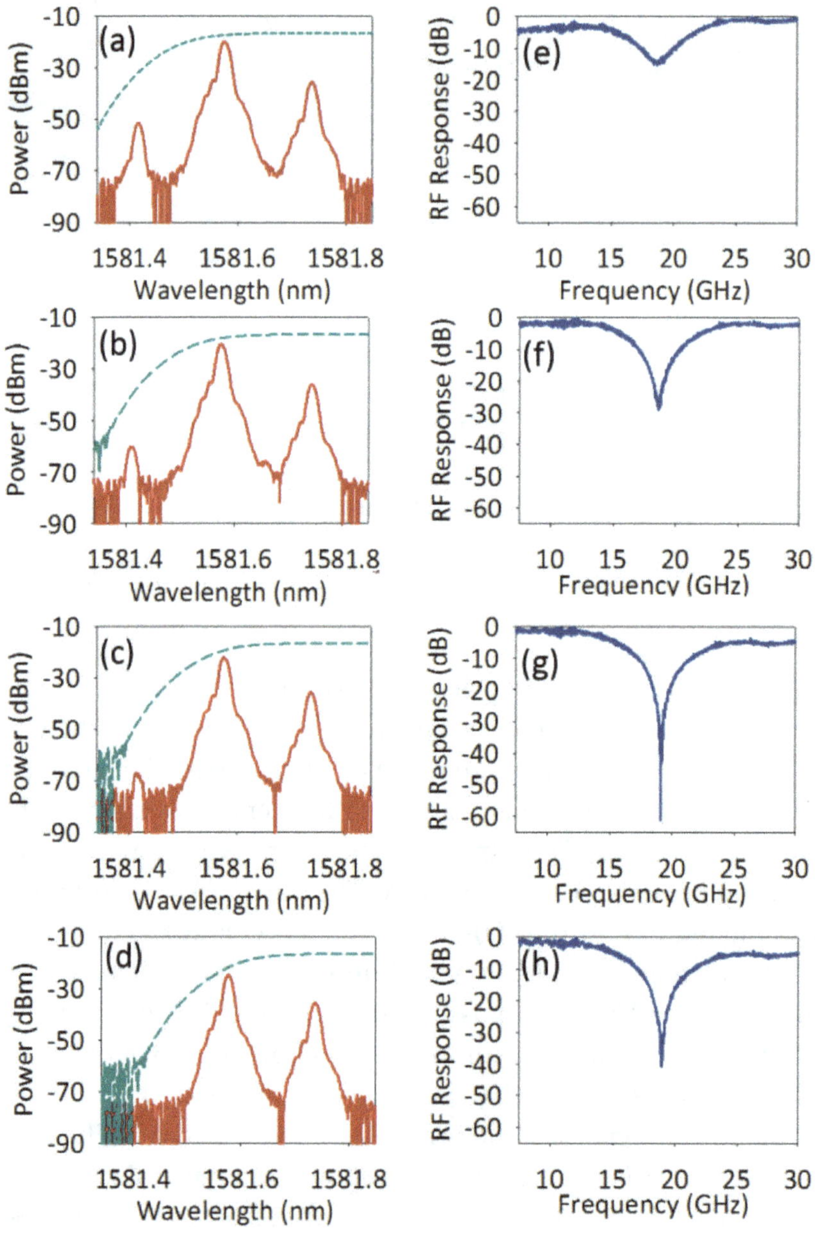

Figure 17. (a)-(d) Optical spectra after the TBF when the central wavelength of the TBF is 1581.746, 1581.776, 1581.806, and 1581.836 nm, respectively. The dashed lines are the corresponding transmission spectrum of TBF. (e)–(h) The corresponding MWP filter responses.

Based on a similar principle, we also demonstrate a rejection ratio tunable notch MWP filter based on a silicon PhC nanocavity. The resonant wavelength of the cavity is around 1554.313 nm. The measured optical spectra after the TBF and the corresponding MWP filter responses are shown in **Figure 19**. The carrier light wavelength is 1554.153 nm. **Figure 19(a)–(d)** depicts the optical spectra when the central wavelength of the TBF is 1554.262, 1554.292, 1554.322, and 1554.352 nm, respectively. **Figure 19(e)–(h)** shows the corresponding MWP filter responses. The experimental results agree well with the simulation. As shown in **Figure 19(a)** and **(e)**, when the USB signal is slightly modified by the TBF, the peak rejection of the obtained MWP filter is very small (~11.8 dB). When we shift the TBF, the power of USB signal decreases, and the peak rejection of the MWP filter increases. A maximum peak rejection of about 62.1 dB is observed when the central wavelength of the TBF is 1551.315 nm (**Figure 19(b)** and **(f)**). When we further increase the central wavelength of the TBF, the USB signal will be further attenuated, thus the peak rejection will decrease again. For example, as shown in **Figure 19(d)** and **(h)**, the peak rejection of the MWP filter decreases to 15 dB when the central wavelength of the TBF is 1551.375 nm. **Figure 19(i)** plots the rejection ratio as a function of the central wavelength of the TBF. The red curve shows the simulation data, while the experimental results are marked by blue circles.

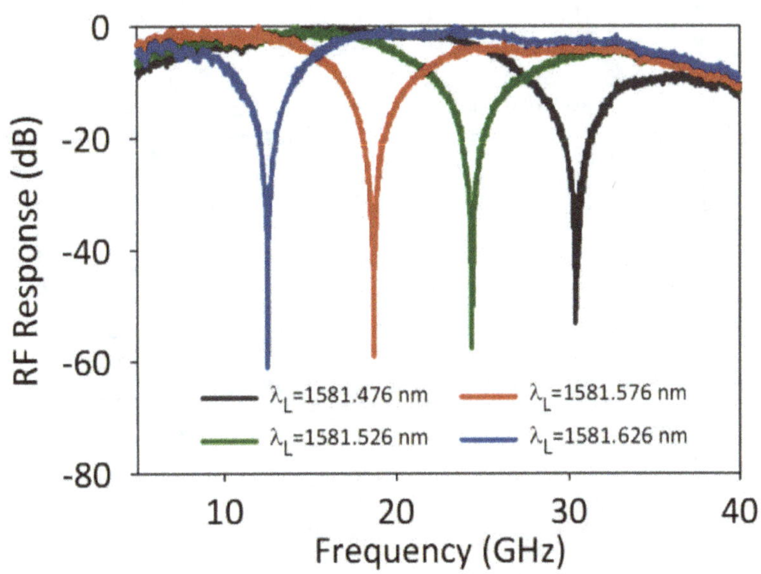

Figure 18. Measured tunable ultra-high peak rejection MWP filter responses with different optical carrier wavelengths.

To evaluate the effect of the filter shape of TBF, we study the performance of the rejection ratio tunable MWP filter with four types of super-gaussian filter shape. We calculate the MWP filter tunability response when the transmission of the TBF is $T_{TBF}(\omega) = e^{-\left(\frac{\omega-\omega_{center}}{2\pi\times 34.89\times 10^{9}}\right)^{N}}$ $(N = 2, 4, 6, 8)$. **Figure 20(a)** shows the rejection ratio of the MWP filter as a function of TBF central wavelength with different filter shapes of TBF. To show the operation stability in high rejection ratio region with different TBF filter shapes, we define the TBF central wavelength tolerance as operation span of TBF central wavelength when the rejection ratio is greater than 40 dB. **Figure 20(b)** plots the TBF tolerance of central wavelength with different filter shapes of TBF. It can be seen that higher-order super-gaussian function leads to worse stability of the system.

Since the high nonlinearities induced by the tight light confinement of silicon photonic devices, the influence of optical nonlinear effects should also be taken into consideration when designing MWP devices. MWP devices such as MWP filters based on silicon waveguide devices might have strong nonlinear microwave responses, which are dependent on the optical carrier power level. Taking the PhC microcavity as an example, we study the nonlinear microwave responses of the MWP filters based on the SOI waveguide devices.

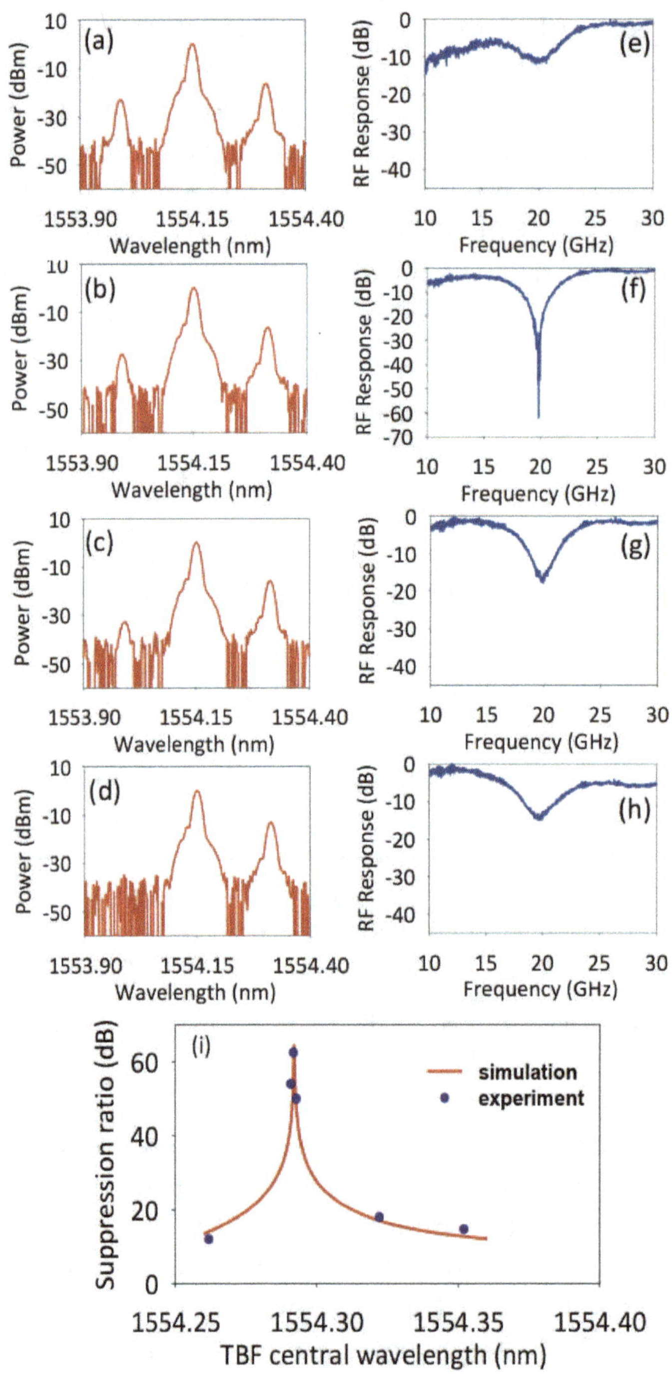

Figure 19. (a)–(d) Optical spectra after the TBF when the central wavelength of the TBF is 1554.262, 1554.292, 1554.322, and 1554.352 nm, respectively. (e)–(h) The corresponding MWP filter response. (i) Rejection ratio as a function of TBF central wavelength.

Figure 21(a) illustrates the typical scheme of an MWP filter based on a PhC microcavity. **Figure 21(b)** and **(c)** summarizes the operation principle of the device. After been phase modulated, the output field is then sent to a PhC microcavity. Assuming that the wavelength of optical carrier is λ_c, locating on the left side of the linear notch resonant wavelength of the PhC microcavity (Case 1), as shown in **Figure 21(b)**, when the optical carrier power is very low and one of the sideband (f_1) is just aligned to the notch resonant wavelength of the PhC microcavity, a bandpass MWP filter with a central frequency of f_1 is obtained. When the optical carrier power is increased, considering the combined nonlinear effects in the PhC microcavity, the notch peak of the microcavity will red shift (e.g., shift by f_2-f_1), resulting in an bandpass MWP filter with a varied central frequency of f_2. When the wavelength of optical carrier is located on the right side of the linear notch resonant wavelength of the PhC microcavity, the situation is a little more complicated. For Case 2, as shown in **Figure 21(c)**, the initial notch peak of the PhC microcavity is aligned to the left sideband corresponding to f_2. When the optical carrier power is increased, the notch peak wavelength will red shift, leading to the decrease of the central frequency of the MWP filter. Remarkably, when we further increase the optical carrier power, the notch peak will go across the optical carrier wavelength. Hence, the central frequency of the MWP filter will first decrease and then increase when gradually increasing the optical carrier power from a very low level to a relatively high level.

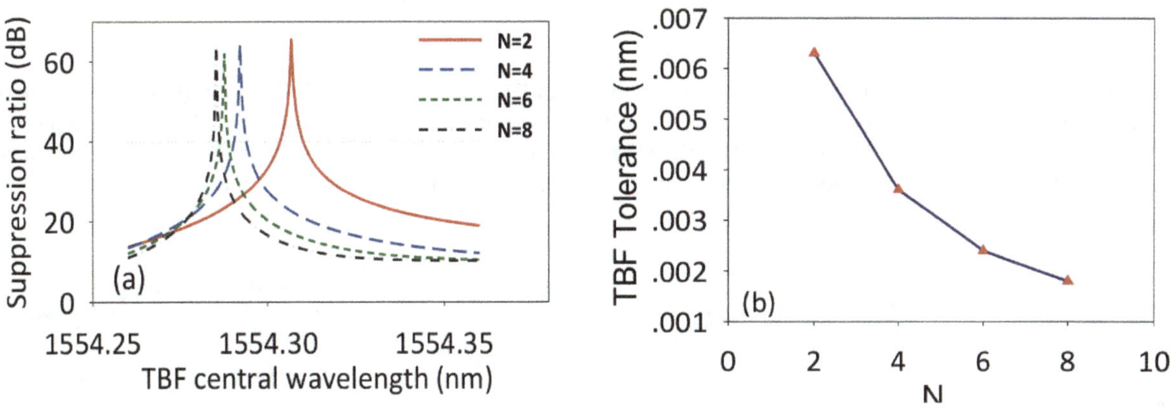

Figure 20. (a) Rejection ratio as functions of TBF central wavelength with different filter shape of TBF. (b) TBF tolerance of central wavelength with different filter shape of TBF.

The resonant wavelength of the cavity is around 1554.152 nm. **Figure 22** shows measured MWP filter responses under different optical carrier power levels for Case 1 and Case 2. The wavelengths of the carrier are 1554.056 and 1554.296 nm, respectively. For Case 1, since the notch resonant wavelength of the PhC microcavity red shifts as the optical carrier power increases, the central frequency of the MWP filter increases. When the optical carrier power increases from −13.5 to 5.5 dBm, the central frequency of the MWP filter is changed from 13.9 to 24 GHz. For Case 2, when the optical carrier is increased from −13.5 to −3.5 dBm, the central frequency of the MWP filter decreases, which indicates that the resonant wavelength of the PhC microcavity is still on the left side of the optical carrier in the process. When we increase the power from −3.5 to −1.5 dBm, the central frequency of the MWP filter suddenly jumps to another frequency at about −2 dBm. When we further increase the power of the carrier light, the central frequency of the MWP filter increases, in contrast with the decrease in low-power region. The obtained results here indicate that the resonance of the

PhC microcavity first jumps to the right side of the carrier, and then red shifts as the carrier further increases.

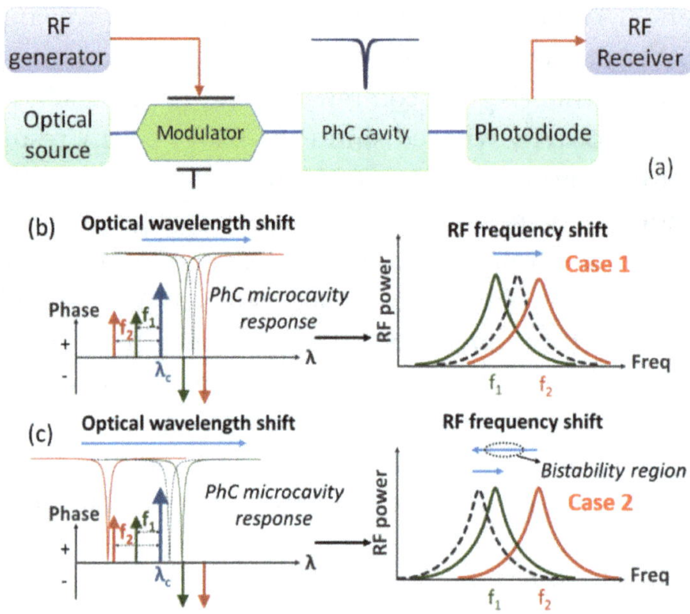

Figure 21. (a) Schematic illustration of the typical scheme of a MWP filter based on PhC cavity. (b) and (c) Working principle of the optical carrier power dependent MWP filter. Case 1: the wavelength of the optical carrier is located on the left side of the linear notch resonant wavelength of the PhC microcavity. Case 2: the wavelength of the optical carrier is located on the right side of the linear notch resonant wavelength of the PhC microcavity.

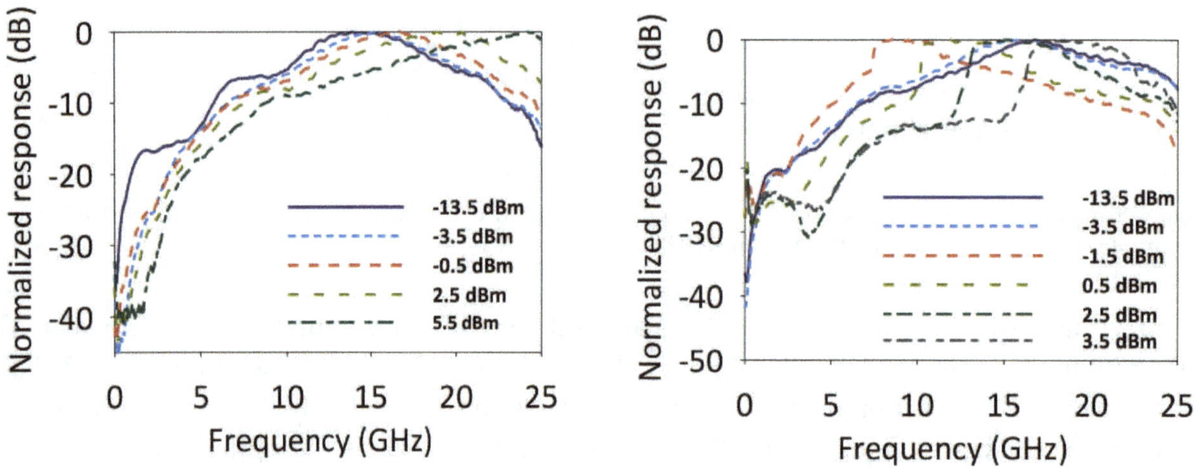

Figure 22. Measured MWP filter responses under different optical carrier power levels for (a) Case 1 and (b) Case 2.

4. Photonic-assisted microwave signal multiplication and ASK modulation

Photonic-assisted generation of microwave signals can find interesting applications in many microwave photonic systems, such as broad-band wireless access networks, software-defined radio, antenna remoting, phased-array antenna, and radar systems [65]. Usually, optical microwave generation is based on heterodyne techniques. Several approaches to generating two phase-correlated optical waves have been proposed and demonstrated based on free space devices [100, 101] and fiber devices [64, 65, 102–111]. Among these approaches, microwave frequency multiplication based on external modulation using MZMs has been considered an effective solution for high frequency and tunable microwave signal generation. Photonic generation of RF binary digital modulation signals is another key technology in microwave photonics. Amplitude-shift keying (ASK), phase-shift keying (PSK), and frequency-shift keying (FSK) are basic modulation formats in wireless communications, which convey information by modulating the amplitude, phase or frequency of a continuous carrier wave. Traditionally, these modulation microwave signals are generated in electrical domain using digital electronic circuits [40]. Due to the electronic bottleneck, the major difficulty of this traditional technique is that the frequency of the generated signals is limited to a few gigahertz. An effective method to generate high frequency RF signals is to generate RF signals in the optical domain [38, 39, 112]. Similar to microwave frequency mulitiplication, the use of MZM is also considered to be a competitive approach to generating binary digital modulation signals. Microwave PSK and FSK signals have been proposed and realized based on commercial MZMs [38, 39, 112]. However, microwave ASK signal as a fundamental digital modulation format in wireless communication has rarely been realized using MZM. In this section, we introduce our recent work on microwave signal multiplication and ASK modulation based on an integrated silicon modulator.

We first propose a simple scheme to obtain frequency-multiplicated microwave signals [113]. **Figure 23** illustrates the schematic illustration of the proposed photonic frequency-multiplicated microwave signal generation system using an integrated silicon MZM. CW light from a tunable laser diode (TLD) is sent to the MZM. The driving signal and the DC-bias are combined by a bias-tee and applied to the MZM. The output light intensity of the MZM can be written as:

$$\begin{aligned} I = \frac{1}{2} I_0 [1 + J_0(m)\cos\varphi_0] + I_0 \cos\varphi_0 \sum_{n=1}^{\infty}(-1)^n J_{2n}(m)\cos(2n\,\omega_{\mathrm{RF}}\,t) \\ + I_0 \sin\varphi_0 \sum_{n=1}^{\infty}(-1)^n J_{2n-1}(m)\cos[(2n-1)\,\omega_{RF}\,t] \end{aligned} \quad (1)$$

where $m = \pi V_{RF}/V_\pi$ is the modulation indice of the MZM. I_0 is the intensity of the input optical carrier. $\phi_0 = \pi V_b/V_\pi$ is the constant phase difference between the two arms determined by the constant DC-bias voltage V_b. V_π is the half-wave voltage of the MZM. V_{RF} and ω_{RF} are the amplitude and angular frequency of the applied electrical drive voltage, respectively. It can be seen from Eq. (1) that if the modulator is biased at V_π, the odd components will be suppressed. Ignoring the high-order harmonic wave, a frequency-doubled microwave signal is obtained. In addition, by adjusting the driving microwave signal amplitude to let $J_2(m) = 0$, the quadruple response will dominate the output, and a frequency-quadrupled signal can be obtained. Similarly, a frequency-tripled signal is generated by suppressing the even components and $J_1(m)$ term of the signal.

We investigate the generation of frequency-doubled microwave signal. **Figure 24(a)** and **(b)** shows the wave-forms of the original 500-MHz driving signal and the generated 1-GHz signal. **Figure 24(c)** and **(d)** shows the obtained 2-GHz signal when the driving signal frequency is 1 GHz.

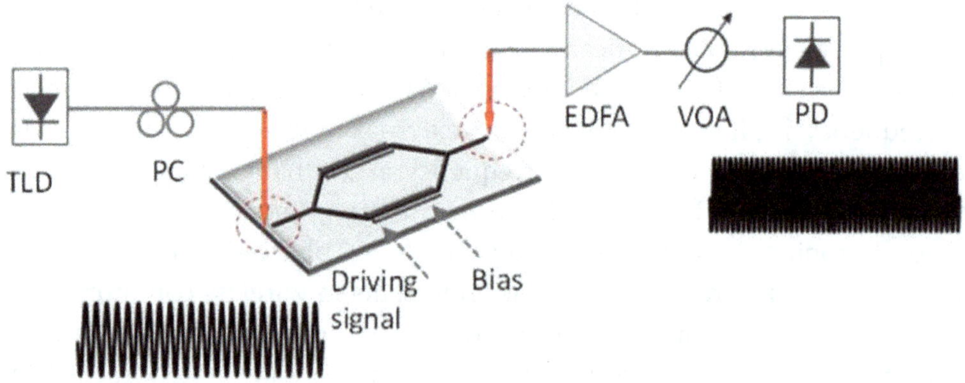

Figure 23. Schematic illustration of the proposed photonic-assisted microwave signal multiplication.

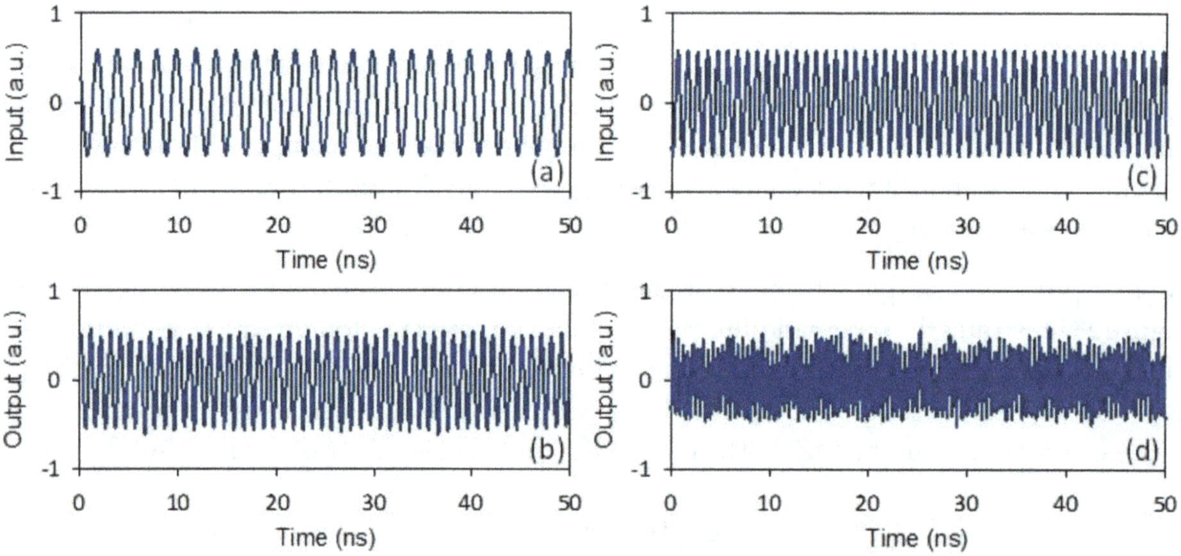

Figure 24. Waveforms of (a) the input 500-MHz driving signal, (b) the generated 1-GHz frequency-doubled signal, (c) the input 1-GHz driving signal, and (d) the generated 2-GHz frequency-doubled signal.

Using the fabricated integrated MZM, we also demonstrate the feasibility of microwave ASK modulation using an integrated photonic approach [113]. **Figure 25** illustrates the schematic illustration of the proposed photonic microwave ASK signal modulation system using an integrated silicon MZM. The microwave carrier signal and the binary coding signal s(t) are applied to the two RF ports of the MZM, respectively. The AC term of the detected signal at the output of the PD can be written as:

$$i_{AC} \propto 2J_1(m)\cos(\omega_{RF}t)\sin(\gamma s(t)-\varphi_0)+J_0(m)\cos(\gamma s(t)-\varphi_0) \tag{2}$$

where ω_{RF}, m, V_{RF}, V_π, and ϕ_0 are defined as previously mentioned. $\gamma=\pi V_s/V_\pi$ is the modulation index in the arm where the binary coding signal is applied. Assuming $\phi_0 = 0$, Eq. (2) can be simplified as:

$$i_{AC} \propto 2J_1(m)\cos(\omega_{RF}t)\sin(\gamma s(t)) + J_0(m)\cos(\gamma s(t)) \tag{3}$$

The second term in Eq. (3) is located in the baseband, which can be easily eliminated by an electric filter. Assuming $\gamma = \pi/2$, the amplitude of the obtained signal is:

$$i = \begin{cases} 2J_1(m)\cos(\omega_{RF}t) & s(t)=1 \\ 0 & s(t)=0 \end{cases} \tag{4}$$

As can be seen, the amplitude of the carrier is "1" with bit "1", and "0" with bit "0". Therefore, two-level microwave ASK signal is generated.

Figure 26(a) shows the waveforms of the original 50-Mb/s baseband signal with a pattern of "110100101101001011". The original 1-GHz microwave carrier is shown in **Figure 26(b)**. **Figure 26(c)** shows the modulated microwave ASK signal.

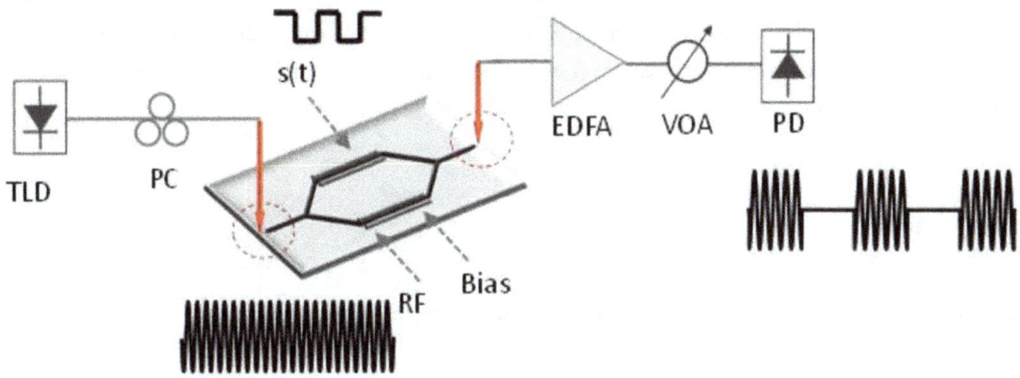

Figure 25. Schematic illustration of the proposed photonic microwave ASK signal generator.

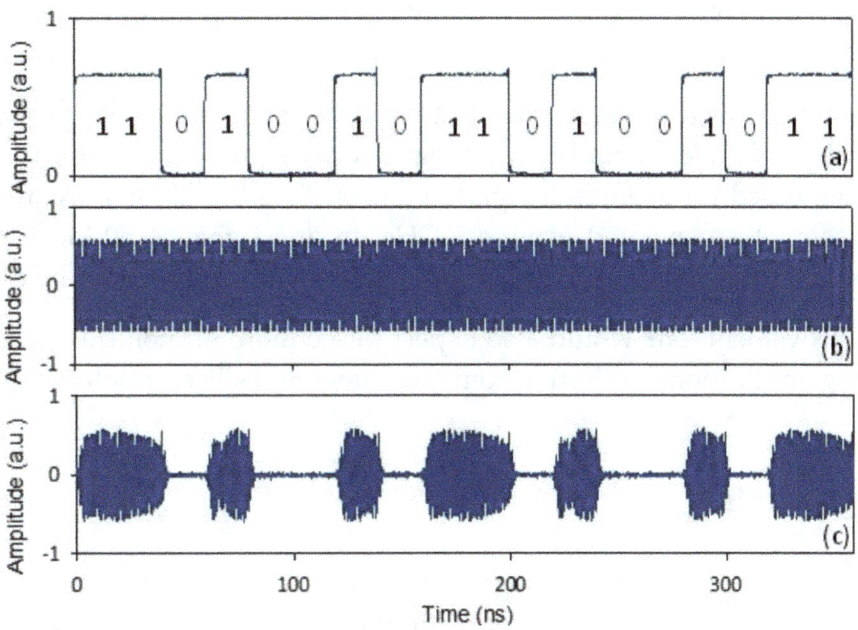

Figure 26. Waveforms of (a) the original 50-Mb/s baseband signal with a pattern of "110100101101001011", (b) original 1-GHz microwave carrier signal, and (c) the output microwave ASK signal.

5. Discussion

In summary, we review our recent works in chip-scale microwave photonic signal processing, including on-chip analog links, MWP filters, and photonic generation of microwave. As a new research topic, integrated MWP has just started to bloom and is set to have a bright future. After evaluating linearity performance of the on-chip analog links, future works may focus on increasing the linearities or decreasing nonlinearities of the on-chip analog links. As to MWP filters, more high performance silicon MWP filters will be demonstrated, involving resonators, photonic crystals, or other silicon photonic devices. In addition, more nonlinear optics can be used for MWP filters. For example, four-wave mixing (FWM) for MWP filtering has been realized [114, 115]. The demonstrated microwave signal multiplication and ASK modulation are still operated in low-frequency region. By optimizing the device design, the operation frequency can be improved and the power consumption can be further reduced. In addition to on-chip analog links, MWP filters, and photonic generation of microwave, silicon photonics can be also adopted to accelerate the success of versatile MWP signal processing functionalities, including arbitrary waveform generation, on-chip pulse shaping, delay line, beamformer, and more.

6. Conclusion

In this chapter, we have reviewed recent research efforts toward on-chip microwave photonic signal processing.

1. Using the fabricated silicon photonic devices, we evaluate the performance of on-chip analog signal transmission. The degradations level of SHD and THD are evaluated.
2. Various types of MWP filters are introduced, and the impact of optical nonlinearities to MWP filters is also discussed.
3. Photonic-assisted microwave signal multiplication/modulation: we propose a simple scheme to obtain frequency-multiplicated microwave signals or amplitude-coded microwave signals based on a single integrated silicon MZM. A 2-GHz frequency-doubled microwave signal is generated using a 1-GHz driving signal. A 50-Mb/s binary amplitude-coded 1-GHz microwave signal is also successfully generated in the experiment.

With future improvement, one would also expect to see more photonic-assisted microwave signal processing applications exploiting compact-integrated silicon photonic devices.

Acknowledgements

This work was supported by the National Program for Support of Top-Notch Young Professionals, the National Natural Science Foundation of China (NSFC) under grants 11574001, 11274131, and 61222502, the Program for New Century Excellent Talents in University under grant NCET-11-0182, the National Basic Research Program of China

(973 Program) under grant 2014CB340004, the Wuhan Science and Technology Plan Project under grant 2014070404010201, the Fundamental Research Funds for the Central Universities (HUST) under grants 2012YQ008 and 2013ZZGH003. The authors thank the Center of Micro-Fabrication and Characterization (CMFC) of WNLO for the support in the manufacturing process of silicon waveguides. The authors also thank to the facility support of the Center for Nanoscale Characterization and Devices of WNLO.

Author details

Jian Wang* and Yun Long

*Address all correspondence to: jwang@hust.edu.cn

Wuhan National Laboratory for Optoelectronics, School of Optical and Electronic Information, Huazhong University of Science and Technology, Wuhan, Hubei, China

References

[1] R. Soref, "The past, present, and future of silicon photonics," IEEE J. Sel. Top. Quantum Electron. 12, 1678–1687 (2006).

[2] B. Jalali, and S. Fathpour, "Silicon photonics," J. Lightwave Technol. 24, 4600–4615 (2006).

[3] M. Hochberg, and T. Baehr-Jones, "Towards fabless silicon photonics," Nat. Photonics. 4, 492–494 (2010).

[4] M. Asghari, and A. V. Krishnamoorthy, "Silicon photonics: Energy-efficient communication," Nat. Photonics. 5, 268–270 (2011).

[5] M. Soljačić, M. Ibanescu, S. G. Johnson, Y. Fink, and J. Joannopoulos, "Optimal bistable switching in nonlinear photonic crystals," Phys. Rev. E. 66, 055601 (2002).

[6] Y. Akahane, T. Asano, B.-S. Song, and S. Noda, "Fine-tuned high-Q photonic-crystal nanocavity," Opt. Express. 13, 1202–1214 (2005).

[7] P. Barclay, K. Srinivasan, and O. Painter, "Nonlinear response of silicon photonic crystal microresonators excited via an integrated waveguide and fiber taper," Opt. Express. 13, 801–820 (2005).

[8] M. Notomi, A. Shinya, S. Mitsugi, G. Kira, E. Kuramochi, and T. Tanabe, "Optical bistable switching action of Si high-Q photonic-crystal nanocavities," Opt. Express. 13, 2678–2687 (2005).

[9] Y. Zhang, D. Li, C. Zeng, Y. Shi, Z. Huang, J. Yu, and J. Xia, "Ultralow power nonlinear response in an Si photonic crystal nanocavity," IEEE Photon. J. 5, 6601409 (2013).

[10] Y. Zhang, D. Li, C. Zeng, Z. Huang, Y. Wang, Q. Huang, Y. Wu, J. Yu, and J. Xia, "Silicon optical diode based on cascaded photonic crystal cavities," Opt. Lett. 39, 1370–1373 (2014).

[11] W. Bogaerts, R. Baets, P. Dumon, V. Wiaux, S. Beckx, D. Taillaert, B. Luyssaert, J. Van Campenhout, P. Bienstman, and D. Van Thourhout, "Nanophotonic waveguides in silicon-on-insulator fabricated with CMOS technology," J. Lightwave Technol. 23, 401–412 (2005).

[12] B. E. Little, S. T. Chu, H. A. Haus, J. Foresi, and J. P. Laine, "Microring resonator channel dropping filters," J. Lightwave Technol. 15, 998–1005 (1997).

[13] B. E. Little, J. Foresi, G. Steinmeyer, E. Thoen, S. Chu, H. Haus, E. Ippen, L. Kimerling, and W. Greene, "Ultra-compact Si-SiO 2 microring resonator optical channel dropping filters," IEEE Photon. Technol. Lett. 10, 549–551 (1998).

[14] B. Little, S. Chu, W. Pan, and Y. Kokubun, "Microring resonator arrays for VLSI photonics," IEEE Photon. Technol. Lett. 12, 323–325 (2000).

[15] G. Priem, P. Dumon, W. Bogaerts, D. Van Thourhout, G. Morthier, and R. Baets, "Optical bistability and pulsating behaviour in Silicon-On-Insulator ring resonator structures," Opt. Express. 13, 9623–9628 (2005).

[16] W. Bogaerts, P. De Heyn, T. Van Vaerenbergh, K. De Vos, S. Kumar Selvaraja, T. Claes, P. Dumon, P. Bienstman, D. Van Thourhout, and R. Baets, "Silicon microring resonators," Laser & Photonics Rev. 6, 47–73 (2012).

[17] T. Barwicz, H. Byun, F. Gan, C. Holzwarth, M. Popovic, P. Rakich, M. Watts, E. Ippen, F. Kärtner, and H. Smith, "Silicon photonics for compact, energy-efficient interconnects [Invited]," J. Opt. Netw. 6, 63–73 (2007).

[18] J. S. Levy, A. Gondarenko, M. A. Foster, A. C. Turner-Foster, A. L. Gaeta, and M. Lipson, "CMOS-compatible multiple-wavelength oscillator for on-chip optical interconnects," Nat. Photonics. 4, 37–40 (2010).

[19] X. Xiao, H. Xu, X. Li, Z. Li, Y. Yu, and J. Yu, "High-speed on-chip photonic link based on ultralow-power microring modulator," in *Optical Fiber Communication Conference* (Optical Society of America, 2014), paper Tu2E. 6.

[20] J. K. Doylend, and A. P. Knights, "The evolution of silicon photonics as an enabling technology for optical interconnection," Laser & Photonics Rev. 6, 504–525 (2012).

[21] Y. Vlasov, "Silicon photonics for next generation computing systems," in *European Conference on Optical Communications*(2008), paper 1–2.

[22] L. Chen, K. Preston, S. Manipatruni, and M. Lipson, "Integrated GHz silicon photonic interconnect with micrometer-scale modulators and detectors," Opt. Express. 17, 15248–15256 (2009).

[23] C. Sun, M. T. Wade, Y. Lee, J. S. Orcutt, L. Alloatti, M. S. Georgas, A. S. Waterman, J. M. Shainline, R. R. Avizienis, and S. Lin, "Single-chip microprocessor that communicates directly using light," Nature. 528, 534–538 (2015).

[24] C. Sun, M. Georgas, J. Orcutt, B. Moss, Y.-H. Chen, J. Shainline, M. Wade, K. Mehta, K. Nammari, and E. Timurdogan, "A monolithically-integrated chip-to-chip optical link in bulk CMOS," IEEE J. Solid-St. Circ. 50, 828–844 (2015).

[25] Q. Li, Y. Liu, K. Padmaraju, R. Ding, D. F. Logan, J. J. Ackert, A. P. Knights, T. Baehr-Jones, M. Hochberg, and K. Bergman, "10-Gb/s BPSK link using silicon microring resonators for modulation and demodulation," in *Optical Fiber Communication Conference* (Optical Society of America, 2014), paper Tu2E. 5.

[26] D. Marris-Morini, L. Virot, C. Baudot, J.-M. Fédéli, G. Rasigade, D. Perez-Galacho, J.-M. Hartmann, S. Olivier, P. Brindel, and P. Crozat, "A 40 Gbit/s optical link on a 300-mm silicon platform," Opt. Express. 22, 6674–6679 (2014).

[27] A. Narasimha, B. Analui, Y. Liang, T. J. Sleboda, S. Abdalla, E. Balmater, S. Gloeckner, D. Guckenberger, M. Harrison, and R. G. Koumans, "A fully integrated 4× 10-Gb/s DWDM optoelectronic transceiver implemented in a standard 0.13 μm CMOS SOI technology," IEEE J. Solid-St. Circ. 42, 2736–2744 (2007).

[28] C. Doerr, L. Chen, L. Buhl, and Y.-K. Chen, "Eight-Channel SiO/Si N/Si/Ge CWDM Receiver," IEEE Photon. Technol. Lett. 23, 1201–1203 (2011).

[29] C. Doerr, and T. Taunay, "Silicon photonics core-, wavelength-, and polarization-diversity receiver," IEEE Photon. Technol. Lett. 9, 597–599 (2011).

[30] C. R. Doerr, N. K. Fontaine, and L. L. Buhl, "PDM-DQPSK silicon receiver with integrated monitor and minimum number of controls," IEEE Photon. Technol. Lett. 24, 697–699 (2012).

[31] J. Klamkin, F. Gambini, S. Faralli, A. Malacarne, G. Meloni, G. Berrettini, G. Contestabile, and L. Potì, "A 100-Gb/s noncoherent silicon receiver for PDM-DBPSK/DQPSK signals," Opt. Express. 22, 2150–2158 (2014).

[32] P. Dong, C. Xie, L. Chen, L. L. Buhl, and Y.-K. Chen, "112-Gb/s monolithic PDM-QPSK modulator in silicon," Opt. Express. 20, B624-B629 (2012).

[33] A. Shastri, M. Webster, G. Jeans, P. Metz, S. Sunder, B. Chattin, B. Dama, and K. Shastri, "Experimental demonstration of ultra-low-power single polarization 56 Gb/s QAM-16 generation without DAC using CMOS photonics," in *2014 The European Conference on Optical Communication (ECOC)* (2014).

[34] C. R. Doerr, L. Chen, D. Vermeulen, T. Nielsen, S. Azemati, S. Stulz, G. McBrien, X.-M. Xu, B. Mikkelsen, and M. Givehchi, "Single-chip silicon photonics 100-Gb/s coherent transceiver," in *Optical Fiber Communication Conference* (Optical Society of America, 2014), paper Th5C. 1.

[35] J. Yao, "Microwave photonics," J. Lightwave Technol. 27, 314–335 (2009).

[36] J. Capmany, J. Mora, I. Gasulla, J. Sancho, J. Lloret, and S. Sales, "Microwave photonic signal processing," J. Lightwave Technol. 31, 571–586 (2013).

[37] F. Liu, T. Wang, Z. Zhang, M. Qiu, and Y. Su, "On-chip photonic generation of ultra-wideband monocycle pulses," Electron. Lett. 45, 1 (2009).

[38] P. Cao, X. Hu, L. Zhang, J. Wu, X. Jiang, and Y. Su, "Photonic generation of microwave frequency shift keying signal using a single-drive Mach–Zehnder modulator," Opt. Express. 22, 14433–14440 (2014).

[39] Z. Tang, T. Zhang, F. Zhang, and S. Pan, "Photonic generation of a phase-coded microwave signal based on a single dual-drive Mach–Zehnder modulator," Opt. Lett. 38, 5365–5368 (2013).

[40] P. Xiang, X. Zheng, H. Zhang, Y. Li, and Y. Chen, "A novel approach to photonic generation of RF binary digital modulation signals," Opt. Express. 21, 631–639 (2013).

[41] D. Zhang, X. Feng, X. Li, K. Cui, F. Liu, and Y. Huang, "Tunable and reconfigurable bandstop microwave photonic filter based on integrated microrings and Mach–Zehnder interferometer," J. Lightwave Technol. 31, 3668–3675 (2013).

[42] Y. Long, and J. Wang, "All-optical tuning of a nonlinear silicon microring assisted microwave photonic filter: theory and experiment," Opt. Express. 23, 17758–17771 (2015).

[43] Y. Long, and J. Wang, "Ultra-high peak rejection notch microwave photonic filter using a single silicon microring resonator," Opt. Express 23, 17739–17750 (2015).

[44] D. Marpaung, B. Morrison, M. Pagani, R. Pant, D.-Y. Choi, B. Luther-Davies, S. J. Madden, and B. J. Eggleton, "Low-power, chip-based stimulated Brillouin scattering microwave photonic filter with ultrahigh selectivity," Optica. 2, 76–83 (2015).

[45] N. Ehteshami, W. Zhang, and J. Yao, "Optically tunable full 360° microwave photonic phase shifter using three cascaded silicon-on-insulator microring resonators," Opt. Commun. 373, 53–58 (2016).

[46] J. Wu, J. Peng, B. Liu, T. Pan, H. Zhou, J. Mao, Y. Yang, C. Qiu, and Y. Su, "Passive silicon photonic devices for microwave photonic signal processing," Opt. Commun. 373, 44–52 (2016).

[47] A. Casas-Bedoya, B. Morrison, M. Pagani, D. Marpaung, and B. J. Eggleton, "Tunable narrowband microwave photonic filter created by stimulated Brillouin scattering from a silicon nanowire," Opt. Lett. 40, 4154–4157 (2015).

[48] A. Zheng, J. Dong, L. Zhou, X. Xiao, Q. Yang, X. Zhang, and J. Chen, "Fractional-order photonic differentiator using an on-chip microring resonator," Opt. Lett. 39, 6355–6358 (2014).

[49] S. Yan, Y. Zhang, J. Dong, A. Zheng, S. Liao, H. Zhou, Z. Wu, J. Xia, and X. Zhang, "Operation bandwidth optimization of photonic differentiators," Opt. Express. 23, 18925–18936 (2015).

[50] Y. Yu, F. Jiang, H. Tang, L. Xu, X. Liu, J. Dong, and X. Zhang, "Reconfigurable photonic temporal differentiator based on a dual-drive Mach-Zehnder modulator," Opt. Express. 24, 11739–11748 (2016).

[51] H. Jiang, D. Marpaung, M. Pagani, K. Vu, D.-Y. Choi, S. J. Madden, L. Yan, and B. J. Eggleton, "Wide-range, high-precision multiple microwave frequency measurement using a chip-based photonic Brillouin filter," Optica. 3, 30–34 (2016).

[52] M. Pagani, B. Morrison, Y. Zhang, A. Casas-Bedoya, T. Aalto, M. Harjanne, M. Kapulainen, B. J. Eggleton, and D. Marpaung, "Low-error and broadband microwave frequency measurement in a silicon chip," Optica. 2, 751–756 (2015).

[53] J. Du, and J. Wang, "Experimental performance evaluation of analog signal transmission in a silicon microring resonator," Opt. Lett. 40, 1181–1184 (2015).

[54] C. Gui, Y. Zhang, J. Du, J. Xia, and J. Wang, "Experimental demonstration of analog signal transmission in a silicon photonic crystal L3 resonator," Opt. Express. 23, 13916–13923 (2015).

[55] W. Liu, M. Li, R. S. Guzzon, E. J. Norberg, J. S. Parker, L. A. Coldren, and J. Yao, "A photonic temporal integrator with an ultra-long integration time window based on an InP-InGaAsP integrated ring resonator," J. Lightwave Technol. 32, 3654–3659 (2014).

[56] J. Zhang, and J. Yao, "Microwave photonic integrator based on a multichannel fiber Bragg grating," Opt. Lett. 41, 273–276 (2016).

[57] H. P. Bazargani, M. Burla, and J. Azaña, "Experimental demonstration of sub-picosecond optical pulse shaping in silicon based on discrete space-to-time mapping," Opt. Lett. 40, 5423–5426 (2015).

[58] H. P. Bazargani, and J. Azaña, "Optical pulse shaping based on discrete space-to-time mapping in cascaded co-directional couplers," Opt. Express. 23, 23450–23461 (2015).

[59] T. Yang, J. Dong, L. Liu, S. Liao, S. Tan, L. Shi, D. Gao, and X. Zhang, "Experimental observation of optical differentiation and optical Hilbert transformation using a single SOI microdisk chip," Sci. Rep. 4, 3960 (2014).

[60] J. Wang, H. Shen, L. Fan, R. Wu, B. Niu, L. T. Varghese, Y. Xuan, D. E. Leaird, X. Wang, and F. Gan, "Reconfigurable radio-frequency arbitrary waveforms synthesized in a silicon photonic chip," Nat. Commun. 6, 5957 (2015).

[61] M. H. Khan, H. Shen, Y. Xuan, L. Zhao, S. Xiao, D. E. Leaird, A. M. Weiner, and M. Qi, "Ultrabroad-bandwidth arbitrary radiofrequency waveform generation with a silicon photonic chip-based spectral shaper," Nat. Photonics. 4, 117–122 (2010).

[62] J. McKinney, D. Leaird, and A. Weiner, "Millimeter-wave arbitrary waveform generation with a direct space-to-time pulse shaper," Opt. Lett. 27, 1345–1347 (2002).

[63] S. Liao, Y. Ding, J. Dong, T. Yang, X. Chen, D. Gao, and X. Zhang, "Arbitrary waveform generator and differentiator employing an integrated optical pulse shaper," Opt. Express. 23, 12161–12173 (2015).

[64] J. O'reilly, P. Lane, R. Heidemann, and R. Hofstetter, "Optical generation of very narrow linewidth millimetre wave signals," Electron. Lett. 28, 2309–2311 (1992).

[65] G. Qi, J. Yao, J. Seregelyi, S. Paquet, and C. Bélisle, "Generation and distribution of a wide-band continuously tunable millimeter-wave signal with an optical external modulation technique," IEEE Trans. Microwave Theory Tech. 53, 3090–3097 (2005).

[66] M. Burla, C. G. Roeloffzen, L. Zhuang, D. Marpaung, M. R. Khan, P. Maat, K. Dijkstra, A. Leinse, M. Hoekman, and R. Heideman, "System integration and radiation pattern measurements of a phased array antenna employing an integrated photonic beamformer for radio astronomy applications," Appl. Optics. 51, 789–802 (2012).

[67] H. Zmuda, R. A. Soref, P. Payson, S. Johns, and E. N. Toughlian, "Photonic beamformer for phased array antennas using a fiber grating prism," IEEE Photon. Technol. Lett. 9, 241–243 (1997).

[68] N.-N. Feng, P. Dong, D. Feng, W. Qian, H. Liang, D. C. Lee, J. B. Luff, A. Agarwal, T. Banwell, and R. Menendez, "Thermally-efficient reconfigurable narrowband RF-photonic filter," Opt. Express. 18, 24648–24653 (2010).

[69] Y. Yue, H. Huang, L. Zhang, J. Wang, J.-Y. Yang, O. F. Yilmaz, J. S. Levy, M. Lipson, and A. E. Willner, "UWB monocycle pulse generation using two-photon absorption in a silicon waveguide," Opt. Lett. 37, 551–553 (2012).

[70] X. Wang, W. Shi, H. Yun, S. Grist, N. A. Jaeger, and L. Chrostowski, "Narrow-band waveguide Bragg gratings on SOI wafers with CMOS-compatible fabrication process," Opt. Express. 20, 15547–15558 (2012).

[71] M. L. Farwell, W. S. Chang, and D. R. Huber, "Increased linear dynamic range by low biasing the Mach-Zehnder modulator," IEEE Photon. Technol. Lett. 5, 779–782 (1993).

[72] A. Khilo, C. M. Sorace, and F. X. Kärtner, "Broadband linearized silicon modulator," Opt. Express. 19, 4485–4500 (2011).

[73] M. Streshinsky, A. Ayazi, Z. Xuan, A. E.-J. Lim, G.-Q. Lo, T. Baehr-Jones, and M. Hochberg, "Highly linear silicon traveling wave Mach-Zehnder carrier depletion modulator based on differential drive," Opt. Express. 21, 3818–3825 (2013).

[74] J. Cardenas, P. A. Morton, J. B. Khurgin, A. Griffith, C. B. Poitras, K. Preston, and M. Lipson, "Linearized silicon modulator based on a ring assisted Mach Zehnder inteferometer," Opt. Express. 21, 22549–22557 (2013).

[75] M. LaGasse, W. Charczenko, M. Hamilton, and S. Thaniyavarn, "Optical carrier filtering for high dynamic range fibre optic links," Electron. Lett. 30, 2157–2158 (1994).

[76] P. Li, L. Yan, T. Zhou, W. Li, Z. Chen, W. Pan, and B. Luo, "Improvement of linearity in phase-modulated analog photonic link," Opt. Lett. 38, 2391–2393 (2013).

[77] H. Huang, Y. Yue, L. Zhang, C. Chase, D. Parekh, F. Sedgwick, M. Wu, C. Chang-Hasnain, M. Tur, and A. Willner, "Analog signal transmission in a high-contrast-gratings-based hollow-core-waveguide," J. Lightwave Technol. 30, 3640–3646 (2012).

[78] M. Song, L. Zhang, R. G. Beausoleil, and A. E. Willner, "Nonlinear distortion in a silicon microring-based electro-optic modulator for analog optical links," IEEE J. Sel. Top. Quantum Electron. 16, 185–191 (2010).

[79] A. Ayazi, T. Baehr-Jones, Y. Liu, A. E.-J. Lim, and M. Hochberg, "Linearity of silicon ring modulators for analog optical links," Opt. Express. 20, 13115–13122 (2012).

[80] S. J. Spector, S. Yegnanarayanan, R. Swint, T. M. Lyszczarz, and P. W. Juodawlkis, "Two-tone measurement of the nonlinear behavior of a silicon-on-insulator (SOI) ring resonator," in *CLEO: Applications and Technology* (Optical Society of America, 2012), paper JW4A. 74.

[81] X. Luo, X. Tu, J. Song, L. Ding, Q. Fang, T.-Y. Liow, M. Yu, and G.-Q. Lo, "Slope efficiency and spurious-free dynamic range of silicon Mach-Zehnder modulator upon carrier depletion and injection effects," Opt. Express. 21, 16570–16577 (2013).

[82] D. Marpaung, C. Roeloffzen, R. Heideman, A. Leinse, S. Sales, and J. Capmany, "Integrated microwave photonics," Laser & Photonics Rev. 7, 506–538 (2013).

[83] R. A. Minasian, "Photonic signal processing of microwave signals," IEEE Trans. Microwave Theory Tech. 54, 832–846 (2006).

[84] J. Capmany, B. Ortega, D. Pastor, and S. Sales, "Discrete-time optical processing of microwave signals," J. Lightwave Technol. 23, 702 (2005).

[85] J. Capmany, B. Ortega, and D. Pastor, "A tutorial on microwave photonic filters," J. Lightwave Technol. 24, 201–229 (2006).

[86] J. Mora, B. Ortega, A. Díez, J. L. Cruz, M. V. Andrés, J. Capmany, and D. Pastor, "Photonic microwave tunable single-bandpass filter based on a Mach-Zehnder interferometer," J. Lightwave Technol. 24, 2500–2509 (2006).

[87] B. Vidal, M. Piqueras, and J. Marti, "Tunable and reconfigurable photonic microwave filter based on stimulated Brillouin scattering," Opt. Lett. 32, 23–25 (2007).

[88] W. Zhang, and R. A. Minasian, "Widely tunable single-passband microwave photonic filter based on stimulated Brillouin scattering," IEEE Photon. Technol. Lett. 23, 1775–1777 (2011).

[89] L. Gao, X. Chen, and J. Yao, "Tunable microwave photonic filter with a narrow and flat-top passband," IEEE Microw. Wireless Compon. Lett. 23, 362–364 (2013).

[90] W. Li, and J. Yao, "A narrow-passband frequency-tunable microwave photonic filter with an improved dynamic range," in *Optical Fiber Communication Conference* (Optical Society of America, Anaheim, California, 2013), paper OTu2H. 3.

[91] P. K. Yu, "A novel digitally tunable microwave-photonic notch filter using differential group-delay module," IEEE Photon. Technol. Lett. 15 (2003).

[92] P. Alipour, A. A. Eftekhar, A. H. Atabaki, Q. Li, S. Yegnanarayanan, C. K. Madsen, and A. Adibi, "Fully reconfigurable compact RF photonic filters using high-Q silicon microdisk resonators," Opt. Express. 19, 15899–15907 (2011).

[93] D. Zhang, X. Feng, and Y. Huang, "Tunable and reconfigurable bandpass microwave photonic filters utilizing integrated optical processor on silicon-on-insulator substrate," IEEE Photon. Technol. Lett. 24, 1502–1505 (2012).

[94] J. Dong, L. Liu, D. Gao, Y. Yu, A. Zheng, T. Yang, and X. Zhang, "Compact notch microwave photonic filters using on-chip integrated microring resonators," IEEE Photon. J. 5, 5500307–5500307 (2013).

[95] M. S. Rasras, K.-Y. Tu, D. M. Gill, Y.-K. Chen, A. E. White, S. S. Patel, A. Pomerene, D. Carothers, J. Beattie, and M. Beals, "Demonstration of a tunable microwave-photonic

notch filter using low-loss silicon ring resonators," J. Lightwave Technol. 27, 2105–2110 (2009).

[96] J. Palací, G. E. Villanueva, J. V. Galán, J. Marti, and B. Vidal, "Single bandpass photonic microwave filter based on a notch ring resonator," IEEE Photon. Technol. Lett. 22, 1276–1278 (2010).

[97] D. Zhang, X. Feng, and Y. Huang, "Reconfigurable microwave photonic filter based on parallel-cascaded microrings assisted with a Mach–Zehnder interferometer," J. Opt. 14, 065502 (2012).

[98] D. Marpaung, B. Morrison, R. Pant, C. Roeloffzen, A. Leinse, M. Hoekman, R. Heideman, and B. J. Eggleton, "Si_3N_4 ring resonator-based microwave photonic notch filter with an ultrahigh peak rejection," Opt. Express 21, 23286–23294 (2013).

[99] D. Marpaung, B. Morrison, R. Pant, C. Roeloffzen, A. Leinse, M. Hoekman, R. Heideman, and B. J. Eggleton, "Ultrahigh suppression and reconfigurable RF photonic notch filter using a silicon nitride ring resonator," in *2014 Conference on Lasers and Electro-Optics (CLEO)*, (2014), paper 1–2.

[100] L. Goldberg, H. Taylor, J. Weller, and D. Bloom, "Microwave signal generation with injection-locked laser diodes," Electron. Lett. 19, 491–493 (1983).

[101] R. Ramos, and A. Seeds, "Fast heterodyne optical phase-lock loop using double quantum well laser diodes," Electron. Lett. 28, 82–83 (1992).

[102] J. O'Reilly, and P. Lane, "Remote delivery of video services using mm-waves and optics," J. Lightwave Technol. 12, 369–375 (1994).

[103] G. Qi, J. Yao, J. Seregelyi, S. Paquet, and C. Bélisle, "Optical generation and distribution of continuously tunable millimeter-wave signals using an optical phase modulator," J. Lightwave Technol. 23, 2687–2695 (2005).

[104] J. Yu, Z. Jia, L. Yi, Y. Su, G.-K. Chang, and T. Wang, "Optical millimeter-wave generation or up-conversion using external modulators," IEEE Photon. Technol. Lett. 18, 265–267 (2006).

[105] W. Li, and J. Yao, "Investigation of photonically assisted microwave frequency multiplication based on external modulation," IEEE Trans. Microwave Theory Tech. 58, 3259–3268 (2010).

[106] L. Gao, W. Liu, X. Chen, and J. Yao, "Photonic-assisted microwave frequency multiplication with a tunable multiplication factor," Opt. Lett. 38, 4487–4490 (2013).

[107] B. Chen, S. Zheng, H. Chi, X. Zhang, and X. Jin, "An optical millimeter-wave generation technique based on phase modulation and Brillouin-assisted notch-filtering," IEEE Photon. Technol. Lett. 20, 2057–2059 (2008).

[108] A. Wiberg, P. Pérez-Millán, M. V. Andrés, and P. O. Hedekvist, "Microwave-photonic frequency multiplication utilizing optical four-wave mixing and fiber Bragg gratings," J. Lightwave Technol. 24, 329 (2006).

[109] C. Zhang, L. Wang, and K. Qiu, "Proposal for all-optical generation of multiple-frequency millimeter-wave signals for RoF system with multiple base stations using FWM in SOA," Opt. Express. 19, 13957–13962 (2011).

[110] B. Vidal, "Photonic millimeter-wave frequency multiplication based on cascaded four-wave mixing and polarization pulling," Opt. Lett. 37, 5055–5057 (2012).

[111] J. Zheng, H. Wang, W. Li, L. Wang, T. Su, J. Liu, and N. Zhu, "Photonic-assisted micro-wave frequency multiplier based on nonlinear polarization rotation," Opt. Lett. 39, 1366–1369 (2014).

[112] J. Chou, H. Yan, and B. Jalali, "Adaptive RF-photonic arbitrary waveform generator," IEEE Photon. Technol. Lett. 15, 581–583 (2003).

[113] Y. Long, L. Zhou, and J. Wang, "Photonic-assisted microwave signal multiplication and modulation using a silicon Mach–Zehnder modulator," Sci. Rep. 6, 20215 (2016).

[114] V. R. Supradeepa, C. M. Long, R. Wu, F. Ferdous, E. Hamidi, D. E. Leaird, and A. M. Weiner, "Comb-based radiofrequency photonic filters with rapid tunability and high selectivity," Nat. Photonics 6, 186–194 (2012).

[115] B. Vidal Rodriguez, J. Palací López, and J. Capmany Francoy, "Reconfigurable photonic microwave filter based on four-wave mixing," in *IEEE Photonics Journal* (Institute of Electrical and Electronics Engineers (IEEE), 2012), paper 759–764.

2

On-Wafer Microwave De-Embedding Techniques

Xi Sung Loo, Kiat Seng Yeo and
Kok Wai Johnny Chew

Abstract

Wireless communication technology has kept evolving into higher frequency regime to take advantage of wider data bandwidth and higher speed performance. Successful RF circuit design requires accurate characterization of on-chip devices. This greatly relies on robust de-embedding technique to completely remove surrounding parasitics of pad and interconnects that connect device to measurement probes. Complex interaction of fixture parasitic at high frequency has imposed extreme challenges to de-embedding particularly for lossy complementary metal oxide semiconductor (CMOS) device. A generalized network de-embedding technique that avoids any inaccurate lumped and transmission line assumptions on the pad and interconnects of the test structure is presented. The de-embedding strategy has been validated by producing negligible de-embedding error (<−50 dB) on the insertion loss of the zero-length THRU device. It demonstrates better accuracy than existing de-embedding techniques that are based on lumped pad assumption. For transistor characterization, the de-embedding reference plane could be further shifted to the metal fingers with additional Finger OPEN-SHORT structures. The resulted de-embedded RF parameters of CMOS transistor show good scalability across geometries and negligible frequency dependency of less than 3% for up to 100 GHz. The results reveal the importance of accounting for the parasitic effect of metal fingers for transistor characterization.

Keywords: CMOS, de-embedding, microwave frequencies, scattering parameters, test structure

1. Introduction

Aggressive scaling of complementary metal oxide semiconductor (CMOS) devices over the past decades has led to tremendous increase in speed, making it suitable to be employed in radio frequency circuits. Current CMOS devices (28 nm) are able to deliver maximum ft (maximum cut-off frequency) of 340 GHz [1] with superior improvement in noise, power, and gain performance. **Table 1** shows the key comparisons of transistor performance characteristics at various technology nodes.

Technology node (nm)	180	130	90	40	28
Supply voltage (V)	1.8	1.5	1.2	1.1	1.05
Peak cut-off frequency, ft (GHz)	60	75	140	260	340
Peak transconductance, gm (μS/μm)	562	809	1030	1128	1377

Table 1. Performance-related characteristics of GLOBALFOUNDRIES' NFET (N-type field-effect transistor) at different technology nodes.

However, the growing complexity of radio frequency-integrated circuits requires highly accurate CMOS device model to predict the circuit behavior correctly for successful design. Therefore, accurate and precision RF measurement data are essential to ensure high quality of RF CMOS model developed. Nevertheless, raw measurement data itself does not represent the high-frequency behavior of intrinsic CMOS device as it includes parasitic effects of test structure. Therefore, additional data processing steps, known as de-embedding, are required to remove the impact of test structure parasitic effects from the raw noise measurement data. De-embedding is challenging for RF characterization of short channel device as parasitic effect of interconnects and lossy Si substrate would appear much larger than the device itself. The reason is that interconnects of test structure do not scale proportionally with CMOS device size as minimum distance between probes has to be maintained to avoid collision. Further, the parasitic effects of test structures (i.e. substrate interaction, distributed effect of interconnects, etc.) become more complex at higher measurement frequency and require more sophisticated de-embedding technique. The aim of this chapter is to present an insight of microwave de-embedding theory and knowledge on various techniques to overcome challenges at high frequencies.

Many de-embedding techniques have been reported up to date. They could be broadly classified as lumped circuit model-based technique [2–6] and network model-based technique [7–12]. As the name implies, the aforementioned techniques model the test fixture parasitics as a combination of parallel-series connections of discrete components. Nevertheless, such a de-embedding technique could not be used to address the distributed effect of metal interconnect. In order to overcome the deficiency mentioned, the network model-based de-embedding techniques [7–12] have been proposed. In particular, the technique proposed in Ref. [7] is renowned for its high generality as it models the surrounding fixture parasitic as a single four-port network without any assumptions made on the network topology. However, it requires five test structures and suffers accuracy degradation at very high frequencies due to ideality assumptions made on the intrinsic standards of dummy test structures. Instead, cascade network model-based de-embedding techniques [8–12] offer alternate way to address the distributed effects of metal interconnects without any precision standards required. Nevertheless, techniques in Refs. [8–11] require lumped assumption on the pad parasitics.

A generalized cascade-based de-embedding technique [12] that addresses the distributed effects of pad and interconnect parasitics will be presented in this chapter. It utilizes unique combination of pad and line de-embedding structures that avoid any lumped assumptions. The description covers parasitic model adopted, de-embedding structures used, and corresponding mathematical de-embedding algorithm in separate subsections. Also, an extension of cascade-based de-embedding technique [13] to further remove the metal finger parasitics of CMOS transistor will be detailed in Section 5. Finally, the validation results of the de-embedding methodology will be presented and supplemented by comparison against existing de-embedding techniques.

2. Test structure

Direct probing on transistor device is impossible due to its minute size. Therefore, test structure is introduced to provide essential electrical interface in between the embedded device and measurement probes through its bond pads and metal interconnections. However, these test fixture components introduce undesirable parasitic effects that lead to error in the device measurements. Particularly, the CMOS test structure is fabricated on silicon substrate that is more lossy than alumina substrate in impedance standard substrate (ISS). One way to reduce the impact of these fixture parasitics is to optimize the design of test structure. **Figure 1(a)** and **(b)** shows the top and cross-sectional view of G-S-G test structure used for characterization of an Negative channel Metal-Oxide Semiconductor (NMOS) device. There are three bond pads with appropriate sizes (70 μm × 60 μm) and separation pitches (50 μm) on both side of test fixture for locating the ground and signal probes at each port. Signal interconnects are used to connect gate and drain lead of NMOS devices to bond pads at input and output ports, respectively. As shown in **Figure 1(b)**, it is actually a metal-via stack that consists of three metal layers, Metal 6 to Metal 8 (M6–M8). Sufficient length of interconnects (>50 μm) is required to avoid possible collision and interference between probes. Also, wide interconnects (10 μm) provide low-resistance paths to gate and drain terminals. The test structure is developed based on shielded design [14] where a wide Metal 1 (M1) ground conductor is included to mitigate the substrate coupling effect. It is connected to the ground bars (M1–M8 metal-via stacks) and is tied to silicon substrate through P+ implants. Since the Metal 1 metal shield is large in size, it provides low-resistive ground connections to all ports. Meanwhile, the NMOS device lies inside the fixture gap of Metal 1 shield where it is directly exposed to silicon substrate. **Figure 2** shows the enlarge view of NMOS transistor. The transistor possesses interdigitated layout whereby the source and drain regions are shared for the reduction of parasitic resistance and capacitance. Nevertheless, layout optimization could only partially suppress the fixture parasitic effect, and further de-embedding is required to remove effects of these fixture parasitics from measurements particularly when it gets worsened at high frequencies.

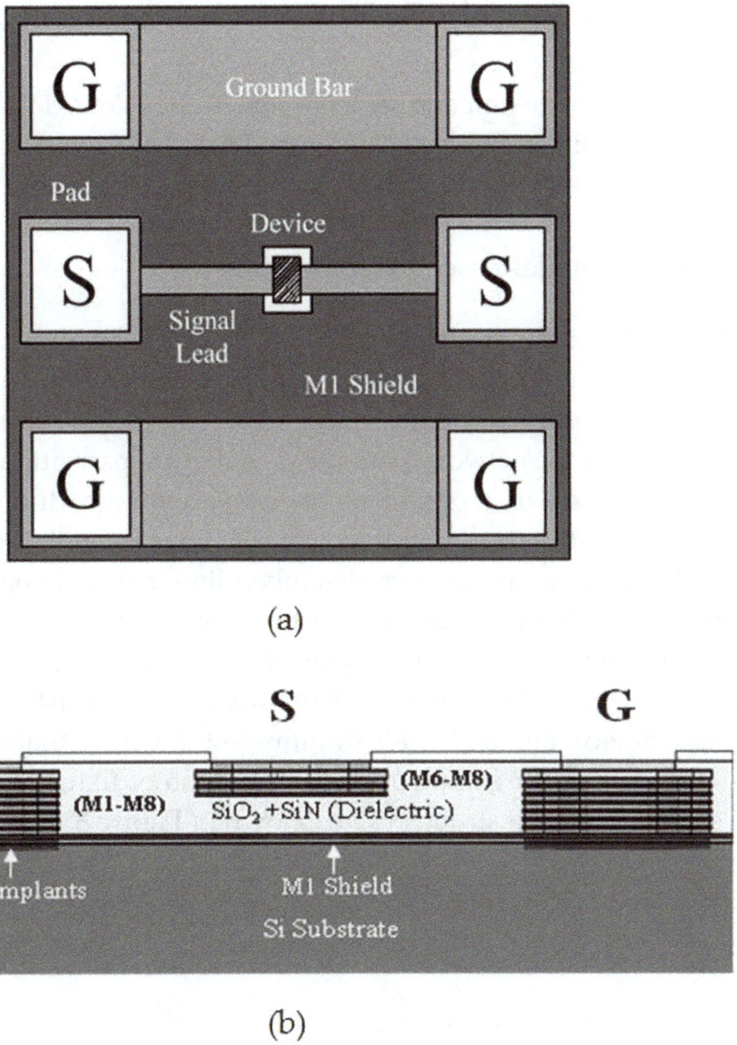

Figure 1. (a) Top schematic and (b) cross-sectional view of transistor test fixture for GHz probing.

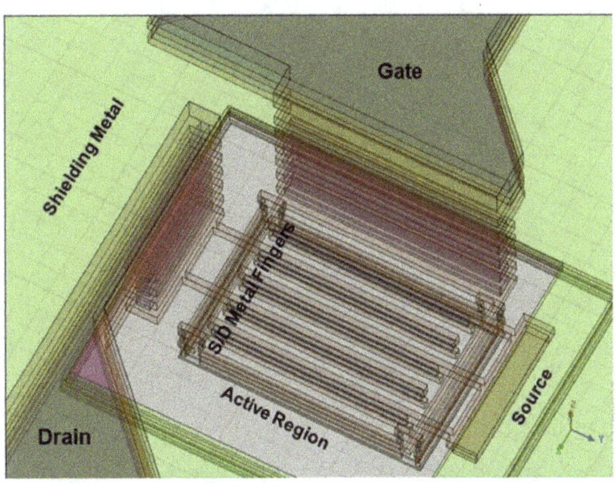

Figure 2. Enlarge view of transistor device embedded in test fixture.

3. Overview of de-embedding techniques

Two popular categories of de-embedding techniques are reviewed theoretically in this section. They are equivalent circuit model-based de-embedding and cascade network model-based de-embedding.

3.1. Equivalent circuit model-based de-embedding

In this method, device (transistor) test fixture is modeled by an equivalent circuit which is basically complex combinations of fixture parasitic components and intrinsic device itself. Network parameters of intrinsic device could be de-embedded from raw S-parameter measurements, provided that the network parameters of fixture parasitic are known. This could be determined through a set of S-parameter measurements on dummy test structures. The number of dummy test structures required typically increase with the complexity of parasitic circuit model. Specifically, open-short de-embedding methodology [2] has been widely adopted for transistor characterization due to its sufficient accurate prediction of fixture parasitic for conventional microwave frequencies of interest at below 30 GHz. As the name implies, it requires an OPEN dummy test structure which consists of only test fixture frame (without intrinsic device) and a SHORT dummy test structure that is associated with short-circuited interconnections for complete characterization of fixture parasitics. The equivalent circuit models of required test structures are shown in **Figure 3**.

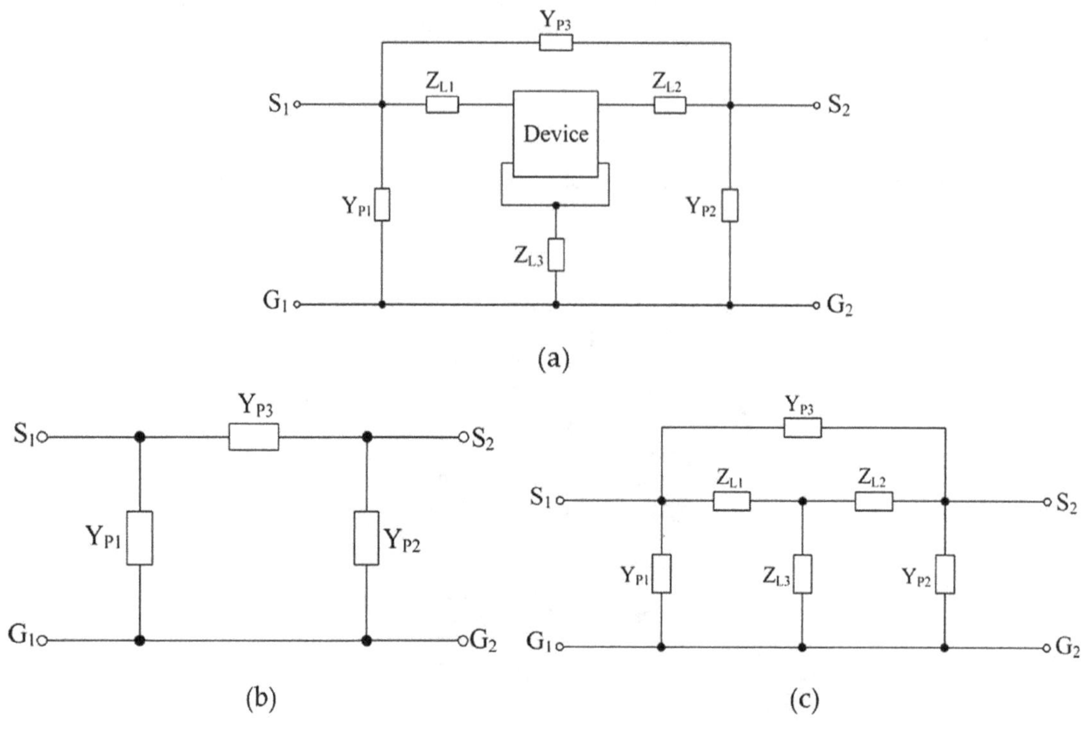

Figure 3. (a) Equivalent circuit model of DUT for open-short de-embedding [6]. (b) π-Circuit model of OPEN dummy test structure. (c) Equivalent circuit model of SHORT dummy test structure for Open-Short de-embedding.

In the device under test (DUT) model (**Figure 3(a)**), admittance component, Y_{p3}, is used to determine the amount of cross talk between port 1 and port 2 due to substrate coupling and fringing capacitances. On the other hand, admittance components, Y_{p1} and Y_{p2}, measure the parasitic capacitances that exist between bond pad and ground for left and right ports. Meanwhile, the metal parasitic components are modeled by a T-network which connected in series with the intrinsic device. Specifically, the series parasitic of metal lines that appear at left and right port is represented by Z_{L1} and Z_{L2}, respectively. Meanwhile, the dangling ground lead parasitic is denoted as Z_{L3}. By employing a two-port network theory for parallel-series network, intrinsic device admittance matrix, Y_{DEV}, could be easily extracted from raw measurements on test structures by Eq. (1).

$$Y_{DEV} = \left[\left(Y_{DUT} - Y_{OPEN}\right)^{-1} - \left(Y_{SHORT} - Y_{OPEN}\right)^{-1}\right]^{-1} \tag{1}$$

3.2. Cascade network model-based de-embedding

At much higher frequency where the length of metal interconnects approaches one-tenth of the frequency wavelength, fixture parasitic could no longer be described by lumped circuit model due to worsening of distributed effect. Fortunately, the problem could be overcome with a network model-based de-embedding technique, which is basically the extension of S-parameter probe-tip calibration techniques (SOLT, TRL, LRRM, etc). The cascade configuration model used to describe S-parameter measurement system is now applied on test fixture model instead. The pad and interconnect parasitics of test fixture are now modeled as error adapters connected in cascade with intrinsic device at left and right port. High generality is achieved with this method as interconnect parasitic is characterized by network parameters instead of inaccurate lumped circuit models. Unlike calibration, fixture parasitic is characterized via measurements of de-embedding structures due to unavailability of accurate on-wafer calibration standards. More recently, on-wafer TRL calibration has been reported [15] to directly remove on-wafer parasitics without extra de-embedding steps and known standards needed. However, it is not suitable for modeling since broadband accuracy is unachievable. Also, it is more recommended to be used with expensive gold pads, low substrate loss process, and high layout symmetry for improved accuracy [15].

In conventional cascade-based de-embedding approaches [8–11], probe pads and interconnects of test fixture are characterized by separate networks that are connected in cascade configuration (**Figure 4(a)**). Similarly, THRU structures are used to extract the two-port network parameters or transmission line parameters of interconnects directly. Specifically, the probe pad network is simplified to lumped circuit model which consists of only one parallel admittance element, Y_{PAD} [8, 9], or encompass additional series pad impedance, Z_{PAD} [10, 11]. PAD OPEN structure is commonly used for finding pad to ground admittance, Y_{PAD}, while PAD SHORT structure is used in [10] to determine the series pad impedance, Z_{PAD}. The major drawback of these techniques is that the pad-line discontinuity effect is not accounted, and extraction of series pad impedance is associated with obvious SHORT interconnection parasitic that span from top metal to bottom ground metal. Although no SHORT structure is used in lumped cascade approach [16], it is unable to account for distributed effects of metallic parasitics. In effort to avoid the deficiencies mentioned, an alternate algorithm has been presented by the authors [11] whereby only THRU structures are used to extract both pad and interconnect parasitics directly. Two set of THRU structures (THRU LL', THRU LH and THRU R'R, THRU RH) are used for determining the network parameters of interconnect at left and right ports, respectively. With each fixture block being characterized using Ref. [11], network parameters of transistor could simply be extracted from raw measurements through chain matrix manipulations as shown in Eq. (2):

$$A_{DEV} = A_{IN}^{-1} A_{DUT} A_{OUT}^{-1} \tag{2}$$

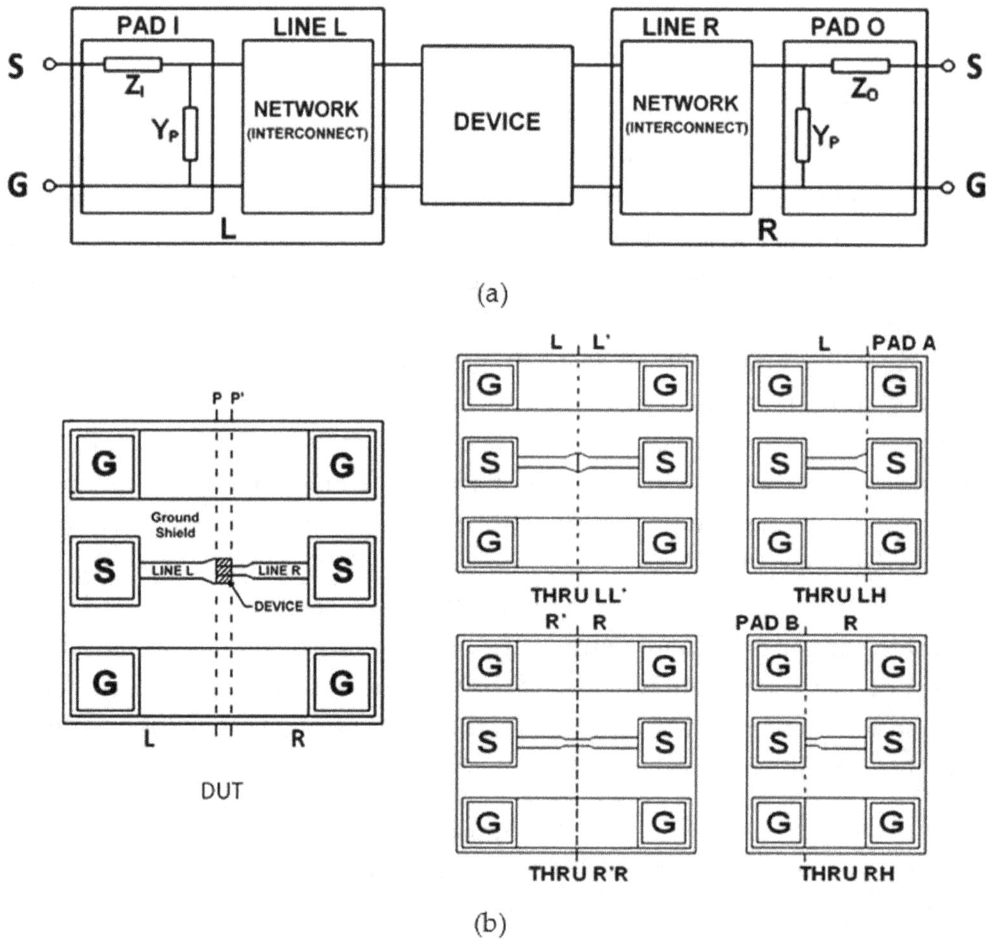

Figure 4. (a) DUT model for author's THRU-based cascade network-based de-embedding approach. (b) Schematic diagram of associated test structures used for author's THRU-based cascade network-based de-embedding technique [11].

For symmetrical DUT structure, the number of THRU structures required for cascade-based de-embedding approach could be reduced further from four to two. Also, the interaction between ports due to leaky substrate and fringing capacitances could be accounted with additional OPEN structure as described in Ref. [17].

4. Generalized cascade-based de-embedding technique

Despite improvement has been made by existing cascade-based de-embedding techniques to address the distributed effects of interconnects, still the pad counterpart is approximated by lumped circuit element. Such issue could potentially be resolved by four-port de-embedding technique in Ref. [7], which avoids any circuit assumption made on the pad and network topology of fixture parasitics. Nevertheless, it requires precision or ideal standards that are unable to be realized on practical CMOS technology. The size of the pad is normally fixed by the dimension of measurement probes and cannot be reduced for optimization of parasitics. As a results, the pad length could become comparable with interconnect length that is usually optimized to be short for parasitic reduction. In effort to overcome the drawback mentioned, a generalized cascade-based de-embedding technique [12] is presented in this section. It utilizes unique combinations of two THRU structures that enable efficient de-embedding of fixture parasitics without any inaccurate lumped pad approximation or requirement of known standards.

4.1. De-embedding concept

Similar to existing cascade de-embedding approaches describe in Section 4, the DUT structure is represented by interconnections of input (L) and output (R) network adapters that appear at both ports of the embedded device (**Figure 5(b)**). Shielded-based test structure is used here to mitigate the forward coupling effect between two ports [14] as described in Section 2. These fixture network adapters (L, R) include the parasitic contribution of probe pads and metal lines of arbitrary lengths (x1, x2). The main advantage of this de-embedding technique is that it does not require any lumped assumption on the pad parasitics since the network parameters of the fixture adapters could be computed directly from measurements on designated de-embedding structures described in the next subsection.

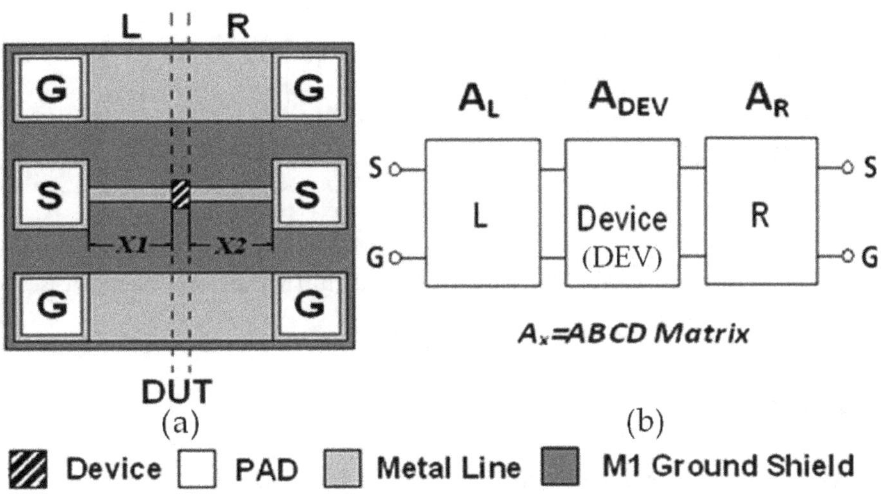

Figure 5. (a) Shielded DUT structure used in device measurement. (b) It is modeled as cascade connections of fixture parasitic networks (L, R) and device [12].

4.2. De-embedding structure

As illustrated in **Figure 6(a)**, two types of THRU de-embedding structures are adopted for extraction of fixture parasitics, namely, THRU LLR and THRU LR structures. The THRU LR structure is simply direct connections of left and right half of test fixture excluding embedded device. Thus, its bond pads and metal line have the same total length as those in DUT structure. Meanwhile, the THRU LLR structure is equivalent to direct connections of left half section of test fixture to the left port of THRU LR structure. Based on the physical layout of the de-embedding structures described, their two-port network models could be determined as illustrated in **Figure 6(b)**. Thus, their cascade matrix (A_{LR}, A_{LLR}) are related to those of fixture parasitic networks (A_L, A_R) by Eq. (3):

$$A_{LR} = A_L A_R \text{ and } A_{LLR} = A_L A_L A_R. \tag{3}$$

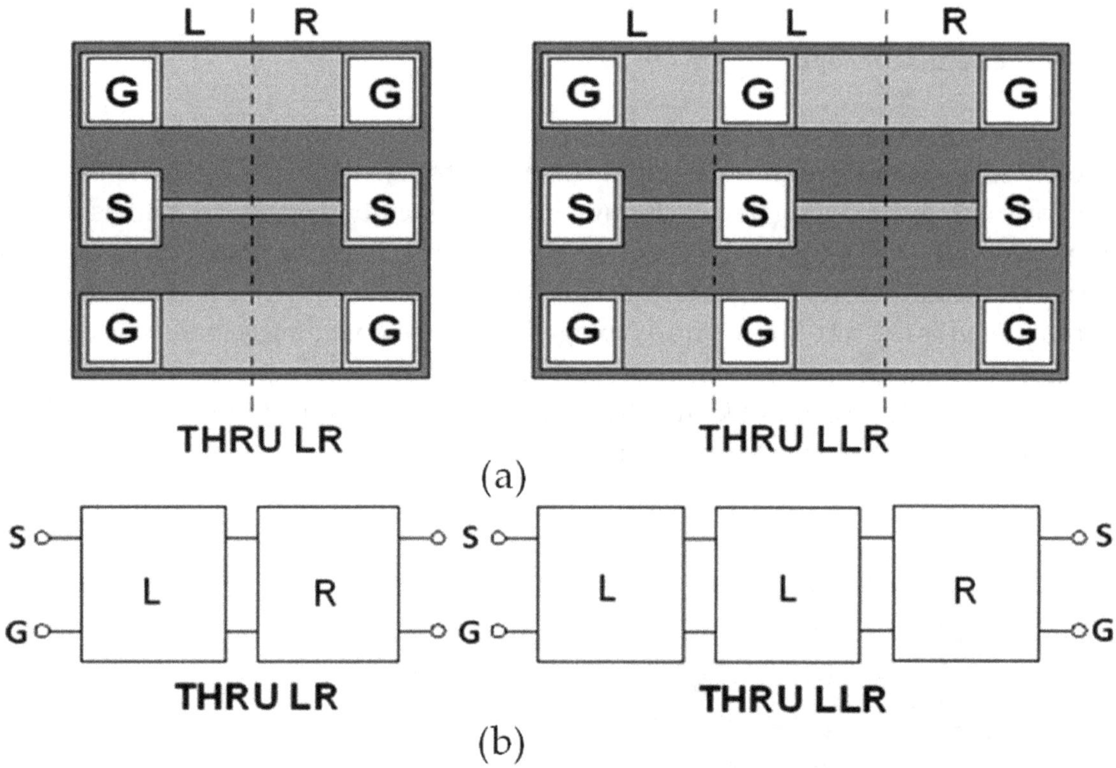

Figure 6. (a) De-embedding structures (THRU LR, THRU LLR) used. (b) They are equivalent to cascade connections of fixture network adapters (L, R) [12].

4.3. De-embedding procedures

The procedure for S-parameter de-embedding is listed as follows:

1. Measure S-parameters of all test structures and convert them into ABCD matrices (A_{DUT}, A_{LR}, and A_{LLR}).

2. Based on the ABCD matrix expression above, determine A_L by $A_{LLR}A_{LR}^{-1}$ and AR by $A_L^{-1}A_{LR}$ with A_L computed.

3. Finally, de-embed both A_L and A_R from measured ABCD matrix of DUT, A_{DUT}, by Eq. (4) to obtain A_{DEV}:

$$A_{DEV} = A_L^{-1} A_{DUT} A_R^{-1} \tag{4}$$

Further correction of forward coupling effect could be achieved with additional OPEN structure as detailed in Ref. [17]. Overall, the generality of the de-embedding technique is improved as the cascade network parameters determined (A_L, A_R) are valid regardless of their internal circuit configuration. Besides that, the de-embedding methodology is greatly simplified as no determination of pad parasitics or transmission line parameters are required.

4.4. De-embedding validation

In order to validate the de-embedding technique presented above, measurements are carried out on test structures using the E8361 PNA and calibrated to probe tips using LRRM technique [18]. The DUT used for validation has symmetrical layout (x1 = x2) whereby both halves of test fixture are mirrored copies of each other. For such condition, the ABCD matrices of L and R parasitic networks are related by $A_L = IA_R^{-1}I$ and vice versa $A_R = IA_L^{-1}I$ where $I = \begin{bmatrix} -1 & 0 \\ 0 & 1 \end{bmatrix}$ is the permutation matrix. Possible extraction errors due to process variations in between the test structures could be minimized by arithmetic averaging of AL (or AR) with $IA_R^{-1}I\left(\text{or}\, IA_L^{-1}I\right)$. Here, the de-embedding validation is performed on 0.13 μm CMOS devices and compared with other techniques [2, 9].

4.4.1. *Verification on zero-length THRU*

The de-embedding accuracy of interconnect and pad parasitics is verified on THRU device of zero electrical length. Theoretically, it exhibits constant transmission coefficient S_{21} of 1 across frequencies since no parasitics associated with its intrinsic behavior. This could be done by applying de-embedding on the THRU LR structure where interconnects at both ports are connected. The deviations of de-embedded S_{21} from theoretical value are shown in **Figure 7**. Note that the de-embedding error by conventional OPEN-SHORT de-embedding technique [2] is relatively higher than the cascade-based methods in [9] and current work due to inaccurate lumped approximations of fixture parasitic. These results show that the proposed de-embedding method is more accurate than [9] at high frequencies (S_{21} error <0.01 at 50 GHz) since no lumped pad assumption is made in the fixture model.

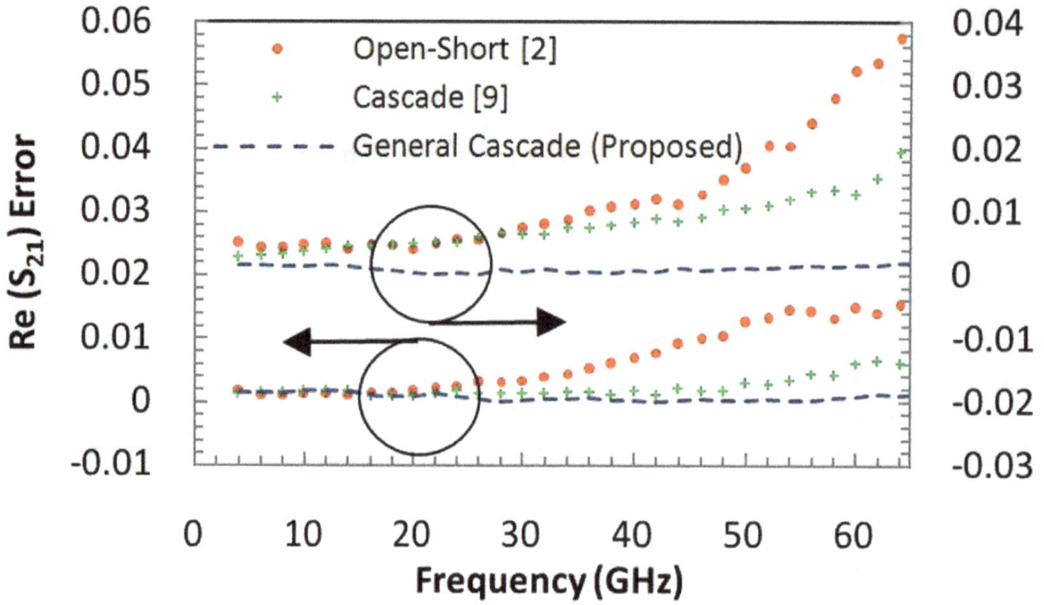

Figure 7. De-embedded S_{21} (real and imaginary) of THRU device versus frequency [12].

4.4.2. Frequency dependencies of transistor parameters

Next, the de-embedding techniques proposed in the current work are validated on frequency dependencies of transistor gate capacitances, C_{gg0}, C_{gd0}, and transconductance, gm. **Figure 8(a)** shows the de-embedded transistor gate capacitances (width = 32 μm, length = 0.13 μm) at zero DC biases where no quasi static effects occur. It could be calculated from $\mathrm{Im}(Y_{11DEV})/\omega$ [19]). Under such circumstances, the transistor gate capacitance exhibits constant behavior across frequencies. The de-embedded C_{gg0} by the de-embedding technique in current work reflects physical behavior described since it is nearly independent of frequencies. It varies only by 2.3% for frequency span of 64 GHz when compared to 10% and 5% by Koolen et al. [2] and Cho et al. [9], respectively. Note that the study by Cho et al. [9] shows more physical results than the study by Koolen et al. [2] as the distributed effects of metal lead are taken into consideration. Still, it demonstrates less physical results than proposed de-embedding technique as the distributed effects of pad are ignored. Meanwhile, the transistor gate drain capacitance could be extracted by $(-\mathrm{Im}(Y_{12DEV})/\omega)$. **Figure 8(a)** demonstrates the de-embedded transistor gate drain capacitances at zero bias. The comparison of de-embedding results shown is consistent with previous on the order of frequency dependency.

The de-embedding validation discussed is further extended to transconductance of transistor in active region (Vgs = Vds = 1.2 V). It could be extracted from real part of Y-parameters, Y_{21}. The comparison results in **Figure 8(b)** show that the de-embedded gm by Koolen et al.[2] and Cho et al. [9] unphysically enhanced over wide range of frequencies despite worsen-ing impact of non-quasi static effects [19]. Meanwhile, the de-embedded gm by the proposed de-embedding technique is more physical as it demonstrates attenuation over 5% at 64 GHz.

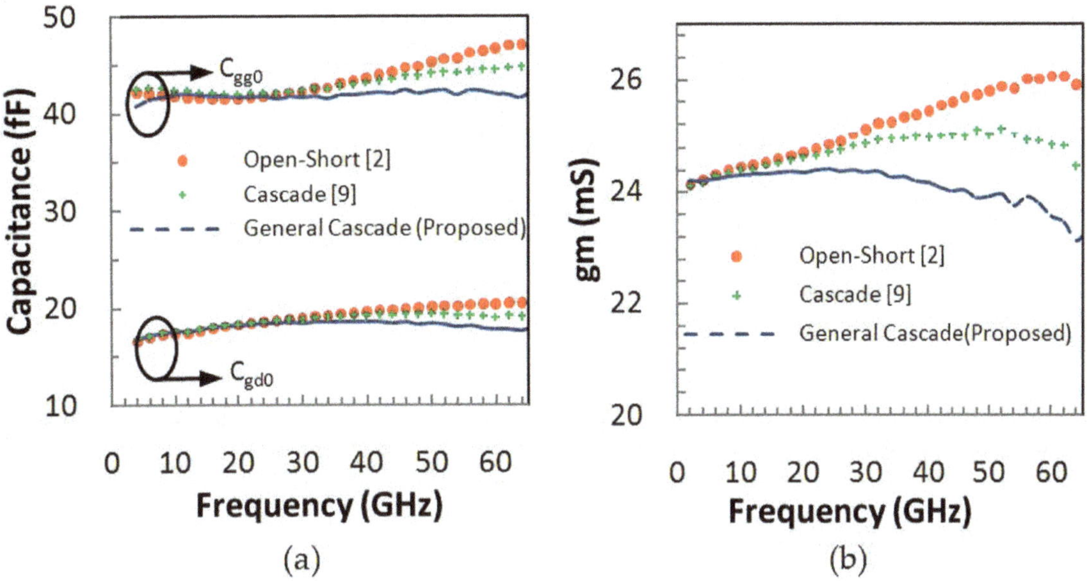

Figure 8. De-embedded parameters of 0.13 μm NMOS transistor versus frequency ((a) C_{gg0}, C_{gd0}, at Vgs = Vds = 0 V and (b) transconductance, gm at Vgs = Vds = 1.2 V) [12].

5. De-embedding metal finger parasitics with hybrid methodology

The de-embedding techniques discussed in previous section focus primarily on removal of pad and interconnect parasitic at top metal level. Nevertheless, it is desirable to establish de-embedding reference plane as close as possible to the device boundary through additional removal of metal fingers parasitic and interconnect via stack. Although several de-embedding techniques [5, 6] have been reported to remove the test fixture parasitics for up to metal fingers, it requires lumped assumptions on interconnects and pad parasitics. Meanwhile, cascade-based de-embedding techniques alone [8–12] are not suitable for removal of metal finger parasitics due to complex inter-couplings between two ports.

In effort to overcome the deficiencies of aforementioned techniques, a hybrid Pad-Line Finger de-embedding technique [13] is presented in this section for mm-wave characterization and modeling of two-port transistor devices. It could simultaneously account for distributed effects of metallic conductors and metal finger parasitics through mix combinations of cascade and lumped series-parallel de-embedding approaches.

5.1. De-embedding concept

Based on the nature of fixture parasitics mentioned, the DUT structure could be described by mix combinations of cascade series-parallel model as shown in **Figure 9**. Similar to de-embedding technique [12] presented in the previous section, metal lines and pads are modeled by generalized cascade network models (PAD, LINE1, 2) to address their transmission line effects. In addition to that, the resistive and coupling characteristics of interdigital fingers are described by series-parallel model (FP, FS).

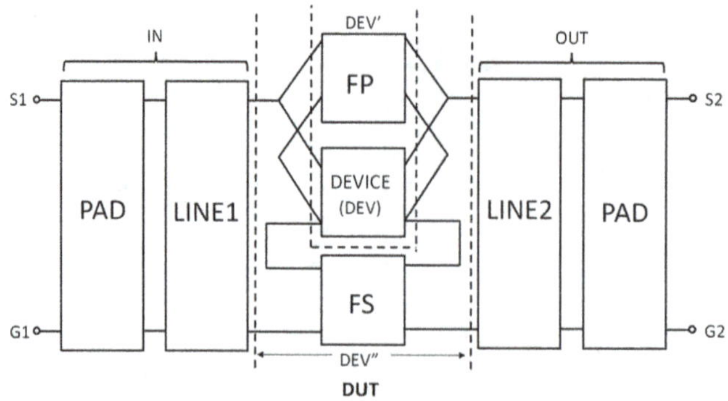

Figure 9. Device under test (DUT) structure modeled as hybrid cascade-series-parallel combinations of two-port device and parasitic networks [13].

5.2. De-embedding structures

Two types of THRU structures (LINE2, PAD-LINE2) are used to extract cascade network parameters of fixture adapters that consist of pad and interconnect parasitics. They consume around 50% less silicon area than those used in the previous section.

As shown in **Figure 10**, the LINE2 structure is equivalent to right half section of DUT structure with pad attached to its left port. On the other hand, the PAD-LINE2 structure differs from LINE2 structure that its left pad is associated with twice of pad length ($2 \times l_{PAD}$). Parasitic extraction for asymmetrical DUT requires additional LINE1 + 2 structure that has total line length of $l_1 + l_2$. It is equivalent to THRU LR structure presented in the previous section. Further extraction of metal finger parasitics is taken care by FINGER OPEN and FINGER SHORT structures. They differ from existing OPEN and SHORT structures that the metal fingers exist in their layouts. Specifically, the FINGER OPEN structure is the same copy of DUT structure but without active region lies underneath the metal fingers. Finally, the FINGER SHORT is similar to FINGER OPEN structure but with source-drain fingers extended and shorted to the gate metals at both ends. Based on the layout configurations, the equivalent network models of the de-embedding structures could be determined as illustrated in **Figure 11**.

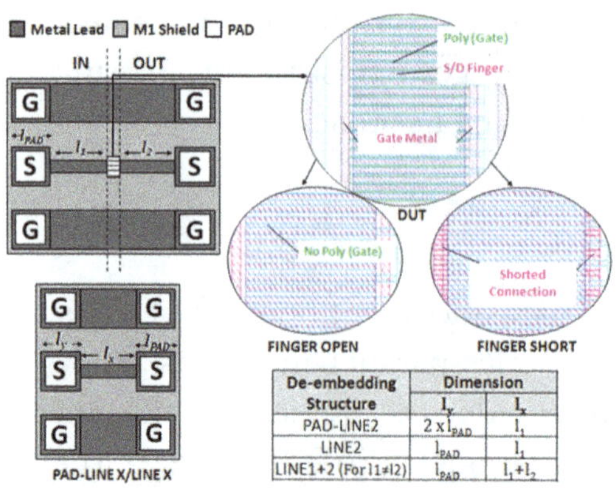

De-embedding Structure	Dimension	
	l_y	l_x
PAD-LINE2	$2 \times l_{PAD}$	l_1
LINE2	l_{PAD}	l_1
LINE1+2 (For l1≠l2)	l_{PAD}	l_1+l_2

Figure 10. Schematic layout of DUT and de-embedding structures used [13].

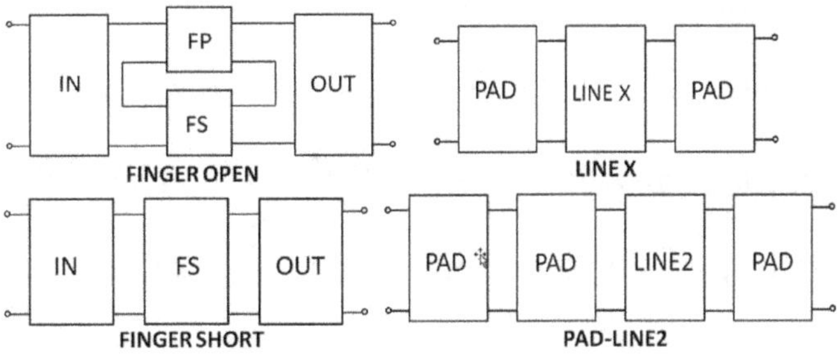

Figure 11. Equivalent parasitic network models of de-embedding structures [13].

5.3. De-embedding procedure

The de-embedding procedure is summarized as follows:

1. Measure S-parameters of all test structures and convert them into cascade ABCD matrices (A_{DUT}, A_{LINE2}, $A_{PAD\text{-}LINE2}$, $A_{FINGER\ OPEN}$, $A_{FINGER\ SHORT}$, and optional $A_{LINE1+2}$).

2. Compute ABCD matrix of input pad by Eq. (5):

$$A_{PAD} = A_{PAD\text{-}LINE2} A^{-1}_{LINE2}. \tag{5}$$

3. Compute ABCD matrix of output half fixture by Eq. (6):

$$A_{OUT} = A^{-1}_{PAD} A_{LINE2} \tag{6}$$

4. Compute ABCD matrix of input half fixture by Eq. (7):

$$A_{IN} = I A^{-1}_{OUT} I \ or A_{IN} = A_{LINE1+2} A^{-1}_{OUT} \text{ for the case of} l1 \neq l2 \text{ where,} I = \begin{bmatrix} -1 & 0 \\ 0 & 1 \end{bmatrix} \tag{7}$$

5. Calculate the resultant ABCD matrix, $A_{DEV''}$ after de-embed the cascade parasitic network components using Eq. (8):

$$A_{DEV''} = A^{-1}_{IN} A_{DUT} A^{-1}_{OUT} \tag{8}$$

Convert the de-embedded results into Z-matrix, $Z_{DEV''}$.

6. Compute ABCD matrix of series parasitic network by Eq. (9):

$$A_{FS} = A^{-1}_{IN} A_{FINGERSHORT} A^{-1}_{OUT} \tag{9}$$

and convert A_{FS} into equivalent Y-matrix, Y_{FS}.

7. Compute Z-matrix of parallel FP network by $Z_{FP} = Z'_{FP} - Z_{FS}$ where ABCD matrix of $Z'_{FP}{}'$ is computed from $A_{IN}^{-1} A_{FINGEROPEN} A_{OUT}^{-1}$. Convert Z_{FP} into equivalent Y-matrix, Y_{FP}.

8. De-embed series metallic parasitic of routing fingers by Eq. (10):

$$Z_{DEV'} = Z_{DEV''} - Z_{FS} \tag{10}$$

9. Finally, de-embed parallel coupling parasitic of fingers by Eq. (11):

$$Y_{DEV} = Y_{DEV'} - Y_{FP} \tag{11}$$

5.4. De-embedding results and discussion

In this section, the de-embedding methodology presented is verified and demonstrated on various performance parameters of the 40 nm CMOS transistor. Specifically, the de-embedding results are supplemented by comparisons with [9] to investigate the impact of metal finger parasitic on transistor de-embedding. The experimental results discussed are based on the same measurement setup as previous with exception that the characterization frequency is further extended up to 100 GHz for benchmarking against [2] and previous work [6].

5.4.1. Verification against electromagnetic simulation

The extracted insertion loss of cascade parasitics at input (IN) and output (OUT) port is validated against electromagnetic simulation (EM) by Integrand's EMX tool. It generates electromagnetic simulations based on boundary element method. As shown in **Figure 12**, the extracted and simulation results agree well with each other. The deviation error of extracted results is within 2% for entire frequency span of 100 GHz.

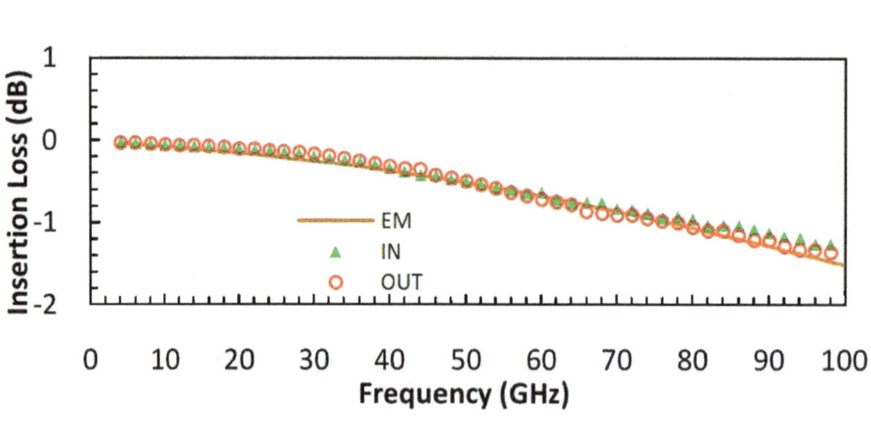

Figure 12. Simulated and extracted insertion loss of input (IN) and output (OUT) network [13].

5.4.2. Scalability of transistor gate capacitance

De-embedding verification on the scalability of transistor gate capacitance provides useful indication on whether the de-embedding technique is correctly applied for up to metal finger. It could be extracted directly from de-embedded Y-parameters ($\mathrm{Im}(Y_{DEV,\,11})/\omega$) [20] as mentioned in previous section. Interestingly, it becomes frequency independent when the non-quasi static effect of the transistor is negligible. This occurs when the transistor is in cut-off mode since no transcapacitance exists in between its gate and drain terminals [21]. **Figure 13** shows the frequency characteristics of de-embedded gate capacitances across different width geometries of transistors at zero bias and frequency of 8 GHz. They are compared against simulation results by GLOBALFOUNDRIES' 40 nm CMOS model. The results clearly show that the de-embedded transistor gate capacitance by the proposed distributed hybrid de-embedding method is scalable and closely agrees with the simulation results. Meanwhile, the de-embedded results by existing cascade-based de-embedding [9] demonstrate larger gap and less scalability. This shows that exclusion of metal finger parasitics in de-embedding has clear impact even at low frequency.

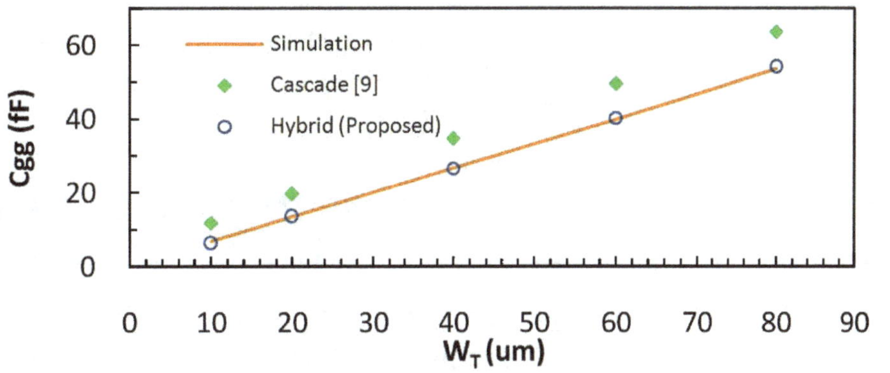

Figure 13. Simulated and de-embedded total gate capacitances of 40 nm NMOS transistor across width geometries at Vgs = 1.1 V, Vds = 1.1 V, and 8 GHz [13].

5.4.3. Frequency variability of de-embedded transistor parameters

In this section, the frequency variability of de-embedded transistor parameters is examined for up to 100 GHz. **Figure 14** shows the comparison of extracted transconductance (Real(Y21)) by different de-embedding techniques for reference plane established underneath metal fingers. As compared to Refs. [2, 6], the extracted transconductance by the proposed de-embedding technique is almost constant with variation of only 1 ms at 100 GHz. The result is physical as non-quasi static effect of 40 nm transistor is negligible for frequencies below 100 GHz and at strong inversion regime [22]. The frequency variability of de-embedded transistor gate capacitance is further examined for drain bias at 0 V. In such case, the characteristic of CMOS transistor could be described by a passive capacitor. As demonstrated in **Figure 15**, the proposed hybrid de-embedding technique has shown to be more robust than previous work by Loo et al. [6] and Koolen et al. [2] as the maximum variation of de-embedded C_{gg} is within 3% for frequency span of 100 GHz. Similar to previous verification results, the larger de-embedding error exhibited by Koolen et al. [2] for frequencies beyond 30 GHz indicates that the behavior of test fixture parasitics could no longer be sufficiently described by single stage of parallel-series lumped equivalent circuit. Although greater improvement has been made by Loo et al. [6] with more comprehensive test fixture model, still it suffers larger error than the proposed de-embedding technique due to lumped approximation of metallic conductors. Finally, the frequency dependency of de-embedded maximum current gain bandwidth product ($H_{21}x\Delta f$) is examined. Theoretically, it is frequency invariant since the maximum current gain of CMOS transistor degrades at constant rate of 20 dB/decade. As illustrated in **Figure 16**, the extracted gain bandwidth product by the proposed hybrid de-embedding technique is more physical since it is almost frequency independent with only 2.1% variation from low frequency value at 100GHz. Comparatively, the de-embedded results by [2, 6] reveal more than 4% error at 100 GHz due to the aforementioned reasons. Note that the transistor gain parameter is more immune to de-embedding error in [2] as it is function of ratio in between transconductance and total gate capacitance.

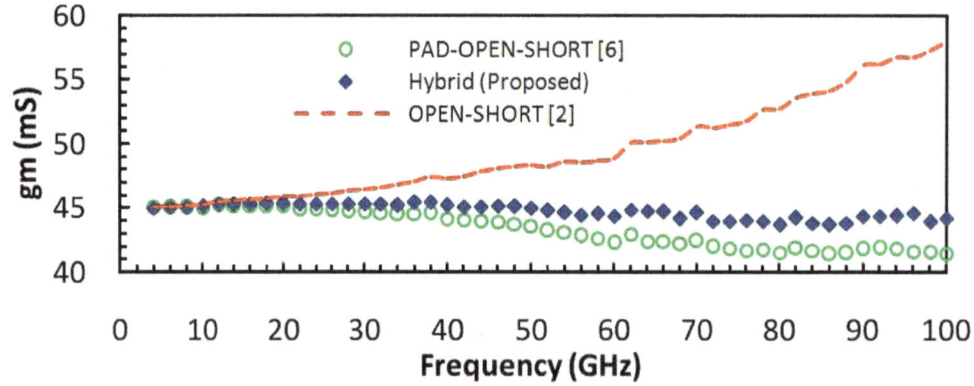

Figure 14. De-embedded transconductance, gm of 40 nm NMOS transistor (W_T = 40 μm) at Vgs = 1.1 V, Vds = 1.1 V [13].

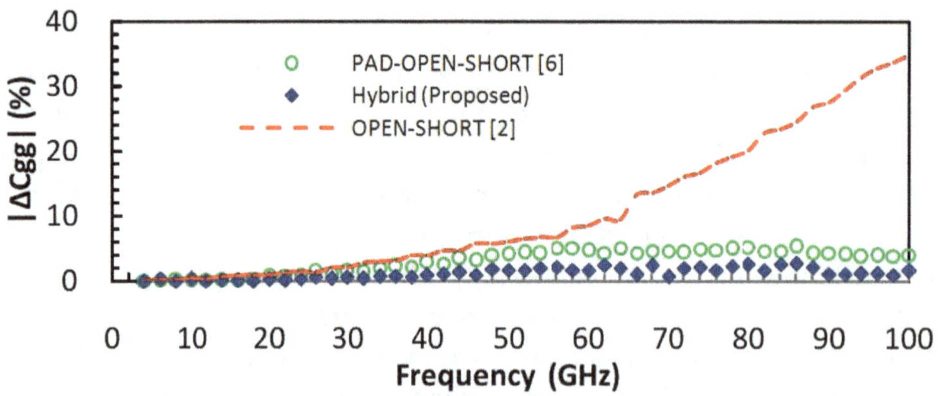

Figure 15. Frequency variability of de-embedded total gate capacitance C_{gg} of 40 nm NMOS transistor (W_T = 40 μm) at Vgs = 1.1 V, Vds = 0 V [13].

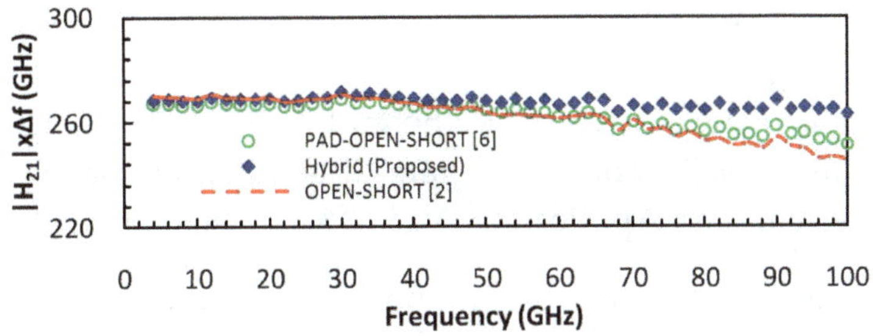

Figure 16. De-embedded maximum current gain bandwidth product, |H21|*Δf of 40 nm NMOS transistor (W_T = 40 μm) at Vds = 1.1 V.

6. Conclusion

This chapter presents a generalized network de-embedding technique that avoids any inaccurate lumped and transmission line assumptions on the pad and interconnects of the test structure. The de-embedding strategy has been validated by producing negligible de-embedding error (<–50 dB) on the insertion loss of the zero-length THRU device. It demonstrates better accuracy than existing de-embedding techniques that are based on lumped pad assumption. For transistor characterization, the de-embedding reference plane could be further shifted to the metal fingers with additional Finger OPEN-SHORT structures. The resulted de-embedded RF parameters of CMOS transistor show good scalability across geometries and negligible frequency dependency of less than 3% for up to 100 GHz. The de-embedding findings suggest that the parasitic effects of metal fingers could not be ignored in modeling of intrinsic transistors. Also, the distributed effect of metallic conductors has to be considered when the line length is comparable to 1/20 of frequency wavelength.

Acknowledgements

The authors like to thank GLOBALFOUNDRIES Singapore Pte Ltd for providing tape-out resources and measurement support.

Author details

Xi Sung Loo[1*], Kiat Seng Yeo[2] and Kok Wai Johnny Chew[1]

*Address all correspondence to: xisung.loo.phd@ieee.org

1 GLOBALFOUNDRIES Singapore Pte Ltd, Singapore

2 Singapore University of Technology & Design, Singapore

References

[1] Verma, P. R.; Shaoqiang, Z.; Wai, C. K.; Boon, T. J. and Nair, R. "Foundry RF technologies," Proceedings of Technical Program–2014 International Symposium on VLSI Technology, Systems and Application (VLSI-TSA), Hsinchu, 2014, pp. 1–2.

[2] M. C. A. M. Koolen, J. A. M. Geelen and M. P. J. G. Versleijen, "An improved de-embedding technique for on-wafer high-frequency characterization," Bipolar Circuits and Technology Meeting, 1991., Proceedings of the 1991, Minneapolis, MN, 1991, pp. 188–191.

[3] Cho, H.; Burk, D. E. "A three-step method for the de-embedding of high-frequency S-parameter measurements," IEEE Transactions on Electron Devices, vol. 38, no. 6, pp. 1371–1375, 1991.

[4] T. E. Kolding, O. K. Jensen and T. Larsen, "Ground-shielded measuring technique for accurate on-wafer characterization of RF CMOS devices," Microelectronic Test Structures, 2000. ICMTS 2000. Proceedings of the 2000 International Conference on, Monterey, CA, 2000, pp. 246–251.

[5] KangI. M. ; Jung, S. J.; Choi, T. H.; Jung, J. H.; Chung, C.; Kim, H. S.; Oh, H.; Lee, H. W.; Jo, G.; Kim, Y. K.; Kim, H. G.; Choi, K. M. "Five-step (Pad CPadShort CPadOpen CShort COpen) de-embedding method and its verification," IEEE Electron Device Letters, vol. 30, no. 4, pp. 398–400, 2009.

[6] Loo, X. S. et al."An accurate two-port de-embedding technique for RF/Millimeter-wave noise characterization and modeling of deep submicrometer transistors," in: IEEE Transactions on Microwave Theory and Techniques, vol. 59, no. 2, pp. 479–487, 2011.

[7] Liang, Q.; Cressler, J. D.; Niu, G.; Lu, Y.; Freeman, G.; Ahlgren, D. C.; Malladi, R. M.; Newton, K.; Harame, D. L. "A simple four-port parasitic deembedding methodology for high-frequency scattering parameter and noise characterization of SiGe HBTs," IEEE Transactions on Microwave Theory and Techniques, vol. 51, no. 11, pp. 2165–2174, 2003.

[8] Chen, C. H.; Deen, M. J. "A general noise and S-parameter deembedding procedure for on-wafer high-frequency noise measurements of MOSFETs," IEEE Transactions on Microwave Theory and Techniques, vol. 49, no. 5, pp. 1004–1005, 2001.

[9] Cho, M. H.; Huang, G. W.; Wang, Y. H.; Wu, L. K. "A scalable noise de-embedding technique for on-wafer microwave device characterization," IEEE Microwave and Wireless Components Letters, vol. 15, no. 10, pp. 649–651, 2005.

[10] Cho, M. H.; Chen, D.; Lee, R.; Peng, A. S.; Wu, L. K. and Yeh, C. S., "Geometry-scalable parasitic deembedding methodology for on-wafer microwave characterization of MOSFETs," IEEE Transactions on Electron Devices, vol. 56, no. 2, pp. 299–305, 2009.

[11] Loo, X. S.; Yeo, K. S. and Chew, K. W. J. "THRU-based cascade de-embedding technique for on-wafer characterization of RF CMOS devices," IEEE Transactions on Electron Devices, vol. 60, no. 9, pp. 2892–2899, 2013.

[12] Loo, X. S.; et al. "A new millimeter-wave fixture deembedding method based on generalized cascade network model," IEEE Electron Device Letters, vol. 34, no. 3, pp. 447–449, 2013.

[13] Loo, X. S.; Chew, K. W. J.; Yeo, K. S.; Win, M. Z. and Boon, C. C. "A hybrid pad-line-finger de-embedding technique for broadband modeling of CMOS transistor," IEEE Microwave and Wireless Components Letters, vol. 26, no. 7, pp. 507–509, 2016.

[14] Kolding, T. E. "Shield-based microwave on-wafer device measurements," IEEE Transactions on Microwave Theory and Techniques, vol. 49, no.6, pp. 1039–1044, 2001.

[15] Williams, D. F.; et al."Calibrations for millimeter-wave silicon transistor characterization," IEEE Transactions on Microwave Theory and Techniques, vol. 62, no. 3, pp. 658–668, 2014.

[16] Song, J.; Ling, F.; Flynn, G.; Blood, W. and Demircan, E. "A de-embedding technique for interconnects," Electrical Performance of Electronic Packaging, Cambridge, MA, pp. 129–132, 2001.

[17] Loo, X. S.; et al."A cascade-parallel based noise de-embedding technique for RF modeling of CMOS device. IEEE Microwave and Wireless Components Letters, vol. 21, no. 8, pp. 448–450, 2011.

[18] Davidson, A., Jones, K. and Strid, E. "LRM and LRRM calibrations with automatic determination of load inductance," ARFTG Conference Digest-Fall, 36th, Monterey, CA, USA, pp. 57–63, 1990.

[19] Cheng, Y. and Matloubian, M., "Frequency-dependent resistive and capacitive components in RF MOSFETs," IEEE Electron Device Letters, vol. 22, no. 7, pp. 333–335, 2001.

[20] Je, M.; Han, J.; Shin, H.; Lee, K. "A simple four-terminal small signal model of RF MOSFETs and its parameter extraction," Microelectronics Reliability., vol. 43, no. 4, pp. 601–609, 2003.

[21] Kwon, I.; Je, M.; Lee, Kwyro; Shin, H. "A simple and analytical parameter-extraction method of a microwave MOSFET," IEEE Transactions on Microwave Theory and Techniques, vol. 50, no. 6, pp. 1503–1509, 2002.

[22] H. Hassan, M. Anis and M. Elmasry, "Impact of technology scaling on RF CMOS," *SOC Conference, 2004. Proceedings. IEEE International*, Santa Clara, CA, 2004, pp. 97–101.

3

LTCC-Based System-in-Package (SiP) Technology for Microwave System Applications

Young Chul Lee

Abstract

Monolithic low-temperature co-fired ceramic (LTCC) SiP modules have been presented for microwave applications. In order to integrate almost passive circuits of a radio system into the LTCC substrate, key technologies such as suppressing parasitic resonant modes, low-loss transitions and compact passive devices have been investigated. Well analyzed mechanisms on the parasitic resonant modes and their suppressing methods have been applied to high-isolation SiP structures. A strip line (SL) to CPW vertical transition using a stepped via structure embedding air cavities has been devised and has been used to design a SL BPF. A surface mount technology (SMT) pad transition has been developed by utilizing a modified coaxial line. A LPF composed of vertical plate capacitors and helical inductors and a 2 × 2 array antenna have been developed. A 61 GHz heterodyne transmitter LTCC SiP module has been implemented by monolithically embedding all passive circuits such as a SL BPF, 2 × 2 array antenna, SMT pads and feeding lines into it. A 60 GHz amplitude shift-keying (ASK) transceiver LTCC SiP module has been implemented as small as 17.8 × 17.9 × 0.6 mm^3 by integrating a high-isolation via fence and a LPF. They have been characterized in terms of an output power, spectrum and link test.

Keywords: LTCC, SiP, transition, BPF, antenna

1. Introduction

In general, the final stage for implementation of microwave systems is system integration. It could be either integrated circuit (IC) die-level integration or package level one. And also, how to integrate microwave systems should be considered at the first stage of its development. Roughly, there are two categories for an integration technology of microwave systems: on-chip technology and in-package one. The progress in these integration technologies has been formidable over the past decade for microwave applications. In these technologies, the main key issue is antenna integration on the chip or in the package, because its integration can provide several advantages such as low cost, compact size, high reliability and high reproducibility.

In the case of on-chip integration technologies based on complementary metal-oxide-semiconductor (CMOS), bipolar CMOS (BiCMOS) and silicon germanium (SiGe) BiCMOS, various researches have been explored up to recently. However, due to high permittivity (~11.7) and low resistivity (~10 Ω cm) [1, 2] of the Si-based substrate, integrated antennas have been suffered from low radiation efficiency and matching bandwidth. In order to modify properties of the substrate, although several novel techniques such as air cavity [3], artificial magnetic conductor [4], suspended membrane [5], ion implantation [6] and meta-surface [7] have been investigated; high-gain and broadband requirements are still below expectations, compared to those of package-based integration technologies [1, 8, 9].

In the package-based integration technologies, aside from highly integrated radio dies, almost passive components such as antenna, BPF and others are monolithically realized in the package substrate, thanks to three-dimensional (3D) stacking capability using via, multilayer and cavity. Very compact and high-performance microwave modules have been recently presented by using multilayer ceramic or printed circuit board (PCB) [10] technologies. Low-temperature co-fired ceramic (LTCC) [11–13] is a representative multilayer ceramic technology. For the past two decades, several LTCC modules involving antennas and other passive circuits with microwave radio chips have been developed and also it is possible to integrate high-gain antennas because of its properties of low-loss substrate and metallization (Ag).

In this chapter, highly integrated monolithic SiP modules have been presented for microwave system applications. In order to integrate radio systems in the single LTCC package module, key technologies such as suppressing parasitic resonant modes, low-loss transitions and compact and high-performance passive devices have been investigated after definition of monolithic SiP module. Finally, a 61 GHz transmitter (Tx) LTCC SiP module and a 60 GHz ASK transceiver one have been implemented in a size of 36 × 12 × 0.9 and 17.8 × 17.9 × 0.6 mm^3, respectively and they have been characterized in terms of an output power, spectrum and link test.

2. Monolithic SiP module

Figure 1 shows a three-dimensional (3D) schematic concept of the monolithic LTCC SiP module integrating a whole radio system consisting of active ICs and passive components in the single LTCC substrate. The filter and antenna are monolithically integrated in the LTCC dielectric and on its top layer, respectively. Active ICs are also mounted on the top of the LTCC multilayers. For miniaturization of the SiP module, passive circuits such as filters, antenna, surface mounted technology (SMT) pads, DC bias feedings and transmission lines are vertically or horizontally deployed by using vertical via interconnections, internal ground plane (I-GND) and signal line transitions in the substrate. Several transmission lines such as a strip line (SL), conductor-backed coplanar waveguide (CB-CPW), or microstrip line (MSL) are utilized within the LTCC block or on its top layer. By considering device and interconnection structures to be integrated in the SiP module, suitable transmission line is designed.

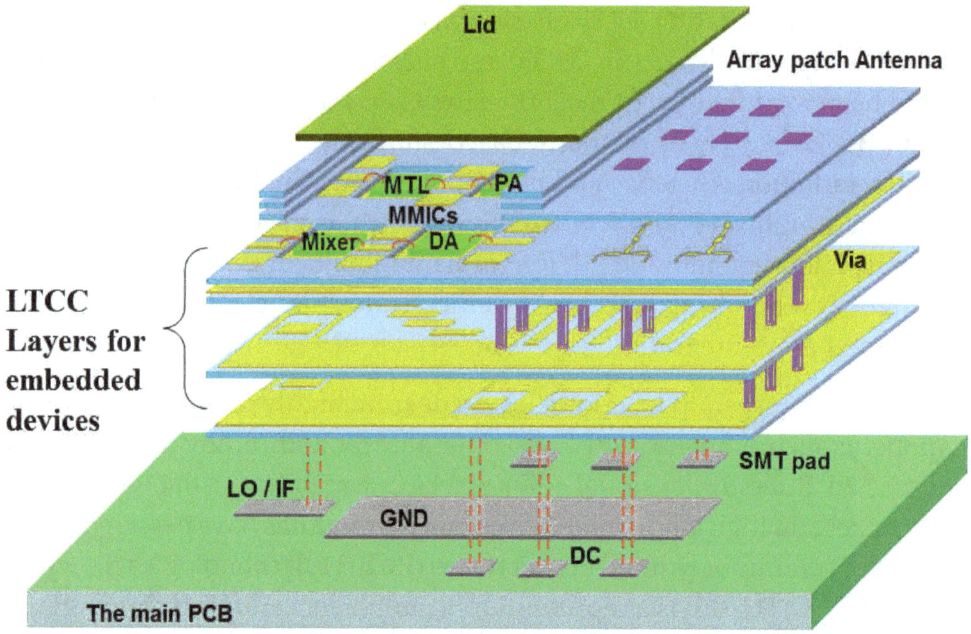

Figure 1. Three-dimensional (3D) schematic concept of a monolithic SiP module.

However, because various integrated transmission lines, passive devices and transitions are compactly integrated in the small area of the monolithic LTCC SiP module, they can make parasitic propagating structures and lead to unwanted cross talk issue. So, suppression of the parasitic resonant modes is one of the key issues in the design of SiP module. In addition, attenuation in interconnections between different signal lines or inter layers should be minimized for power efficiency and noise performance of the SiP module. And also, high-performance and compact passive devices should be designed.

3. Suppression of parasitic resonant modes

A CB-CPW and SL are in general used as signal lines of a SiP module because of their low dispersion and radiation. The CB-CPW consists of a lower and upper ground plane, embedded vias and a signal line (W) and the SL as shown in **Figure 2A**. However, these ground planes and vias can make parasitic resonant circuits such as rectangular waveguide (R-W/G), a parallel-plate waveguide and a patch antenna [14, 15] and they cause undesired resonant modes. An input signal is coupled by the gap of the CB-CPW, propagates through the parallel plate and finally radiates due to a parasitic patch resonator. In the case of the SL as shown in **Figure 2B**, it is basically a buried device and its structure is also composed of a lower and upper ground plane, vias and a strip. The ground planes and vias generate the parasitic R-W/G, which is analogous to that of the CB-CPW.

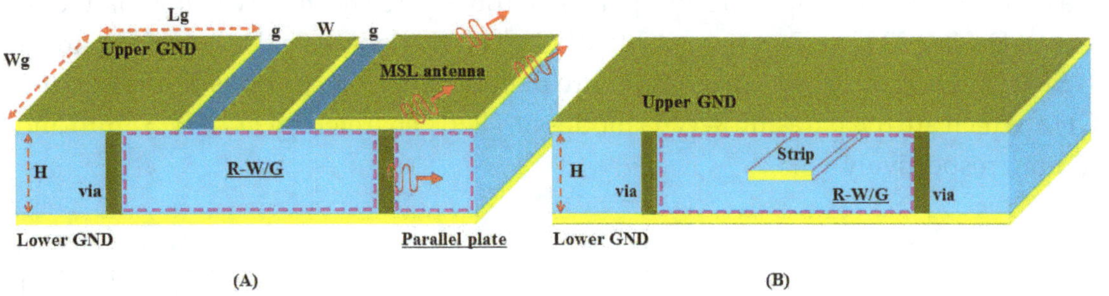

Figure 2. Structure of a CB-CPW (A) and SL (B) involving parasitic resonant circuits.

The resonant frequency (f_{WG}) due to the R-W/G is given by Pozar [16],

$$c_{p\ sample} = \frac{Q_{sample}}{Q_{ref}} \frac{m_{ref}}{m_{sample}} C_{p\ ref} \tag{1}$$

where D is the spacing between vias and it is the same as a horizontal dimension of a rectangular waveguide, ε_o is the permittivity of free space, μ_o is the permeability of free space and ε_r is the relative dielectric constant of the substrate.

The resonant frequency (f_{pr}) due to the surface ground planes is similar to that of the simple patch antenna [14, 15],

$$f_{pr} = \frac{c}{2W_g} \left(\frac{\epsilon\_r + 1}{2} \right)^{-1/2} [\text{GHz}] \tag{2}$$

where W_g is a width of a rectangular patch and c is the speed of light.

Figure 3 shows measured insertion loss characteristics of a 50 Ω CB-CPW line (A) and a SL BPF (B) fabricated in the LTCC substrate. Unwanted resonant modes are distinctly observed in the CB-CPW and SL BPF. In the case of the CB-CPW as shown in **Figure 3A**, its width, gap and length are 300, 150 and 3260 μm, respectively and its substrate height is 400 μm. Vias in order to short the lower and upper ground planes are placed at the both sides with a distance (D) of 1620 μm. The parasitic R-W/G is generated in the CB-CPW due to the ground planes and vias. In these dimensions of the CB-CPW, a propagation mode of the parasitic R-W/G is generated at 34 GHz. Resonances at 19.5 and 39 GHz are due to a parasitic patch antenna mode and its harmonic, respectively. Some input signal in the CB-CPW propagates to the parasitic R-W/G by coupling through its gap, passes through a parallel-plate W/G and finally radiates from the parasitic patch antenna. **Figure 3B** presents tested insertion losses of the 40 GHz SL BPF. The BPF was realized with six-stacked LTCC layers. Among the six-stacked layers, the SL filter was placed on the third layer and the CPW pads are on the top layer for on-wafer probing.

Both top and bottom ground planes were connected to each other through ground vias to equalize the electric potential. The inset of **Figure 3B** shows the critical two dimensions (D_{CPW} and D_{SL}), which cause resonant phenomena. D_{CPW} of 1.15 mm is the distance between embedded vias. The lower and upper ground planes of the SL are shorted by using via blocks. The distance of D_{SL} is 2.32 mm. Two parasitic R-W/Gs with the lengths of D_{CPW} and D_{SL} are created in this SL BPF. Therefore, its performances are degraded due to spurious responses at 36 and 47.6 GHz, which are the parasitic R-W/G modes of TE20 generated by D_{SL} of 2.32 mm and D_{CPW} of 1.15 mm, respectively.

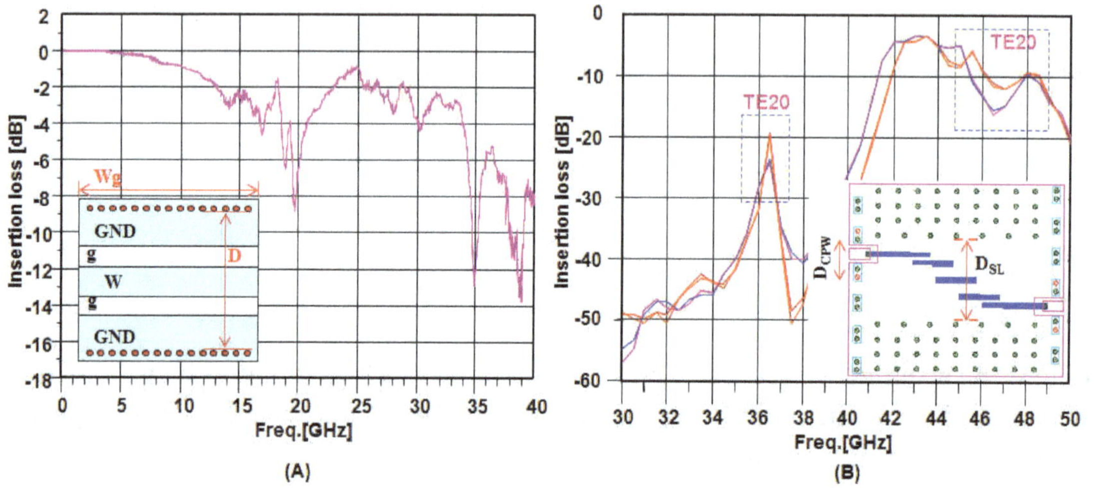

Figure 3. Unwanted resonance modes due to parasitic structures in the fabricated CB-CPW (A) [an inset: the top view of the CB-CPW] and the fabricated SL BPF (B) [the inset: the layout of the BPF].

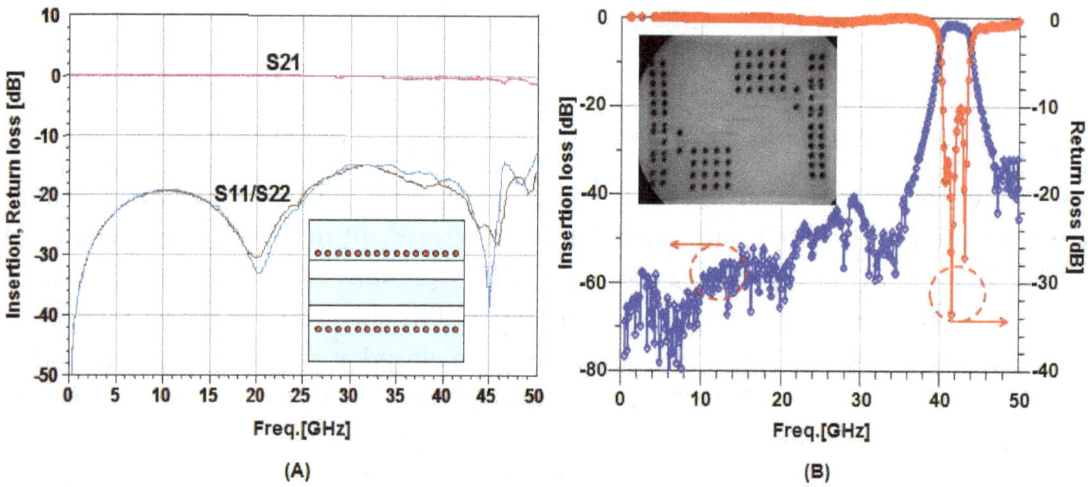

Figure 4. Resonance-free CB-CPW (the inset: the top view of the CB-CPW) (A) and SL BPF (the inset: X-ray photo of the fabricated SL BPF) (B).

The parasitic R-W/G modes (f_{WG}) in operating frequency band can be successfully suppressed by placing the vias within shorter distance than the calculated value from Eq. (1), because parasitic propagating modes are generated in the higher frequency than that of an interesting band. By using this method, the CB-CPW and SL BPF were modified and fabricated. Their measured characteristics are indicated in **Figure 4**. The modified via placement for the CB-CPW and the SL BPF is presented in each inset of **Figure 4A** and **B**, respectively. The parasitic resonant modes are clearly suppressed in the operation frequency region [17]. In the CB-CPW, the distance (D) of 1620 μm is shortened to 700 μm and its layout and measured results are shown in **Figure 4A**. For the SL BPF, D_{CPW} of 1.15 mm is modified to 760 μm, which corresponds to the parasitic rectangular WG mode of TE10 at 76 GHz. One of the two GND-via blocks facing each other is deleted in order not to make the parasitic R-W/G in the SL structure. The X-ray photo of the fabricated SL BPF and the measured loss characteristics are presented in **Figure 4B**.

4. Low-loss transitions for 3D integration

For the 3D integration of microwave radio systems, several vertical transitions such as CPW-to-SL transition [18, 19], MSL-to-SL transition [20, 21], CPW-to-CPW transition [18, 22] and coaxial-like surface mount technology (SMT) pad transition [23] have been developed. These transitions allow the integration of passive and active circuits to be placed in inner layers or mounted on the top layer. The main issue for **3D** interconnection using transitions is to reduce attenuation and discontinuity. In particular, radiation due to structural discontinuity causes cross talk issue. Therefore, several attempts have been tried in order to remedy problems due to discontinuities. In order to improve impedance matching or compensate parasitics, the coaxial-like transition and intermediate ground planes have been utilized [18–23].

4.1. Strip line (SL)-to-CPW vertical via transition

In the vertical via transitions, the total physical height of the directly stacked vias has a decisive effect on their RF performance besides capacitive or inductive effects in the transition region. In the case of its height over one tenth of the wavelength (>0.1 λ) in the mm-wave frequencies, input signals can be significantly radiated or reflected.

Figure 5 shows proposed SL-to-CPW transition and its simulated performances [24]. A cross-sectional structure of the proposed SL-to-CPW via transition is shown in **Figure 5A**. Each LTCC layer is 100 μm high. The CPW on the top layer is connected with the embedded SL on the 4th layer (L4) by using vertical vias, which are subdivided into three-stepped one. Because of the proposed three-stepped via structure, the critical dimension, which mainly causes the physical discontinuity, is decreased from 300 to 100 μm. The wavelength (λ) on the LTCC CPW with relative dielectric constant of 7.0 is 2.56 mm at 60 GH. Its rate of the critical dimension to λ is decreased from 11 to 3.9%, respectively. However, it leads to the increase in the shunt capacitance between the vias and SL ground planes. For reduction in the increased shunt capacitance, the embedded air cavities are inserted below the stepped vias. In order to evaluate the proposed vertical via transition, the SL-to-CPW vertical transition is designed in back-to-back type as shown in **Figure 5B**. The ground planes of the CPWs and SL are connected by shielding vias. In order to design the 50 Ω CPW and SL, the CPW with the width of 250 μm and gap of

99 μm is designed and the width of the SL is 135 μm. For comparison purposes, the conventional transition using the directly stacked vias is also designed. For the conventional transition, two 526 μm-long CPW lines are connected with the 2650 μm-long SL. In the proposed one, two 526 μm-long CPW lines and 2050 μm-long SL are used. The air cavities are embedded through the 2nd to 5th layer below the 7th layer via and through the 2nd to 3rd layer below the 5th layer via. By using a 3-D finite integration technique (FIT) simulator [25], all transitions have been designed and analyzed. Calculated results of the proposed transition in comparison with the conventional one are presented in **Figure 5C** and **D**. The proposed transition shows better performance in terms of return and insertion loss than that of the conventional one because of reduced via discontinuities. For quantitative analysis of improved performance, radiation losses ($1-|S_{11}|^2-|S_{21}|^2$) of the SL-to-CPW vertical transitions are calculated by using the simulated insertion and return loss and are illustrated in **Figure 5D**. At 60 GHz, radiation of the proposed transition is reduced by 23 and 62% by using the three-stepped via structure and embedded air cavities, respectively, compared to the conventional one.

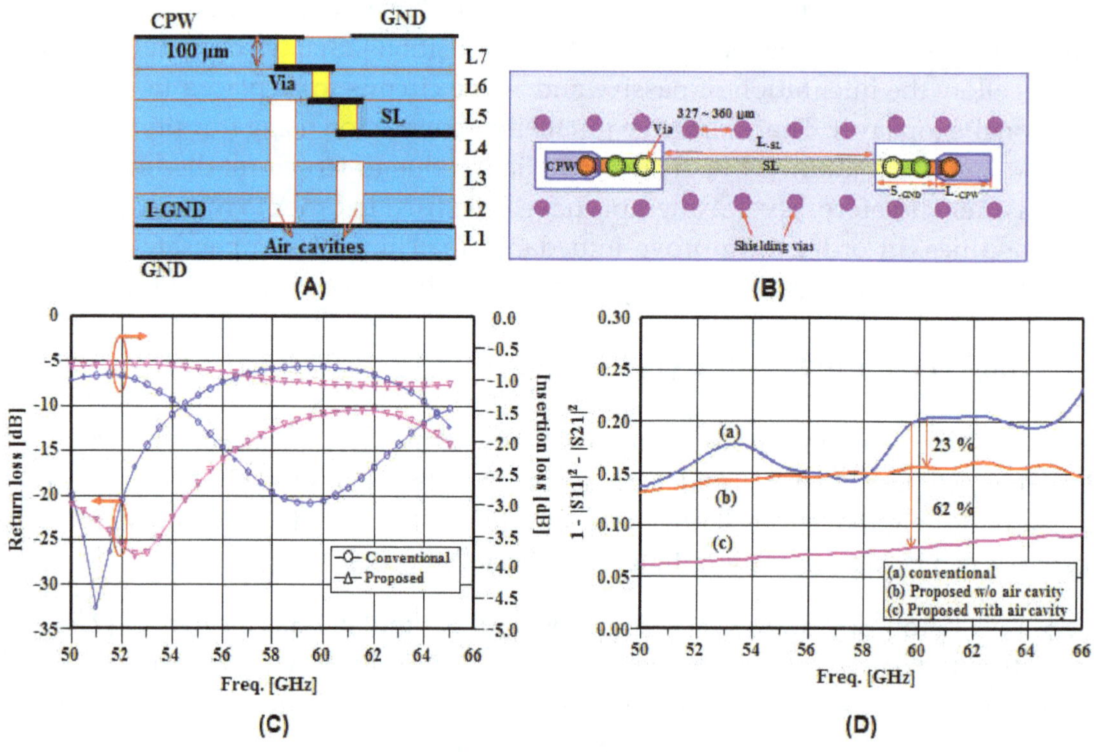

Figure 5. Proposed SL-to-CPW via transition and its simulated performances [(A): its cross-sectional view (Lx: the number of LTCC layers), (B): its layout in a back-to-back structure, (C): return and insertion loss characteristics and (D): radiation loss ones].

The designed transitions in the back-to-back structure were fabricated using seven-layered LTCC substrate and the fabricated ones were characterized by using a probing method as shown in **Figure 6**. For the proposed transition, embedded air cavities are clearly formed below the 5th and 7th layer via as shown in an inset of **Figure 6B** and also its S_{11} and S_{21} characteristics are improved compared to the conventional one. The measured S_{11} and S_{21} of the CPW-SL-

CPW are less than −10 dB and −2.0 dB, respectively, from 50 to 65 GHz. In particular, its low S_{21} of −1.6 dB is achieved at 60 GHz. These values represent all losses along the three-segment transmission lines and the two vertical via transitions. Considering the total loss of transmission lines with −0.19 dB, which is calculated by using a conventional line calculator, the transition loss per a ST transition is 0.7 dB at 60 GHz.

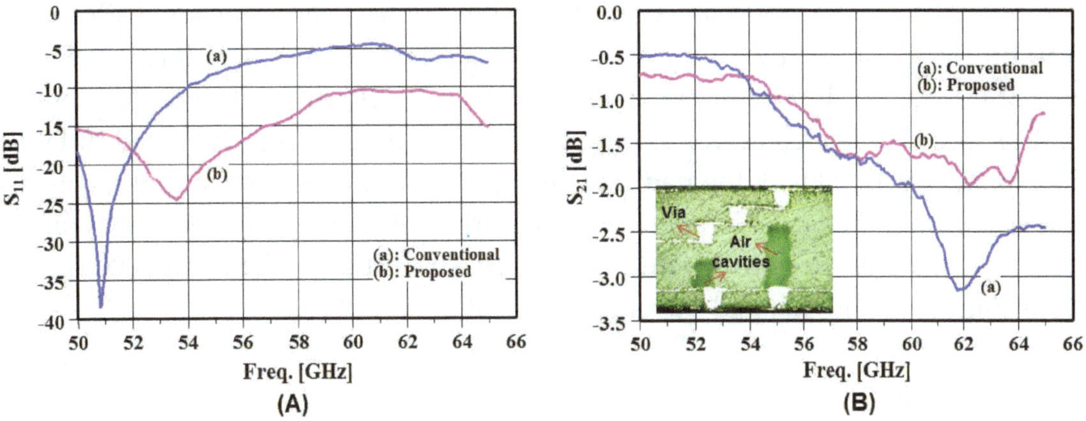

Figure 6. Measured return losses (A) and insertion ones (B) of the fabricated vertical via transitions in the back-to-back type using the proposed transition and the conventional one (an inset: cross-sectional views of the fabricated proposed one).

4.2. SMT pad using a coaxial-line transition

In general, LTCC SiP modules have been developed in type of SMT package because of low cost and easy assembly. In order to mount LTCC SiP modules on the main board, the low-loss transition between the signal port and SMT pad is required. In addition, the excitation of package modes [26] should be investigated for millimeter-wave applications. Typical transitions from I/O (input and output) port to the SMT pad in the SiP module have been designed by using vertical via structures [18–24]. In the case of long transition with several stacked vias over seven layers (>0.7 mm), it is difficult to control discontinuities and radiation. Therefore, by using a coaxial-line structure, the SMT pad has been implemented for the SMT LTCC SiP applications.

In order to suppress radiation in the long transition with several stacked vias, a SMT pad transition using a coaxial-line structure is proposed. By using a commercial tool [25], a CPW-to-coaxial line-to-SMT pad transition in a nine-layer LTCC substrate has been designed and its structure and designed results are presented in **Figure 7A**, **B** and **C**, respectively. The relative dielectric constant of the LTCC substrate is 7.8 at 20 GHz. Each layer is 100 μm thick. The total height of the vertical vias in the transition region is 700 μm. Because of a bulky structure of a 50 Ω coaxial line, an inner and outer diameter is optimized in terms of its transition loss and size. Considering losses in the transition, the impedance of the coaxial line of 37 Ω is determined. The diameter of the inner conductor (via) is fixed in 135 μm. The diameter and width of the outer conductor are 695 and 235 μm, respectively. The CPW line is designed in the cavity on L7 for interconnection with other devices or measurement. An embedded CPW (ECPW) between the coaxial line and CPW is designed on the same layer for their interconnection. The CPW line is with a 144 μm wide strip and a gap of 83 μm. The width and gap of the ECPW are 90 and 95 μm, respectively. Because of the overlapped part between the outer conductors of the coaxial line and the CPW line on the PCB board, E-fields are concentrated and consequently, a significant amount of reflection can be generated. The overlapped part, which is coaxial-line outer conductor to the left of the SMT pad, is cut off. In addition, in order to suppress the radiation at the bending part of interconnection between the coaxial line and CPW, a semicircular cap on the 9th layer (L9) of the LTCC substrate is designed. It radius is 928 μm. The modified SMT pad transition is shown in **Figure 7A**. Its E-field distribution at its cross section and its designed characteristics, comparing the effect of the cap are illustrated in **Figure 7B** and **C**, respectively. E-fields of the pad transition with a cap (WC) are confined between the cap and coaxial cable in the LTCC substrate in **Figure 7B**. From 8 to 20 GHz, the return losses (S_{11} and S_{22}) and insertion losses (S_{21} and S_{12}) of the pad transition with the cap (WC) are clearly improved compared with the pad transition without the cap (WOC).

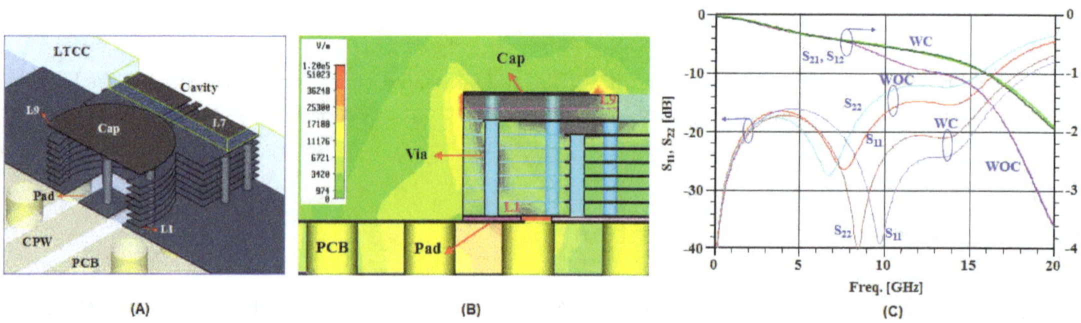

Figure 7. Perspective view (A) of the SMT pad transition using a modified coaxial-line structure, E-field distribution at its cross section (B) and its designed characteristics, comparing the effect of the cap on the top layer (C).

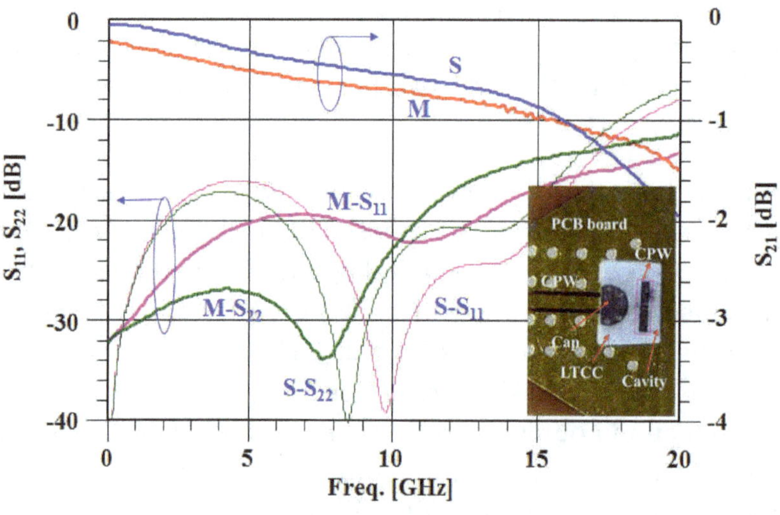

Figure 8. Measured results of the fabricated SMT pad transition, compared to the simulated ones (the inset: the fabricated SMT pad transition on the PCB board, M: measurement and S: simulation).

The modified SMT pad transition was fabricated by using LTCC standard fabrication process. Its measured characteristics are presented in **Figure 8**, compared with simulated ones. Poles of both return (S_{11} and S_{22}) and insertion loss (S_{21}) make a difference between the simulated and measured results. This difference comes from parasitic components due to soldering works. The measured insertion loss of 0.9 dB is achieved at 15 GHz. The return losses of S_{11} and S_{22} are below −14 dB at the same frequency. This SMT pad can be used for X- or Ku-band applications or LO-frequency ports for millimeter-wave SiP applications.

5. Compact LTCC passive devices—SL BPF, LPF and 2 × 2 array antenna

For compact radio system SiP module applications, the key components are the band-pass filter (BPF) and antenna, because they cover significant space and are difficult to be integrated in the RF ICs by using semiconductor technology. In general, they have been implemented in types of planar structures. However, planar structured circuits are usually bulky and are prone to unwanted radiation. Therefore, compact size, shielded electromagnetic cross talk and low-loss **3D** interconnection have been considered as the most important issues for passive device integration in the RF system SiP module.

5.1. A dual-mode four-pole 60 GHz LTCC SL BPF

A SL structured BPF using a dual-mode patch resonator can satisfy key issues for filter integration such as miniaturization and suppressed cross talk, because the dual-mode patch resonator offers a very compact structure [27] and radiation of the SL structure buried between upper and lower ground planes is negligible. However, in order to interconnect it with other circuits on the surface, the low-loss vertical via transition is required.

Figure 9A, **B** and **C** shows a fully embedded SL dual-mode BPF, its perspective view and a layout of two CPW-to-CPW planar transitions, respectively. The BPF is designed on seven-layer LTCC substrate with a relative dielectric constant of 7.0. The dual mode can be generated by adding a perturbation (cut) at a point that is 45° from the axes of coupling to the patch resonator (P-R). Two resonators are used on the 3rd and 5th layer for wide bandwidth characteristics and two P-R blocks are 684 μm away. The feed lines, external coupling between the resonators on the 3rd and 5th layers and internal coupling between their two blocks are on the 4th layer. Its center frequency (fc) and bandwidth (BW) are 61 GHz and 4.5%, respectively. The side length of the resonator is about half a wavelength (613 μm). The widths of the feed lines are 135 μm. By changing the depth of the cut, the coupling coefficients can be controlled and the calculated optimum cut length is 150 μm. The external coupling distances on the 4th layer are 140 μm and the internal coupling is realized by an overlap of 40 μm between two resonators on the 3rd and 5th layer. By using the low-loss CPW-to-SL transitions described in Section 4.1, the 600 μm-thick SL BPF is interconnected with 100 μm-thick RFICs mounted in the SiP module. However, the steep height difference between their GND planes can cause radiation problems. Therefore, by using CPW planar transition, GND planes are gradually transited from the 1st layer to the 3rd and 5th layers as shown in **Figure 9B**. This CPW planar transition consists of three CPW lines and two transitions (TR1 and TR2). Their width and the gap for 50 Ω impedance CPW lines are designed. The width of a CPW1 (=CPW2) and CPW3 is 244 and 100 μm, respectively. Its corresponding gap is 90, 140 and 90 μm, respectively. The transition length of TR1 and TR2 is 40 and 144 μm, respectively. LCPW1, LCPW2 and LCPW3 for each CPW length are 500, 430 and 500 μm, respectively.

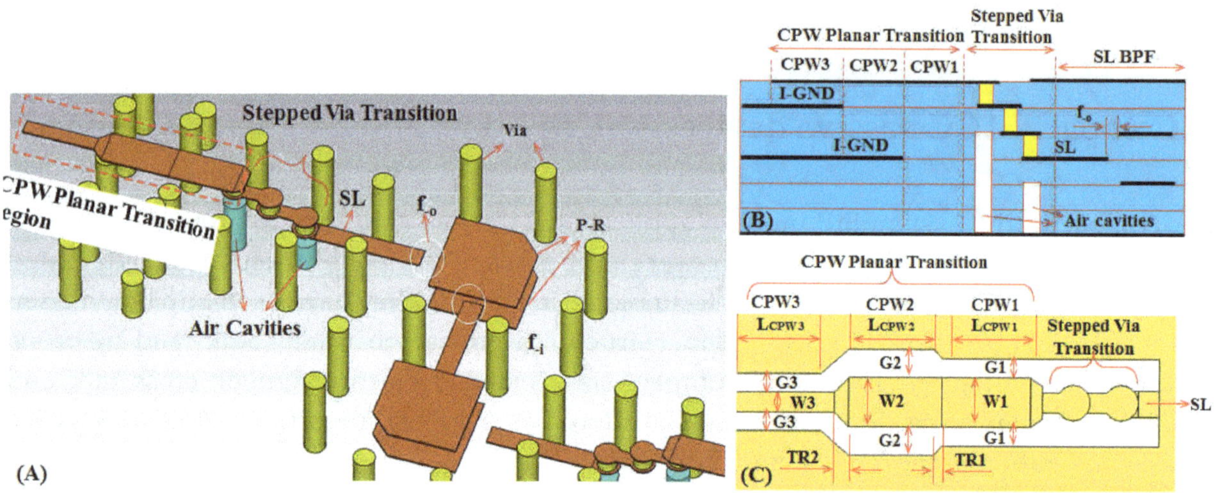

Figure 9. Perspective view (A) of a 60 GHz SL LTCC BPF involving CPW-to-SL stepped via transitions embedding air cavities, its cross-sectional view (B) and CPW-to-CPW planar transitions (C) [P-R: a patch resonator, f_o: an input and output feed line, f_i: an inter coupling feed line, I-GND: an internal ground plane].

The SL 60 GHz BPF fabricated in seven LTCC dielectric layers and its measured results are presented in **Figure 10**. Its total size including the entire transitions is 3.2 × 6.5 × 0.7 mm^3. By using the on-wafer probing method, the implemented BPF was tested. The comparison of the simulated and measured results is presented. While the measurement shows a lower center frequency and narrower BW than the simulation results, two results coincide rather well in the pass band from 60.075 to 61.925 GHz. The misalignment among feed lines and resonators results in different coupling coefficients, compared to the designed ones. Therefore, its frequency characteristics are a little different from the simulated results. The measured fc and fractional BW are 60.8 GHz and 4.1%, respectively. The return loss is less than −10.0 dB at the pass band. Its insertion loss including two vertical and four planar transitions is 4.98 dB. Considering the insertion loss of the transitions, its insertion loss is 3.74 dB.

5.2. A 5th order low-pass filter (LPF)

In order to eliminate harmonics or analog components among output signals, in general, a LPF has been used. The LPF based on the Chebyshev LPF prototype [16] as shown in **Figure 11** is designed in order to fully embed in this LTCC SiP module [28]. It has a cutoff frequency of 1.5 GHz, ripple of 0.05 dB and order of 5. In order to improve its return loss characteristics, values of capacitance (C1–C3) and inductance (L1, L2) in the designed basic LPF circuit are optimized.

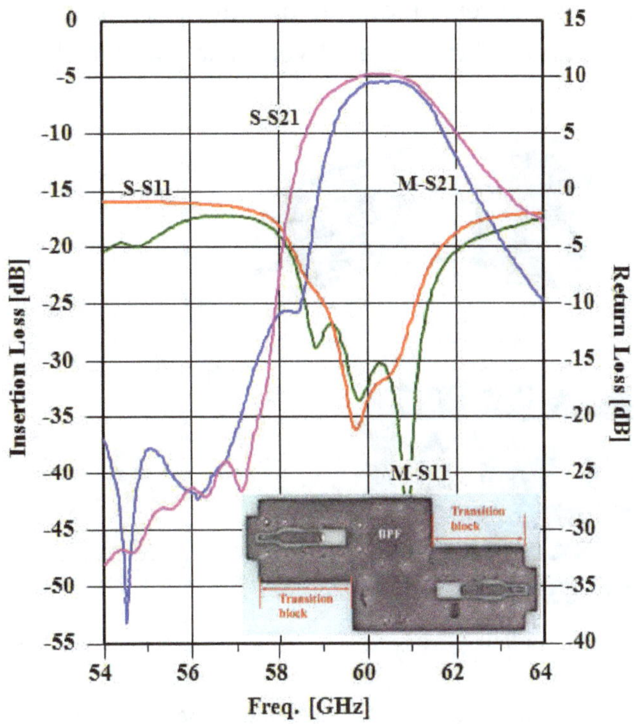

Figure 10. Measured performance of the fabricated SL BPF, compared to the simulated one (the inset: the photo of the fabricated BPF, S: simulation and M: measurement).

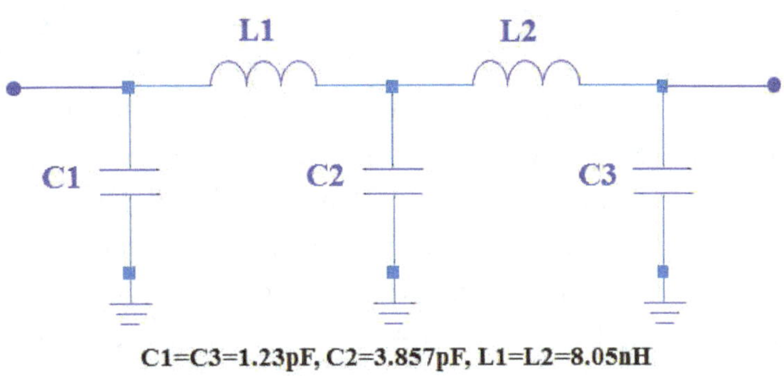

Figure 11. A prototype circuit of the 5th order Chebyshev LPF.

These capacitors and inductors are vertically designed within the six-layered LTCC dielectric whose permittivity is 7.2 at 2 GHz. In the case of the capacitor, three parallel-plate capacitors are interconnected in parallel by using vertical vias. Its capacitance value (C1 and C2) is controlled by an overlapped area among plates. The overlapped area of the C1 (=C3) and C2 is 400 × 500 and 600 × 1100 μm^2, respectively. In the case of the inductor, a 10,175 um-long line is coiled in the six-layer LTCC dielectric. The number of turn is 4.5. Its width of a metal strip is 170 μm. The number of turn and inner opening area of the helical inductor is 4.5 and 600 × 330 μm^2, respectively. The 5th order LPF is designed by integrating these elements in the size of 4.0 × 3.2 × 0.68 mm^3. It was fabricated using a LTCC commercial foundry. In **Figure 12A**, **B** and **C**, **3D** structures of the LPF elements (C1, C2 and L1), the designed LPF and measured results of the fabricated LPF are presented. The measured insertion loss (S_{21}) and return losses (S_{11}, S_{22}) are less than −0.46 and −11 dB, respectively. They are similar to simulated results.

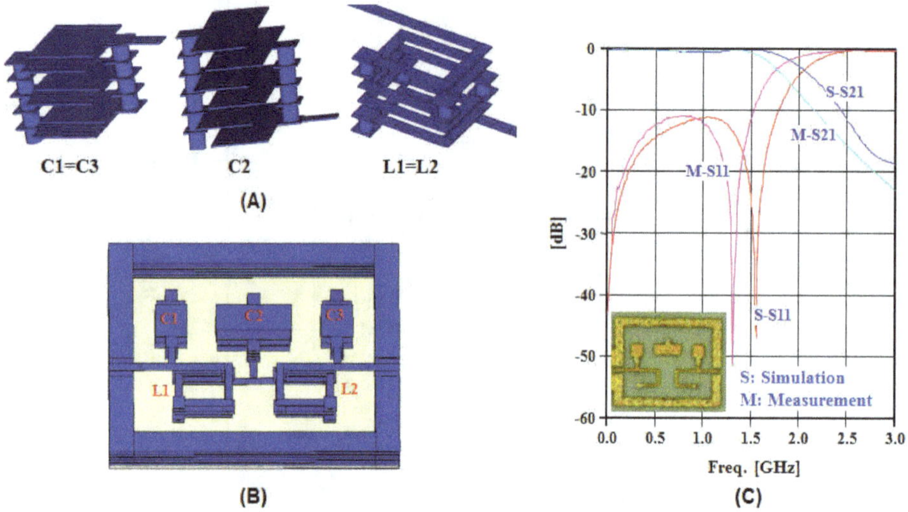

Figure 12. Perspective views of the elements (A) and LPF integrating them (B) and measured results compared to the simulated ones (C) [the inset: the fabricated LPF].

5.3. A compact 2 × 2 array patch antenna

The 2 × 2 patch array antenna is designed with a LTCC MSL structure. **Figure 13A** and **B** shows the **3-D** structure of the LTCC antenna and an embedded MSL (EMSL) power divider, respectively. Three layers from L6 to L8 are for the antenna and additional layers from L1 to L5 are used for internal and outer ground planes. The radiating patches are placed on the 8th layer (L8) and their size is the same as 645 × 1299 μm^2. The EMSL structured feeding network is designed on the 7th layer (L7) using a T-divider (power divider). In this structure, 70.7 Ω-quarter-wavelength ($\lambda_g/4$) transformers are required. However, it is impossible to implement them because limitation of the line width is 90 μm in the LTCC design rule. Therefore, the additional $\lambda_g/4$ transformers with low impedance (Z) are designed at the common port as shown in **Figure 16B**. The width of high-Z lines is 90 μm and their impedance is 47 Ω. For the low-Z lines, their width and length are optimized considering overall characteristics. The optimized width and its impedance are 130 μm and 40 Ω, respectively. The GND plane is on the 5th layer. The antenna size is as small as 10 × 10 × 0.3 mm^3.

Figure 14A and **B** shows the measured return loss characteristic and beam patterns of the fabricated antenna, respectively. Its X-ray photo is in the left inset of **Figure 14A**. In order to test a return loss and beam patterns, a WR15 waveguide (WG)-to-MSL transition was used as shown in the right insets of (A). A −10 dB bandwidth is 6.3 GHz from 56.5 to 62.8 GHz. At 61 GHz, the measured E- and H-plane radiation patterns are presented in **Figure 14B**. A gain of 7 dBi and a 3-dB beam width of 36° in H-plane pattern are obtained. The E-plane pattern is wider because of spurious generated in the E-plane direction of the feeding network.

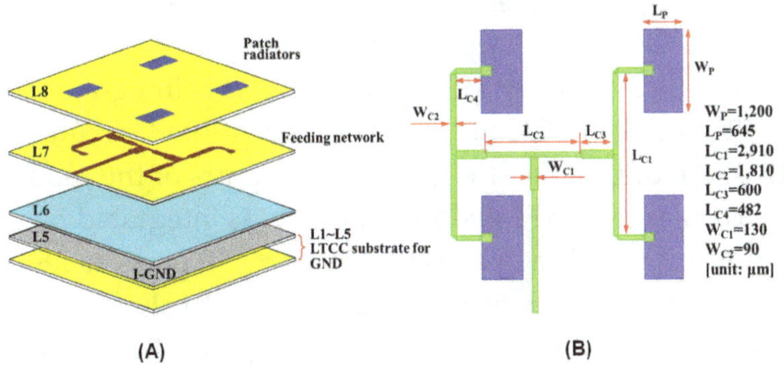

Figure 13. Layer structure of the 2 × 2 array LTCC antenna (A) and its feeding network and radiating patches (B).

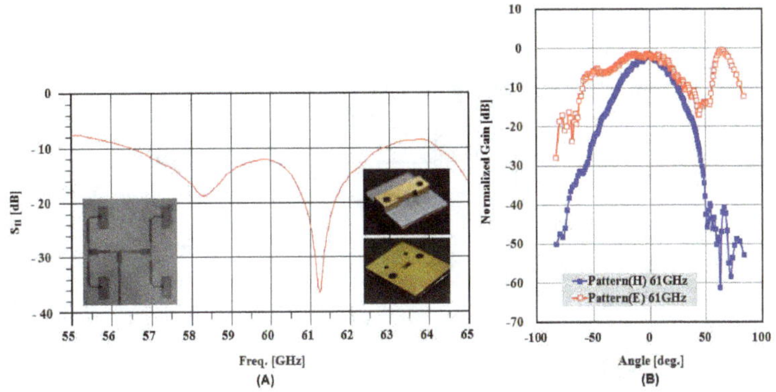

Figure 14. Measured return loss characteristic of the fabricated 2 × 2 array LTCC antenna (the left and right inset: its X-ray photo and the antenna assembled with a WR15 WG-to-MSL transition, respectively) (A) and its measured beam patterns (B).

6. Compact LTCC SiP modules for V-band applications

In this chapter, a typical heterodyne 61 GHz transmitter (Tx) and a highly integrated 60 GHz amplitude shift-keying (ASK) transceiver (TRx) SiP module are presented in detail [29–31]. They have been designed and implemented by using the key technologies such as suppression unwanted resonant modes, low-loss vertical transitions and compact passive devices presented in the previous chapters.

6.1. Monolithic 61 GHz transmitter LTCC SiP module

A block diagram of a typical heterodyne 61 GHz transmitter (Tx) is shown in **Figure 15**. This Tx is comprised of a BPF, a antenna, a up-converting mixer, two frequency multipliers (MTLs), a drive amplifier (DA) and a power amplifier (PA). The local oscillation (LO) signal (59.15 GHz) of the mixer is supplied by multiplying the external LO source of 14.79 GHz by 4.

The 61 GHz Tx is monolithically integrated into the single SiP module as shown in **Figure 16**. This SiP module consists of a nine-layer LTCC dielectric. The BPF, which is implemented in the previous subchapter 5.1, is fully embedded through L2 to L7 by using the CPW-to-SL vertical transition and the CPW-to-CPW planar transition. The BPF is connected with a driver amplifier and mixer IC. The 2 × 2 array MSL patch antenna is integrated in L6 through L8. In order to mount the 61 GHz Tx LTCC SiP module on a printed circuit board (PCB), a SMT package is adopted. Therefore, using SMT pads, all the ports of the module are designed on its bottom side. In particular, the SMT pad for a LO port is integrated by using the transition implemented in Section 4.2. Pad dimensions for DC ports, IF ports of 1.85 GHz and LO ports of 14.79 GHz are 700 × 700, 320 × 550, 560 × 560 μm^2, respectively. Five active chips mounted in the cavity of the L7 are isolated from each other using isolation cavities structure, which consists of L8 and L9. The DC bias lines and long IF feed lines are shielded using isolating ground planes and vias.

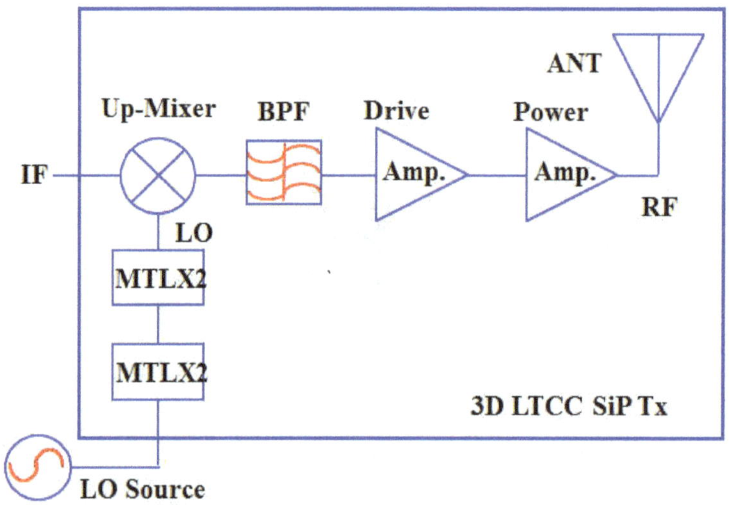

Figure 15. Block diagram of the 61 GHz heterodyne transmitter.

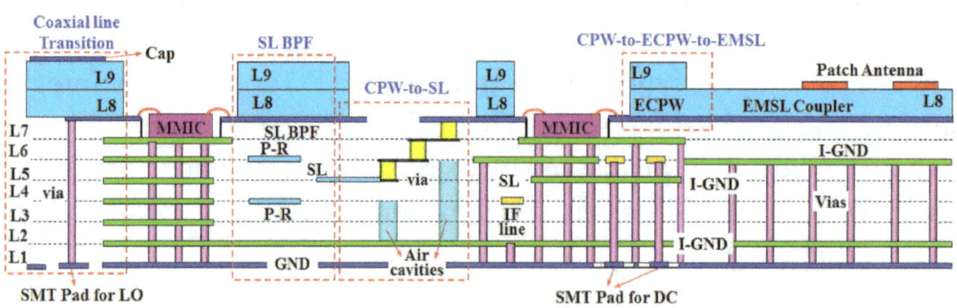

Figure 16. A cross-sectional view of the monolithic 61 GHz Tx LTCC SiP module integrating a BPF, an antenna, MMICs and DC bias circuits (ECPW: the embedded CPW, EMSL: the embedded MSL, P_R: the patch resonator and Lx: the layer number of the LTCC multilayer).

Designed SiP module was implemented in nine-LTCC multilayers using the standard LTCC process. **Figure 17A** shows the fabricated monolithic LTCC SiP module of the 61 GHz Tx. The whole size of the transmitter is as small as 36 × 12 × 0.9 mm^3. **Figure 17B** shows its bottom side with SMT pads.

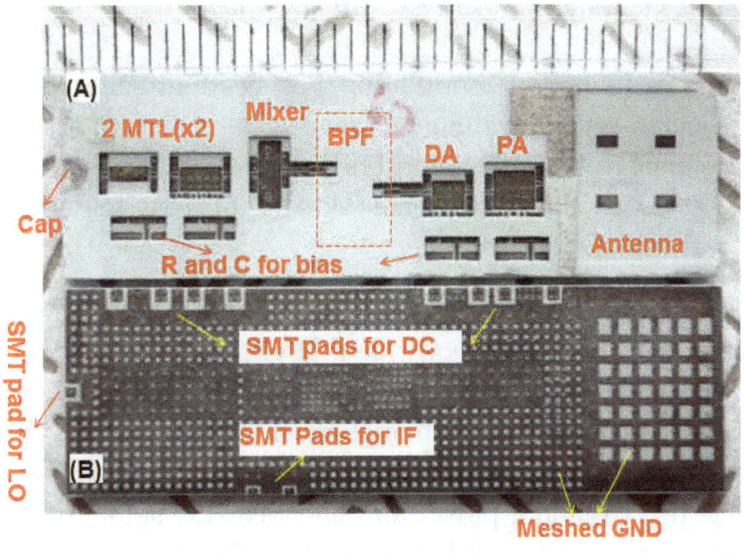

Figure 17. Implemented monolithic 61 GHz Tx LTCC SiP module [36 × 12 × 0.9 mm³, (A) and (B): top and bottom side].

The fabricated LTCC Tx module was mounted on the PCB. At the output port of the power amplifier of the module, the output power and frequency spectrum were measured using on-wafer probing. **Figure 18A** plots the RF output power and the power gain as a function of the IF input power of 1.85 GHz. A measured output power at a 1-dB gain compression point (P1dB) and up-conversion gain is 10.2 dBm and 7.3 dB, respectively, at 61 GHz. Output spectrums such as a LO, RF and spurious signals are shown in **Figure 18B**. The isolation level between the LO and RF and the spurious one are less than 26.4 and 22.4 dBc, respectively. The measured output performance demonstrates that the integrated BPF suppresses effectively the LO and spurious signal.

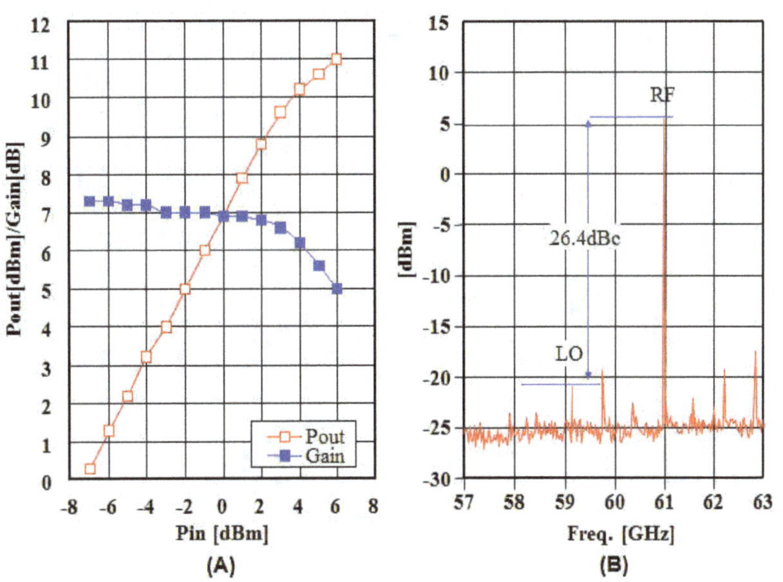

Figure 18. Measured output performance of the fabricated module (A) output power and conversion gain and (B) output frequency spectrum.

6.2. Compact 60 GHz ASK transceiver LTCC SiP module

The amplitude shift-keying (ASK) modulation has been utilized in various microwave systems [32]. In particular, several millimeter-wave systems have adopted it for high-speed applications, because of circuit simplicity and high power efficiency. In addition, an analog-digital converter (ADC) is hard to be implemented and it is easy to demodulate ASK noncoherently by using an envelope detector.

In this work, the 60-GHz ASK transceiver (TRx) is designed and implemented. **Figure 19A, B** and **C** shows its block diagram, its layout for a LTCC SiP module and its cross-sectional structure, respectively. The Rx part consists of a high-gain and low-noise amplifier (LNA) block, detector, low-pass filter (LPF) and attenuator (ATT), which is inserted for the impedance buffering in the high-gain budget. The Tx part is composed of an up-converting mixer, two frequency multipliers (MTLs) and a power amplifier. The LO signal of the Tx is supplied to the mixer by multiplying the external LO source of 7.78 GHz by eight times. The carrier frequencies of the Tx and Rx link are 62.24 and 58.75 GHz, respectively. The whole 60-GHz ASK TRx is integrated into the six-layered LTCC SiP module in the size of 17.8 × 17.9 × 0.6 mm^3 as shown **Figure 19B**. In the conceptual vertical structure in **Figure 19C**, a LPF, isolation via fence and DC bias components are embedded. RFICs are mounted on cavities in L6. Each LTCC layer is 84 μm high. A via diameter is 120 μm before co-firing process. The main signal line is a 50 Ω CPW line, whose width and gap are 123 and 100 μm, respectively. The five-order Chebyshev LPF with a cutoff frequency of 1.5 GHz inserted in the ASK de-modulator (Rx part) in order to eliminate harmonics and analog components at its output. The previously implemented LPF in Section 5.2 was utilized. The ground plane in the bottom side of the Tx and Rx part is also separated for eliminating return path.

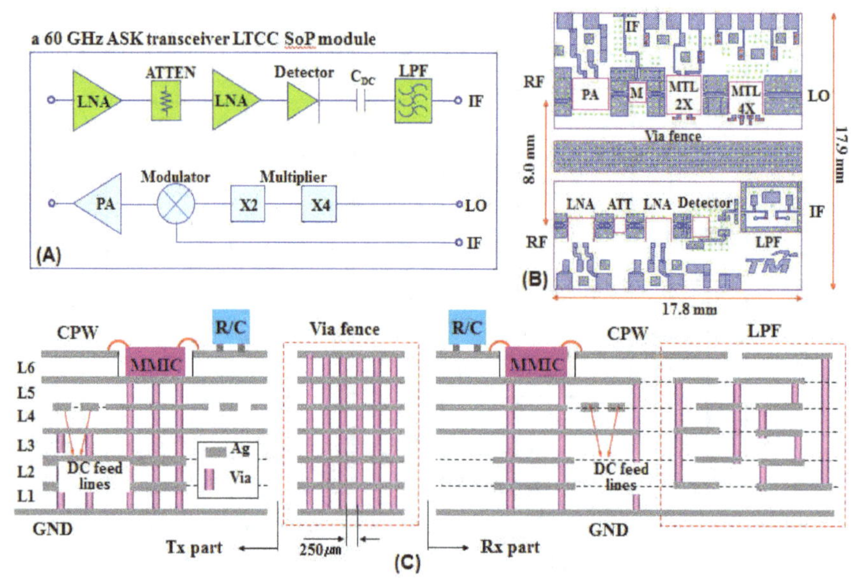

Figure 19. A block diagram of a 60 GHz ASK transceiver (TRx) (A), the layout of the ASK TRx LTCC SiP module (B) and its conceptual vertical structure (C) embedding a LPF, via fence, RFICs and DC bias components in the 6-layer LTCC dielectrics [Lx: the number of the LTCC dielectric layer].

The high-isolation via fence between Tx and Rx part is investigated by using a 3-D electromagnetic (EM) tool [25] as shown in **Figure 20A**. In general, the isolation of over 80 dBc is required between the ASK modulator and demodulator. In order to confine the EM-fields within each Tx or Rx area, the via fence is designed between 50 Ω CPW lines, which can be assumed as the signal path of the Tx (P1–P2) and Rx (P3–P4) part. The spacing between them is 8 mm, which is the same that between the Tx and Rx port in the diplexer of the main system. The diameter of vias and spacing between their edges are 120 and 250 μm, respectively. The E-field distribution of the designed model at 60 GHz is presented in **Figure 20A**. This result shows that E-fields inputted from a port 1 (P1) can be effectively confined within one path (Tx) due to the via fence. Simulated isolation characteristics between two lines from DC to 100 GHz are presented in **Figure 20B**. It clearly shows that the isolation better than 80 dBc is obtained.

The designed ASK TRx LTCC SiP module was fabricated using a six-layer LTCC substrate in the commercial Foundry [33]. The implemented ASK TRx LTCC SiP module is as small as 17.8 × 17.9 × 0.6 mm^3 as shown in the inset of **Figure 21A** and it was assembled into a metal housing with DC bias boards. **Figure 21B** and C shows the measured RF and IF spectrum in the Tx and Rx part, respectively, of the ASK TRx LTCC SiP module. In **Figure 21B**, the measured output power (Pout) is 12.8 dBm at the LO of 62.24 GHz and IF frequency swept from 10 MHz to 1.5 GHz. By inserting a 20 dB attenuator between the Tx and Rx parts considering the free space path loss, the IF spectrum of the Rx part was tested. By changing the IF signal from 10 MHz to 1.5 GHz at the Tx part with the LO signal of 58.752 GHz, the flat IF output signal less than −2 dBm is obtained up to 1.25 GHz as shown in **Figure 21C**. The conversion gain is 38 dB at the IF output of 1.25 GHz.

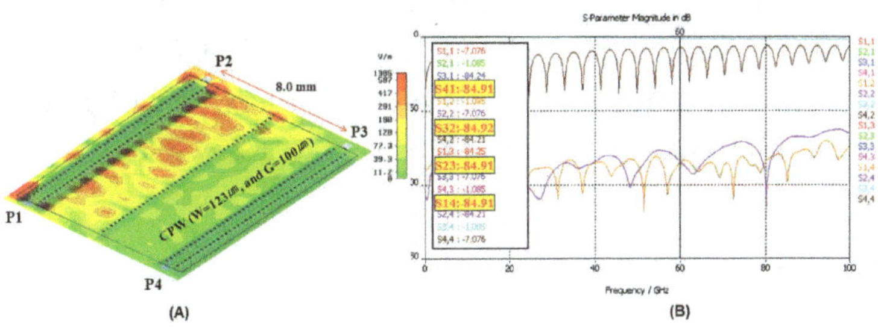

Figure 20. An assumed model for evaluation of isolation between the Tx and Rx and its E-field distribution (A) and simulated results (B).

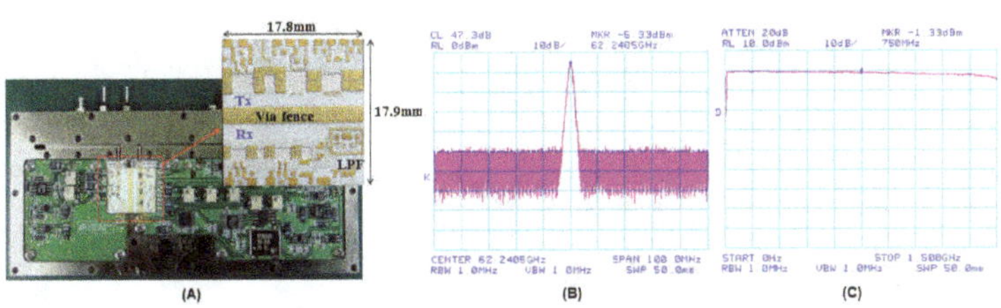

Figure 21. Fabricated ASK TRx LTCC SiP module (A), its measured spectrum at the output port of the Tx part (B) and the IF port of the Rx one (C).

7. Conclusion

In this chapter, highly integrated monolithic LTCC SiP modules have been presented for microwave applications. Almost passive circuits of the whole radio system have been monolithically embedded in the LTCC multilayer dielectric substrate. The main key technologies for the monolithic SiP module are suppressing parasitic resonant modes, low-loss transitions and compact and high-performance passive devices. In general, the parasitic rectangular waveguide consisting of via, upper ground plane and lower one is easily and frequently formed in the SiP module and also effectively suppressed by reducing its horizontal dimension, spacing between vias. The low-loss SL-to-CPW vertical via transition using the stepped via structure and embedded air cavity achieves −0.7 dB transition loss at 60 GHz. By using the modified coaxial line, the SMT pad transition is developed and demonstrates 0.9 dB loss at 15 GHz. By using the developed SL-to-CPW transition, the dual-mode four-pole 60 GHz SL BF is fully embedded in the LTCC substrate in the size of $3.2 \times 6.5 \times 0.7$ mm^3 and the insertion loss of 3.74 dB and the BW of 4.1% are obtained. The fully embedded 5th order LPF composed of vertical plate capacitors and helical inductors is implemented as small as $4.0 \times 3.2 \times 0.6$ mm^3. Its measured insertion and return losses are −0.46 dB and less than −11 dB, respectively. The 2×2 array patch antenna with the gain of 7 dBi and beam width of 36° has been developed. By utilizing the well analyzed and developed key technologies, 61 GHz transmitter and 60 GHz ASK transceiver LTCC SiP modules have been implemented. The 61 GHz Tx LTCC SiP module achieves an output power of 10.2 dBm at 61 GHz and the conversion gain of 7.3 dB. Because of the integrated SL BPF, the LO and spurious signals are suppressed below 26.4 dBc and 22.4 dBc, respectively. Using the off-shelf receiver, the wireless link is verified. In the case of the 60 GHz ASK LTCC SiP module, in order to achieve 80 dBc isolation between the Tx and Rx part, the high isolated substrate using the via fence is proposed and used in the SiP design. The 60 GHz ASK TRx LTCC SiP module is fabricated as small as $17.8 \times 17.9 \times 0.6$ mm^3 and it achieves the output power of 12.8 dBm at LO of 62.24 GHz and the flat IF output signal less than −2 dBm up to 1.25 GHz.

Acknowledgements

These works were financially supported by the Ministry of Science and Technology of Korea and KISTEP from 2002 to 2006 and by Telecom Malaysia Research & Development (TMRND) from 2008 to 2009.

Author details

Young Chul Lee

Address all correspondence to: leeyc@mmu.ac.kr

Department of electronic engineering, Mokpo National Maritime University (MMU), Mokpo, South Korea

References

[1] Bouayadi O.E., Lamy Y., Dussopt L. and Simon G. 3D integration and packaging of mmWave circuits and antennas: opportunities and challenges. Microwave Journal. 2016;59:22–28.

[2] Zhang Y. P. and Liu D. Antenna-on-chip and antenna-in-package solutions to highly integrated millimeter-wave devices for wireless communications. IEEE Transactions on Antennas and Propagation. 2009;57:2830–2841. doi:10.1109/TAP.2009.2029295

[3] Papapolymerou I., Franklin Drayton R. and Katehi L.P.B. Micromachined patch antennas. IEEE Transactions on Antennas and Propagation. 1998;46:275–283. doi:10.1109/8.660973

[4] Kuo H.-C., Yue H.-L., Ou Y.-W., Lin C.-C. and Chuang H.-R. A 60-GHz CMOS sub-harmonic RF receiver with integrated on-chip artificial-magnetic-conductor Yagi antenna and balun bandpass filter for very-short-range gigabit communications. IEEE Transactions on Microwave Theory and Techniques. 2013;61:1681–1691. doi:10.1109/TMTT.2013.2247622

[5] Muller A., Saadaoui M., Pons P., Neculoiu D., Bary L., Dubuc D., Grenier K., Vasilache D., Buiculescu C., Petrini I. and Plana R. Membrane supported endfire antennas manufactured using silicon micromachining. IEEE International Workshop on Antenna Technology: Small Antennas and Novel Metamaterials. 2005;113–116. doi:10.1109/IWAT.2005.1461017

[6] Chan K.T., Chin A., Lin Y.D., Chang C.Y., Zhu C.X., Li M.F., Kwong D.L., McAlister S., Duh D.S. and Lin W.J. Integrated antennas on Si with over 100 GHz performance, fabricated using an optimized proton implantation process. IEEE Microwave and Wireless Components Letters. 2003;13:487–489. doi:10.1109/LMWC.2003.817146

[7] Pan S., Caster F., Heydari P. and Capolino F. A 94-GHz extremely thin metasurface-based BiCMOS on-chip antenna. IEEE Transactions on Antennas and Propagation. 2014;62:4439–4451. doi:10.1109/TAP.2014.2330575

[8] Swaminathan M., Bavisi A., Yun W., Sundaram V., Govind V. and Monajemi P. Design and fabrication of integrated RF modules in liquid crystalline polymer (LCP) substrates. 31st Annual Conference of IEEE Industrial Electronics Society. 2005;2346–2351. doi:10.1109/IECON.2005.1569270

[9] Li R., DeJean G., Maeng M., Lim K., Pinel S., Tentzeris M. M. and Laskar J. Design of compact stacked-patch antennas in LTCC multilayer packaging modules for wireless applications. IEEE Transactions on Advanced Packaging. 2004;27(4):581–589. doi:10.1109/TADVP.2004.831866

[10] Pilard R., et al. HDI organic technology integrating built-in antennas dedicated to 60 GHz SiP solution. Proceedings of the 2012 IEEE International Symposium on Antennas and Propagation. 2012. doi:10.1109/APS.2012.6348633

[11] Lamminen A. E., Saily J. and Vimpari A. R. 60-GHz patch antennas and arrays on LTCC with embedded-cavity substrates. IEEE Transactions on Antennas and Propagation. 2008;56(9):2865–2874. doi:10.1109/TAP.2008.927560

[12] Ohata K., et al. 1.25 Gbps wireless gigabit ethernet link at 60GHz-band. 2003 IEEE MTT-S International Microwave Symposium Digest. 2003. doi:10.1109/MWSYM.2003.1210955

[13] Kuo J.-L., et al. 60-GHz four-element phased-array transmit/receive system-in-package using phase compensation techniques in 65-nm flip-chip CMOS process. IEEE Transactions on Microwave Theory and Techniques. 2012;60(3):743–756. doi:10.1109/TMTT.2011.2176508

[14] Haydl W. H. Resonance phenomena and power loss in conductor-backed coplanar structures. IEEE Microwave and Guided Wave Letter. 2000;20(12):514–516. doi:10.1109/75.895086

[15] Das N. K. Methods of suppression or avoidance of parallel-plate power leakage from conductor-backed transmission lines. IEEE Transactions on Microwave Theory and Techniques. 1996;44(2):169–181. doi:10.1109/22.481565

[16] Pozar D. M.. Microwave Engineering. John Wiley & Sons, Inc: 1998.

[17] Lee Y. C., Cho Y. H. and Park C. S. BPF integrated amplifier module using LTCC technology for millimeter-wave applications. Microwave and Optical Technology Letters. 2015;57(8):1821–1825. doi:10.1002/mop.29206

[18] Panther C. G., Stubbs M. G. and Wight J. S. Vertical transitions in low temperature co-fired ceramics for LMDS applications. IEEE MTT-S International Microwave Symposium Digest. 2001;3:1907–1910. doi:10.1109/MWSYM.2001.967281

[19] Lei S., Guo Y. X. and Ong L. C. CPW to stripline transitions in LTCC for millimeter-wave applications. IEEE Asia Pacific Microwave Conference (APMC) Proceedings. 2005. doi:10.1109/APMC.2005.1606430

[20] Schmuckle F. J., Jentzch A., Heinrich W., Butz J. and Spinnler M. LTCC as MCM substrate: design of strip-line structures and flip-chip interconnections. IEEE MTT-S International Microwave Symposium Digest. 2001;3:1093–1096. doi:10.1109/MWSYM.2001.967280

[21] Yang T.-H., Chen C.-F., Huang T.-Y., Wang C.-L. and Wu R.-B. A 60GHz LTCC transition between microstrip line and substrate integrated waveguide. IEEE Asia Pacific Microwave Conference (APMC) Proceedings. 2005. doi:10.1109/APMC.2005.1606294

[22] Heyen J., Gordiyenko A., Heide P. and Jacob A.F. Vertical feedthroughs for millimeter-wave LTCC modules. 33rd European Microwave Conference Proceedings. 2003;411–414. doi:10.1109/EUMC.2003.1262309

[23] Lee Y. C. SMT pad using a coaxial line structure for LTCC SoP applications. Microwave and Optical Technology Letters. 2009;51(7):1769–1772. doi:10.1002/mop.24447

[24] Lee Y. C. CPW-to-stripline vertical via transitions for 60GHz LTCC SOP applications. Progress In Electromagnetics Research Letters. 2008;2:37–44. doi:10.2528/PIERL07122805

[25] CST MICROWAVE STUDIO, CST Inc., [Internet] Available: http://www.cst.com

[26] Ziroff A., Nalezinski M., Menzel, Wolfgang A. Novel approach for LTCC packaging using a PBG structure for shielding and package mode suppression. European Microwave Conference. 2003;419–422. doi:10.1109/EUMC.2003.1262311

[27] Lee Y. C. and Park C. S. A fully embedded 60-GHz novel BPF for LTCC system-in-package applications. IEEE Transactions on Advanced Packaging. 2006;29(4):804–809. doi:10.1109/TADVP.2006.884807

[28] Lee Y. C., Kim T. W., Ariffin A. B. and Myoung N. G. 60-GHz amplitude shift-keying receiver LTCC system-on-package module. Microwave and Optical Technology Letters. 2011;53(4):758–761. doi:10.1002/mop.25824

[29] Lee Y. C., Chang W.-I. and Park C. S. Monolithic LTCC SiP transmitter for 60GHz wireless communication terminals. IEEE MTT-S International Microwave Symposium Digest. 2005. doi:10.1109/MWSYM.2005.1516839

[30] Lee Y. C. and Park C. S. LTCC-based monolithic system-in-package (SiP) module for millimeter-wave applications. International Journal of RF and Microwave Computer-Aided Engineering. 2016; 26(9): 803–811. dio:10.1002/mmce.21032

[31] Lee Y. C. 60-GHz amplitude shift-keying transceiver LTCC system-on-package module. Microwave and Optical Technology Letters. 2013;55(1):48–51. doi:10.1002/mop.27264

[32] Tarusawa Y., Ogawa H. and Hirota T. A new constant resistance ASK modulator using double-sided MIC. IEEE Transactions on Microwave Theory and Techniques. 1987;35:819–822. doi:10.1109/TMTT.1987.1133758

[33] RN2-LTCC [Internet]. Available: http://WWW.rn2-ltcc.com

4

Materials Characterization using Microwave Waveguide Systems

Kok Yeow You

Abstract

This chapter reviews the application and characterization of material that uses the microwave waveguide systems. For macroscopic characterization, three properties of the material are often tested: complex permittivity, complex permeability and conductivity. Based on the experimental setup and sub-principle of measurements, microwave measurement techniques can be categorized into either resonant technique or nonresonant technique. In this chapter, calibration procedures for non-resonant technique are described. The aperture of open-ended coaxial waveguide has been calibrated using Open-Short-Load procedures. On the other hand, the apertures of rectangular waveguides have been calibrated by using Short-Offset-Offset Short procedures and Through-Reflect-Line calibration kits. Besides, the extraction process of complex permittivity and complex permeability of the material which use the waveguide systems is discussed. For one-port measurement, direct and inverse solutions have been utilized to derive complex permittivity and complex permeability from measured reflection coefficient. For two-port measurement, in general, the material filled in the waveguide has been conventional practice to measure the reflection coefficient and the transmission coefficient by using Nicholson-Ross-Weir (NRW) routines and convert these measurements to relative permittivity, ε_r and relative permeability, μ_r. In addition, this chapter also presents the calculation of dielectric properties based on the difference in the phase shifts for the measured transmission coefficients between the air and the material.

Keywords: microwave waveguides, relative permittivity, relative permeability, conductivity, resonant methods, nonresonant methods, materials characterization

1. Introduction

For macroscopic material characteristic investigations, three properties of the material are often measured: relative permittivity ε_r, relative permeability μ_r and conductivity σ. Normally, many microwave measurements only focus on the properties of relative permittivity, ε_r rather than the permeability μ_r and the conductivity σ. Recently, there has been an increased interest in the determination of dielectric properties of materials at microwave frequencies range. This is because those properties were played the important roles in the construction of high-frequency electronic components, the superconducting material properties, the quality of printed circuit board (PCB) substrate, the efficiency of microwave absorption materials, metamaterial characterizations and the performance of dielectric antenna design. Based on the setup and sub-principle of measurements, measurement techniques can be categorized into either resonant methods or nonresonant methods. In practice, the prime considerations in measuring the dielectric properties of the materials are the thickness required of the material, the size of the waveguide, limitations of the operating frequency and the accuracy of the measurements.

For material characterizing using resonant methods, a resonator is filled with a material or sample as shown in **Figure 1** [1, 2]. This produces a resonance frequency shift and also a broadening of the resonance curve compared to the resonator without filled with any sample. From measurements of shifting resonance frequency, the properties of the sample can then be characterized. The particular resonance frequency for the resonator without filled with the sample is depends on its shape and dimensions. The resonance measurement techniques are good choices for determining low-loss tangent, tanδ values for the low-loss sample, but such techniques cannot be used for the measurement of swept frequency.

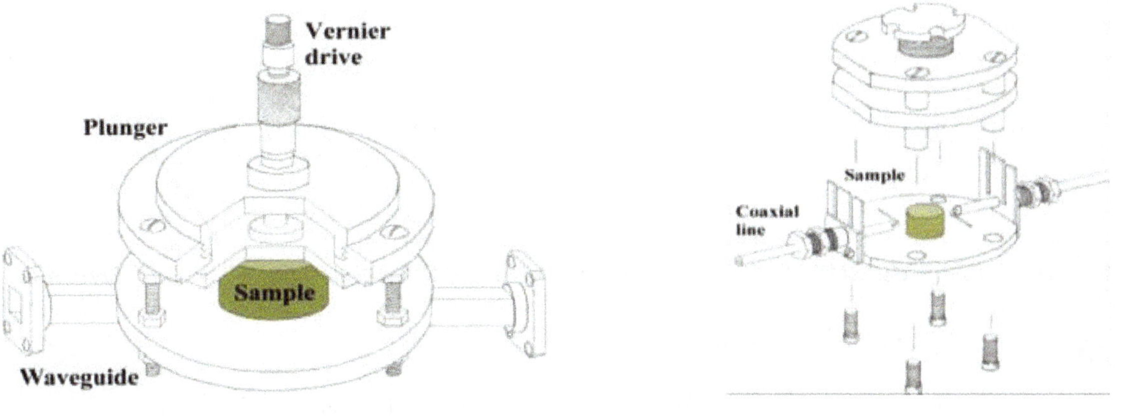

Figure 1. The resonator cavity filled with sample under test [1, 2].

A free-space and transmission/reflection measurement techniques are grouped in the category of nonresonant methods. The free-space technique is a far-field measurement, and a horn antenna is used as the radiator as shown in **Figure 2** [3–5]. The free-space method is suitable for the measurement for thin film sample with high temperature because horn radiators do not come into direct contact with the sample, and thus, the RF circuits of the instrument are safer from heat damage. However, this method provides a less precise measurement because the sensing field is highly dispersed. Furthermore, the distance between the sample surface and the horn aperture is difficult to gauge precisely. The coaxial, circular or rectangular waveguides are implemented in transmission/reflection measurement techniques which are directly in contact with the sample. Although various measurement techniques are available to be used, when choosing the appropriate technique, some other factors are required to be considered in the selection of technique, such as accuracy, cost, samples shape and operating frequency. This chapter is focused only on coaxial and rectangular waveguides.

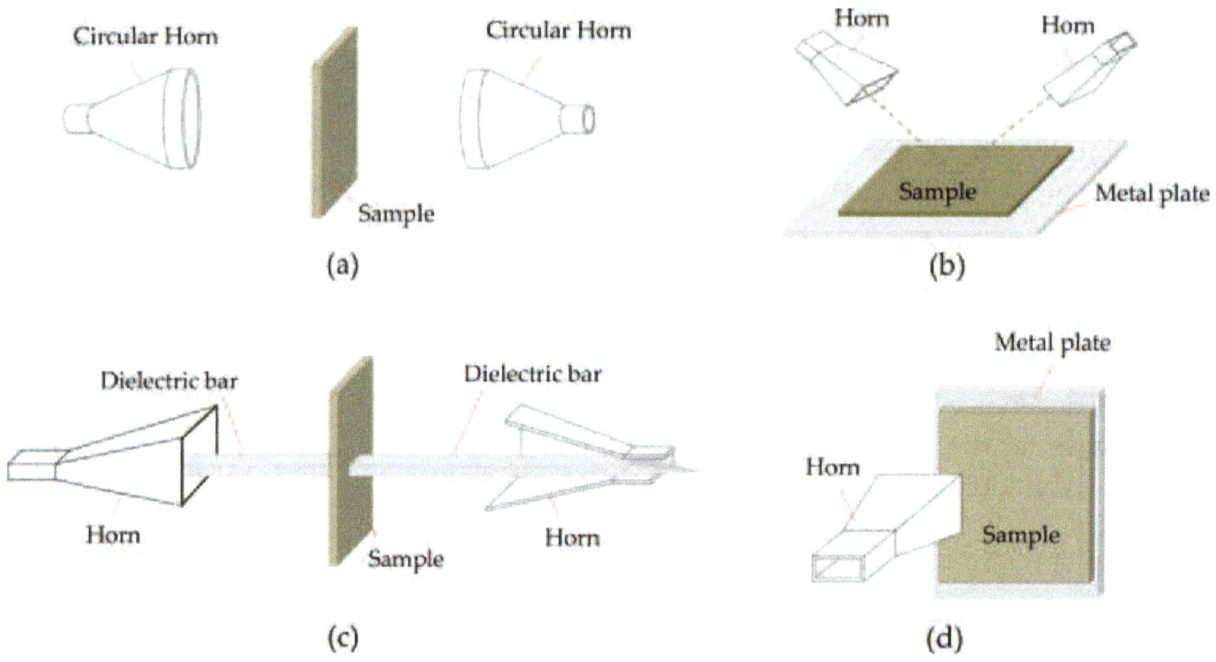

Figure 2. Free-space measurement setup for dielectric measurement of the thin sample [3–5].

2. Microwave measurement using coaxial and rectangular waveguides

2.1. Coaxial and rectangular waveguides

There are various sizes of the coaxial probes and rectangular waveguides, which are dependent on the operating frequency and its application. The coaxial probe is a waveguide consisting of inner and outer conductors, with radii a and b, respectively, as shown in **Figure 3**. On the other hand, the rectangular waveguide is a rectangular metal pipe with width, b and height, a, which guides high-frequency electromagnetic waves from one place to another without significant loss in intensity. There are several commercial rectangular waveguides, such as WR510, WR90, WR75 and WR62 waveguides, which covering a broad measurement range for L-band, X-band and Ku-band, respectively, as shown in **Figure 4**. Generally, the material characterization using waveguide discontinuity methods can be categorized into one-port and two-port measurements. The measurements assume that only the dominant transverse electric, TE_{10} mode propagates in the rectangular waveguide. On the other hand, only transverse electromagnetic mode (TEM) is assumed to be propagated in the coaxial line waveguide.

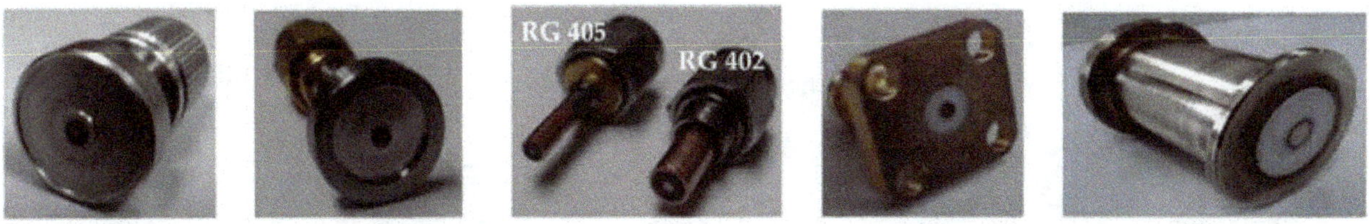

(a) (b) (c) (d) (e)

Figure 3. (a) Keysight dielectric probe kit with inner radius of outer conductor, b = 1.5 mm and radius of inner, a = 0.33 mm. (b) Customized small coaxial probe with b = 0.33 mm and a = 0.1 mm [6]. (c) RG402 and RG 405 semi-rigid coaxial probe [7]. (d) SMA stub coaxial probe with b = 2.05 mm and a = 0.65 mm [8]. (e) Customized large coaxial probe with b = 24 mm and a = 7.5 mm.

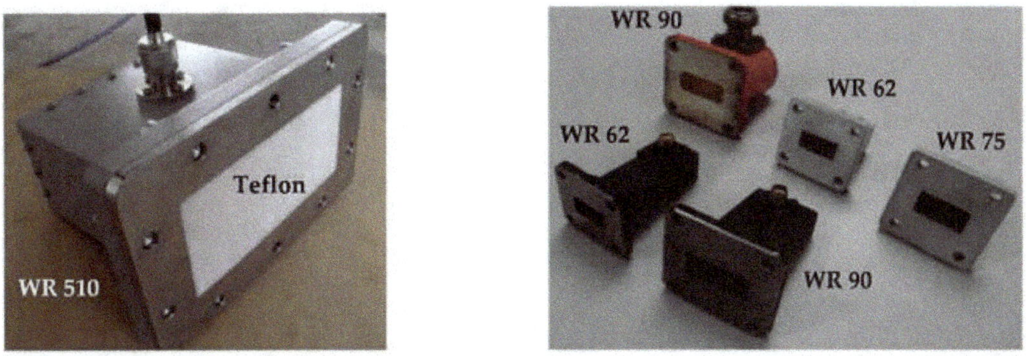

(a) (b)

Figure 4. (a) WR510 waveguide-to-coaxial adapter and (b) WR 90, WR 75 and WR 62 waveguide-to-coaxial adapters [9].

2.2. Measurements principles

The one-port measurement is based on the principle that a reflected signal (reflection coefficient, S_{11}) through the waveguide, which end aperture is contacting firmly with the material under test (sample), will obtain the desired information about the material as shown in **Figure 5**. The main advantage of using one-port reflection technique is that the method is the simplest, broadband, nondestructive way to measure the dielectric properties of a material. However, one-port measurement is suitable only for measuring the relative permittivity, ε_r, of the dielectric material (nonmagnetic material, $\mu_r = 1$). This is due to insufficient information to predict the permeability, μ_r, if only obtained the measured reflection coefficient, S_{11} without transmission coefficient, S_{21}.

For **Figure 5a** and **c** measurements, the sample is considered infinite, as long as the sample thickness d is greater than the radius of the outer conductor b. However, the radiation, or sensing area, for an aperture rectangular waveguide is much greater than that of a coaxial probe. For instance, the WR90 waveguide has a radiation distance up to 20 cm in the air from the aperture waveguide. Hence, the sample under test must be much thicker when the rectangular waveguide is utilized in the measurement. Besides, the coaxial probe and rectangular waveguide are also capable of testing the thin film sample as shown in **Figure 5b** and **d**. The measurements required that the thin sample is backed by a metallic plate.

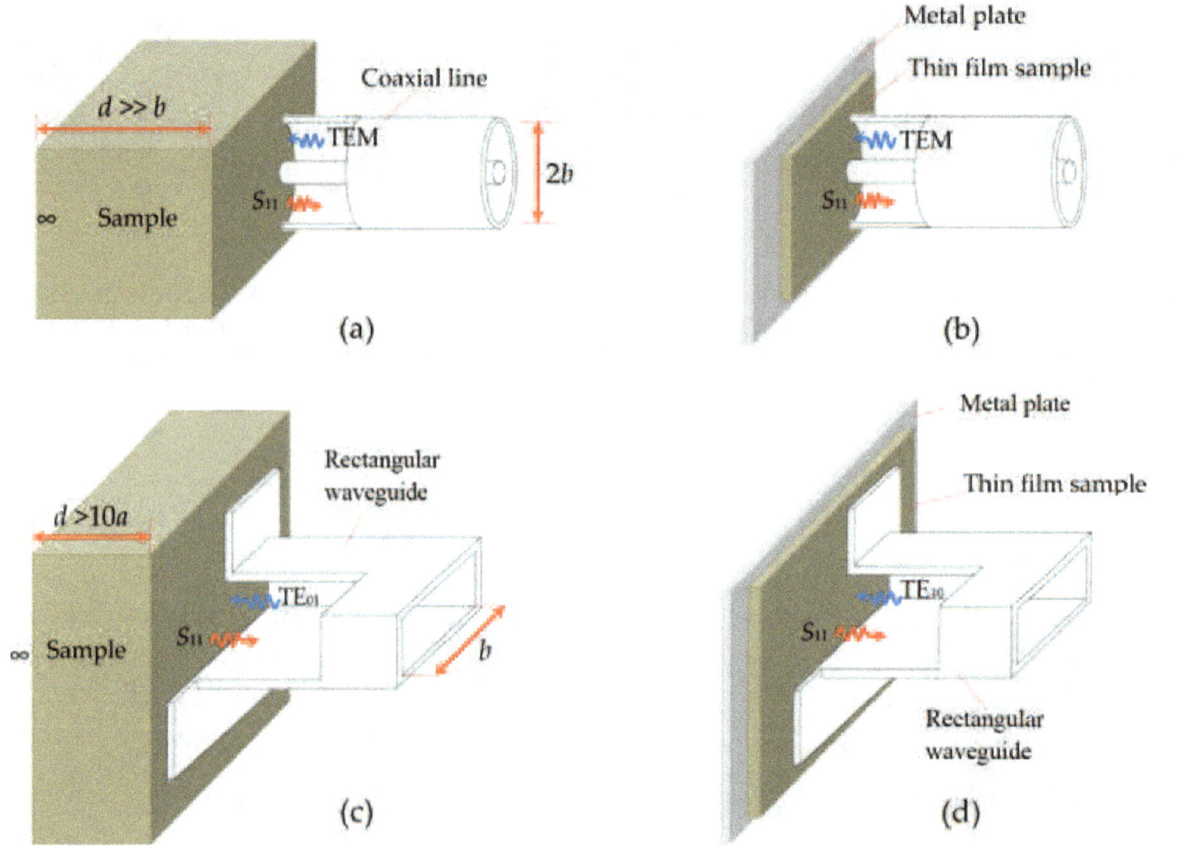

Figure 5. Open-port measurements using the (a), (b) coaxial probe and the (c), (d) rectangular waveguide.

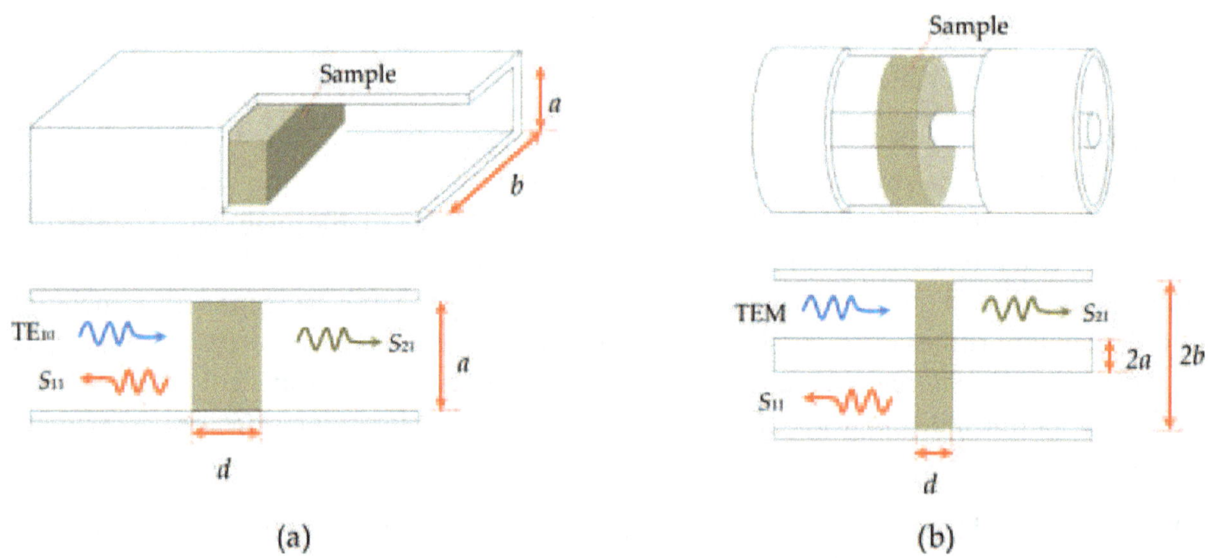

Figure 6. Two-port measurements using the (a) rectangular waveguide and the (b) coaxial transmission line.

The two-port measurement uses both reflection and transmission methods. Here, the material under test is placed between waveguide transmission lines or segments of the coaxial line as shown in **Figure 6**. The two-port measurement using coaxial or rectangular waveguides became popular due to the convenient formulations derived by Nicholson and Ross [10] in

1970, who introduced a broadband determination of the complex relative permittivity, ε_r and permeability, μ_r of materials from reflection and transmission coefficients (S_{11} and S_{21}). For measurements in **Figure 6**, the sample must be solid and carefully machined with parallel interfaces, and must perfectly fill in the whole cross section of the coaxial line or waveguide transmission line. The main advantage of using two-port Nicholson-Ross-Weir (NRW) technique [10, 11] is that the both relative permittivity, ε_r and relative permeability, μ_r of the sample can be predicted simultaneously. When using NRW method for thin samples, the thickness of the sample must be less than $\lambda/4$.

2.3. Measurements setup

In this chapter, the dimensions of the used coaxial probe and the rectangular waveguide as examples of the one-port measurement are shown in **Figure 7a** and **b**, respectively. The coaxial probe is capable of measuring the reflection coefficients covered the frequency range between 0.5 and 7 GHz. On the other hand, the rectangular waveguide adapter covers frequency from 8.2 to 12.4 GHz. For two-port measurement, a 5 cm length of the coaxial and the rectangular transmission lines is implemented. The experiment setup of the waveguides with an Agilent E5071C vector network analyzer (VNA) is shown in **Figure 8**.

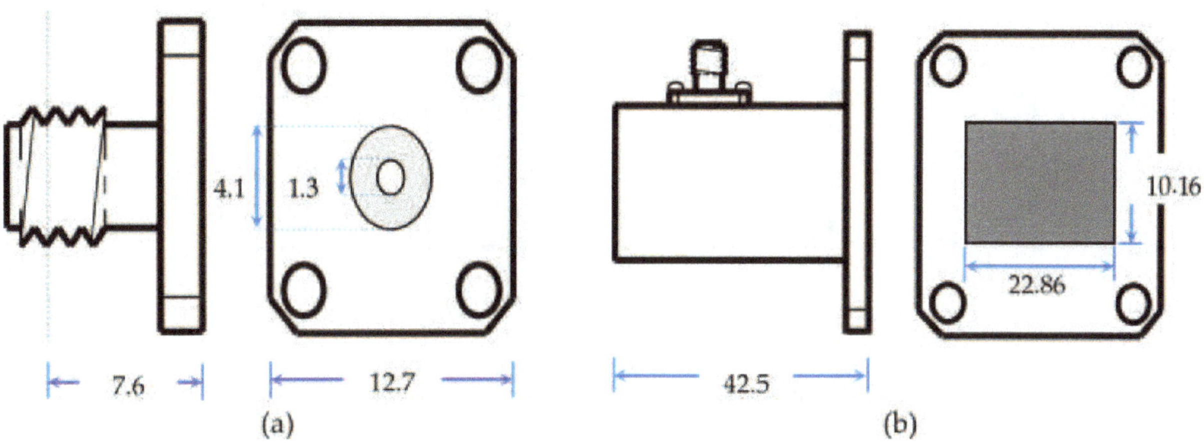

Figure 7. Cross-sectional and front views for the dimensions (in millimeter) of the (a) coaxial probe and the (b) rectangular waveguide.

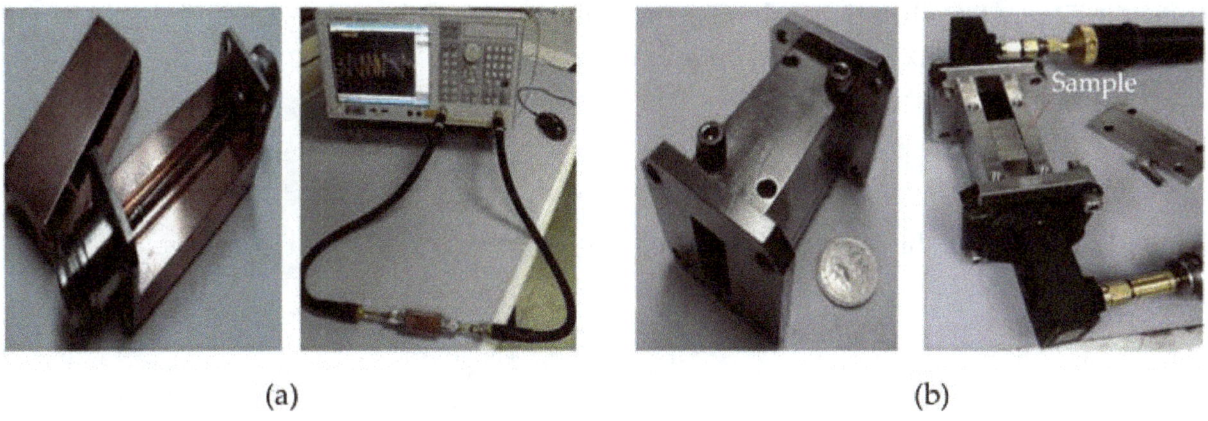

Figure 8. The experiment setup for the (a) coaxial cavity and the (b) rectangular waveguide cavity.

3. Waveguide calibrations

3.1. One-port calibrations

3.1.1. Open standard calibration

The reflection coefficient S_{11a_sample} of the sample at the probe aperture (at the BB' plane) should be measured as shown in **Figure 9b** [8]. However, during the measurement process, only the reflection coefficient S_{11m_sample} at the end of the coaxial line (at the AA' plane) is measured. The measured S_{11m_sample} must be calibrated due to the reflection at the AA' plane, which is separated from the sample (at the BB' plane) by a coaxial line. Thus, a de-embedding process should be done to remove the effects of the coaxial line.

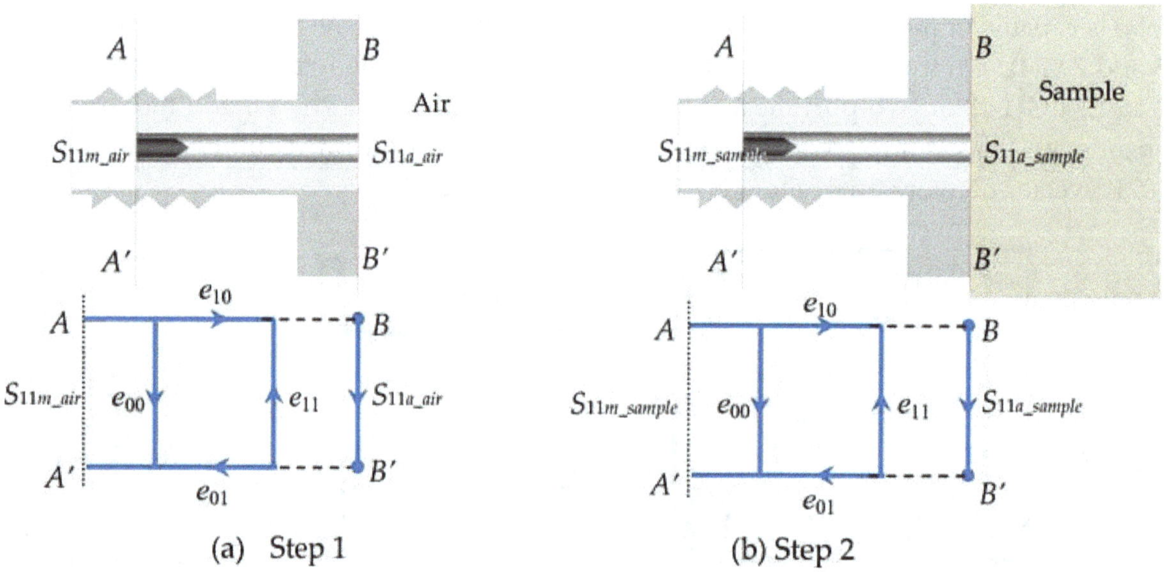

Figure 9. Error network and finite coaxial line: (a) terminated by air; (b) terminated by a sample under test.

In this subsection, a simple open standard calibration is introduced which requires the probe aperture open to the air as shown in **Figure 9a**. Firstly, the S_{11m_air} for the air is measured. Later, the probe aperture is contacted with the sample under test, and its S_{11m_sample} is measured as shown in **Figure 9b**. The relationship between the S_{11m_air} at the plane AA' and S_{11a_air} at the probe aperture BB' is expressed in a bilinear equation as:

$$S_{11a\_air} = \frac{S_{11m\_air} - e_{00}}{e_{11} S_{11m\_air} + e_{10}e_{01} - e_{00}e_{11}} \tag{1}$$

Similarly, the relationship between measured S_{11m_sample} and S_{11a_sample} is given as:

$$S_{11a\_sample} = \frac{S_{11m\_sample} - e_{00}}{e_{11} S_{11m\_sample} + e_{10}e_{01} - e_{00}e_{11}} \tag{2}$$

The e_{00}, e_{11} and $e_{10}e_{01}$ are the unknown scattering parameters of the error network for the coaxial line.

The e_{00} is the directivity error that causes the failure to receive the measured reflection signal completely from the sample being tested at plane BB'. The e_{11} is the source matching error due to the fact that the impedance of the aperture probe at plane BB' is not exactly the characteristic impedance (Z_o = 50 Ω). The $e_{10}e_{01}$ is the frequency tracking imperfections (or phase shift) between plane AA' and sample test plane BB'. For this calibration, the e_{00} and e_{11} terms in Eqs. (1) and (2) are assumed to be vanished ($e_{00} = e_{11} = 0$). By dividing Eq. (2) into Eq. (1), yields

$$\frac{S_{11a\_sample}}{S_{11a\_air}} = \frac{S_{11m\_sample}}{S_{11m\_air}} \tag{3}$$

Once the S_{11m_air} and S_{11m_sample} are obtained, the actual reflection coefficient, S_{11a_sample}, of the sample at the probe aperture, BB' can be found as:

$$S_{11a\_sample} = \frac{S_{11a\_air}}{S_{11m\_air}} \times S_{11m\_sample} \tag{4}$$

The standard values of the reflection coefficient, S_{11a_air} in (4), can be calculated from Eq. (5) that satisfying conditions: (DC < f < 24) GHz.

$$S_{11a\_air} = \frac{1 - j(\omega/Y_o)(C_of^{-1} + C_1 + C_2f + C_3f^2)}{1 + j(\omega/Y_o)(C_of^{-1} + C_1 + C_2f + C_3f^2)} \tag{5}$$

Symbol $\omega = 2\pi f$ and $Y_o = [(2\pi)/\ln(b/a)]\sqrt{(\varepsilon_o\varepsilon_c/\mu_o\mu_r)}$ are the angular frequency (in rad/s) and characteristic admittance (in siemens), respectively. For instance, the complex values of the C_o, C_1, C_2 and C_3 in (5) for Teflon-filled coaxial probe with $2a$ = 1.3 mm, $2b$ = 4.1 mm and ε_c = 2.06 are given as [12]:

$$C_o = 5.368082994761808 \times 10^{-7} + j\,2.320071598550666 \times 10^{-7} \quad (\mathrm{F} \times \mathrm{Hz})$$

$$C_1 = 3.002820660256831 \times 10^{-14} - j\,2.988971515445163 \times 10^{-16} \quad (\mathrm{F})$$

$$C_2 = 1.112989441958266 \times 10^{-25} + j\,7.730261500907114 \times 10^{-26} \quad (\mathrm{F/Hz})$$

$$C_3 = 3.140652268416283 \times 10^{-36} - j\,6.786433840933426 \times 10^{-36} \quad (\mathrm{F/Hz^2})$$

It should be noted that this simple calibration technique will not eliminate the standing wave effects in the coaxial line.

3.1.2. *Open-short-load (OSL) standard calibrations*

In this subsection, the three-standard calibration is reviewed in which open, short and load standards are used in the de-embedding process [13]. Let S_{11a_a}, S_{11a_s} and S_{11a_w} represent the known reflection coefficients for the open, short and load (water) standards which are terminated at the aperture plane BB', while S_{11m_a}, S_{11m_s} and S_{11m_w} are the measured reflection coefficients for open, short and load standards at plane AA'. The S_{11m_a} is measured by leaving the open end of the probe in the air as shown in **Figure 10a**. Later, the measurement is repeated to obtain the S_{11m_s} by terminating the probe aperture with a metal plate as shown in **Figure 10b**. Finally, the S_{11m_w} is obtained by immersing the probe in water as shown in **Figure 10c**.

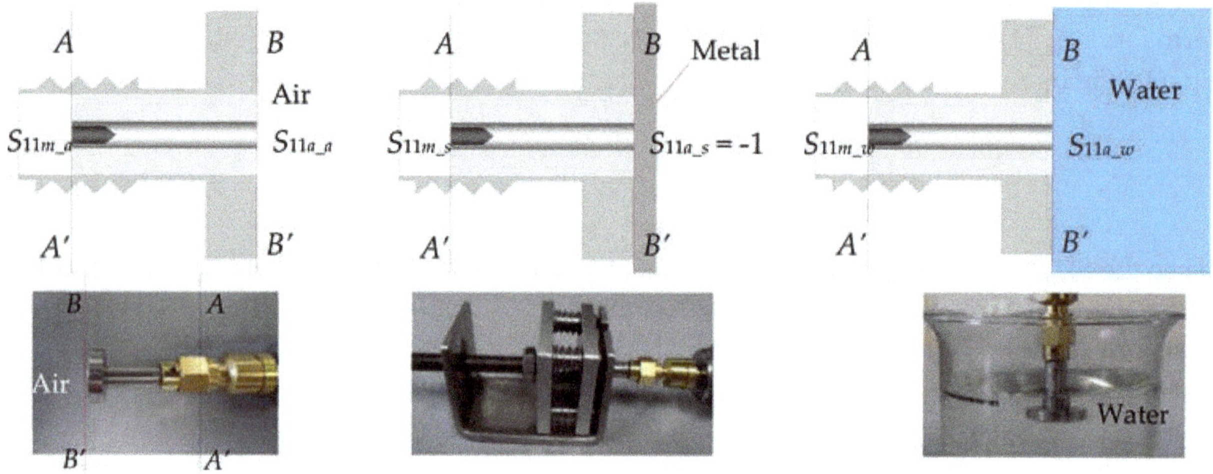

(a) Step 1 (b) Step 2 (c) Step 3

Figure 10. Finite coaxial line: (a) terminated by free space; (b) shorted by a metal plate; (c) immersed in water.

Once the complex values of S_{11a_air}, S_{11a_short}, S_{11a_water}, S_{11m_a}, S_{11m_s} and S_{11m_w} are known, the three unknown complex coefficients (e_{00}, e_{11}, and $e_{10}e_{01}$) values in Eq. (2) can be found as:

$$e_{00} = \frac{S_{11a\_s}S_{11a\_w}S_{11m\_a}\Delta_{w\_s} + S_{11a\_o}S_{11a\_s}S_{11m\_w}\Delta_{s\_a} + S_{11a\_a}S_{11a\_w}S_{11m\_s}\Delta_{a\_w}}{S_{11a\_w}S_{11a\_s}\Delta_{w\_s} + S_{11a\_a}S_{11a\_s}\Delta_{s\_a} + S_{11a\_w}S_{11a\_a}\Delta_{a\_w}} \tag{6a}$$

$$e_{11} = \frac{-(S_{11a\_a}\Delta_{w\_s} + S_{11a\_w}\Delta_{s\_a} + S_{11a\_s}\Delta_{a\_w})}{S_{11a\_w}S_{11a\_s}\Delta_{w\_s} + S_{11a\_a}S_{11a\_s}\Delta_{s\_a} + S_{11a\_w}S_{11a\_a}\Delta_{a\_w}} \tag{6b}$$

$$e_{10}e_{01} = (e_{00} \times e_{11}) + \frac{S_{11a\_a}S_{11m\_a}\Delta_{w\_s} + S_{11a\_w}S_{11m\_w}\Delta_{s\_a} + S_{11a\_s}S_{11m\_s}\Delta_{a\_w}}{S_{11a\_w}S_{11a\_s}\Delta_{w\_s} + S_{11a\_a}S_{11a\_s}\Delta_{s\_a} + S_{11a\_w}S_{11a\_a}\Delta_{a\_w}} \tag{6c}$$

where

$\Delta_{a\_w} = S_{11m\_a} - S_{11m\_w}$, $\Delta_{s\_a} = S_{11m\_s} - S_{11m\_a}$, and $\Delta_{w\_s} = S_{11m\_w} - S_{11m\_s}$

3.1.3. Short-offset-offset short (SOO) standard calibrations

The open-short-load (OSL) technique is rarely used in the one-port rectangular waveguide calibration due to unavailable commercial open kit for the rectangular waveguide. In this subsection, the short-offset-offset short (SOO) calibration [14, 15] is introduced for waveguide calibration by using waveguide adjustable sliding shorts as shown in **Figure 11**. The calibration procedures are shown in **Figure 12**.

In this calibration method, the measured reflection coefficients for one shorted aperture and two different lengths, l of offset short are required. Let S_{11m_1}, S_{11m_2} and S_{11m_3} represent the known measured reflection coefficients at plane AA' for the shorted aperture and the two offset shorts at location l_1 and l_2 from the waveguide aperture, respectively. Before calibration, the selection of the appropriate offset short length, l_1 and l_2 will be an issue. The lengths of the offset shorts can be determined by conditions:

1. The three phase shift between the S_{11m_1}, S_{11m_2} and S_{11m_3} must not be equal:

$$\angle(S_{11m\_2})-\angle(S_{11m\_1})\neq\angle(S_{11m\_3})-\angle(S_{11m\_1})\neq\angle(S_{11m\_3})-\angle(S_{11m\_2})$$

2. The resolution degree between any three phase shift must be significant large (>100°) as shown in **Figure 13**. In this work, the distance l_1 and l_2 for the offset shorts from the X-band waveguide aperture are equal to 0.007 m and 0.013 m, respectively.

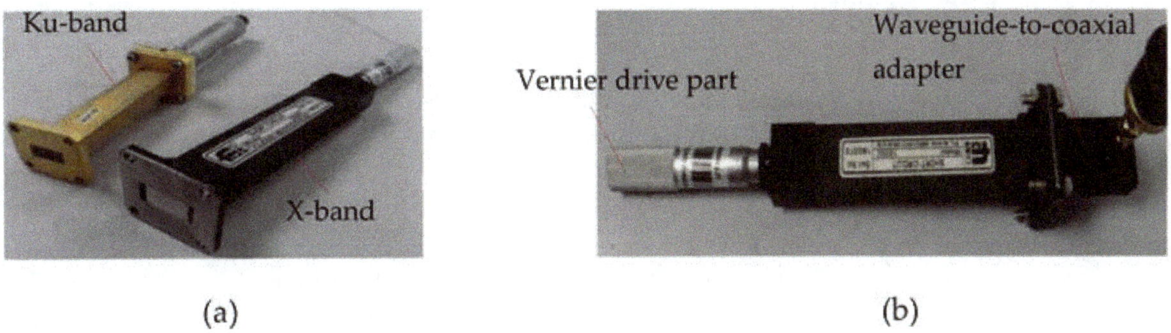

Figure 11. (a) Ku-band and X-band waveguide adjustable sliding shorter. (b) Connection between sliding short with waveguide-to-coaxial adapter.

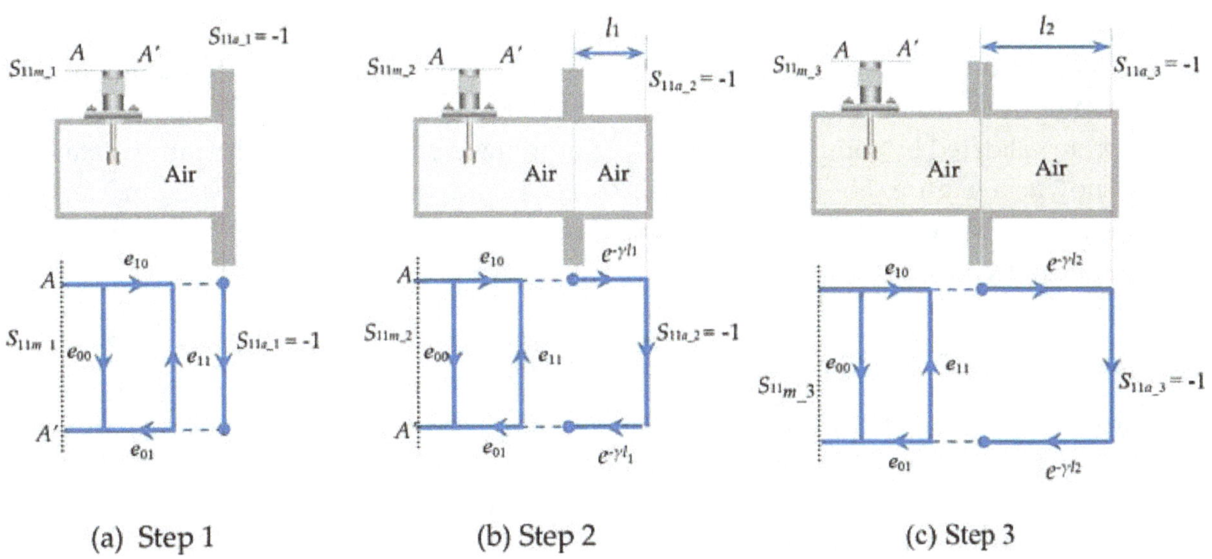

Figure 12. Calibration procedures of the aperture rectangular waveguide using an adjustable shorter. (a) Step 1; (b) Step 2; (c) Step 3

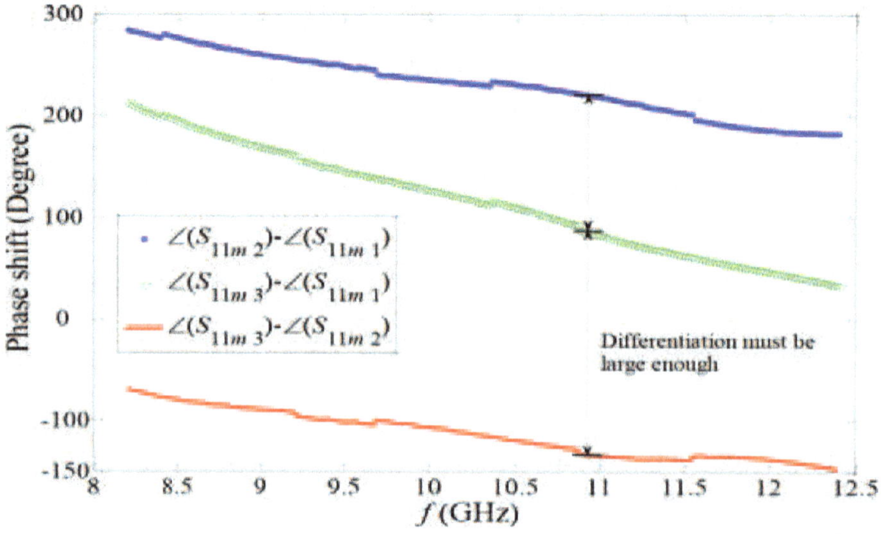

Figure 13. The three phase shift of the measured reflection coefficients for the shorted aperture and two offset shorts with l_1 = 0.7 cm and l_2 = 1.3 cm, respectively.

Once the S_{11m_1}, S_{11m_2}, S_{11m_3}, l_1 and l_2 are obtained, the three unknown complex coefficients (e_{00}, e_{11}, and $e_{10}e_{01}$) values in Eq. (2) can be found as:

$$e_{00} = \frac{S_{11m\_1}S_{11m\_2}(e^{-2\gamma l_1}-1)-S_{11m\_2}S_{11m\_3}\left(e^{2\gamma(l_2-l_1)}-1\right)-S_{11m\_1}S_{11m\_3}\left(e^{-2\gamma l_1}-e^{2\gamma(l_2-l_1)}\right)}{(e^{-2\gamma l_1}-1)(S_{11m\_2}-S_{11m\_3})-\left(e^{2\gamma(l_2-l_1)}-1\right)(S_{11m\_2}-S_{11m\_1})} \tag{7a}$$

$$e_{11} = \frac{e^{2\gamma l_1}(S_{11m\_2}-e_{00})+e_{00}-S_{11m\_1}}{S_{11m\_1}-S_{11m\_2}} \tag{7b}$$

$$e_{10}e_{01} = (e_{00}-S_{11m\_1})(1+e_{11}) \tag{7c}$$

The complex reflection coefficient, S_{11a_sample}, at the waveguide aperture which is open to the air was measured. Then, the measured S_{11a_sample} was converted to normalized admittance, Y/Y_o parameter by a formula: $Y/Y_o = (1 - S_{11a\_sample})/(1 + S_{11a\_sample})$. The SOO calibration techniques were validated by comparing normalized admittance, Y/Y_o with the literature data [15–22] as shown in **Figure 14**. The real part, Re(Y/Y_o), and the imaginary part, Im(Y/Y_o), of admittance results were found to be in good agreement with literature data over the operational range of frequencies.

3.2. Two-port calibrations [through-reflect-line (TRL)]

The through-short-line (TRL) calibration model [23] is used for two-port rectangular waveguide measurement. The TRL technique requires three standards, which are *through, short* and *line* measurements at the *CC′* and *DD′* planes so-called reference planes (at the front surface of the sample under test) as shown in **Figure 15**.

The error coefficients (e_{00}, e_{11}, $e_{10}e_{01}$, e_{33}, e_{22}, e_{32}, and e_{23}) in **Figure 15** can be obtained by solving the matrix equation of Eq. (8).

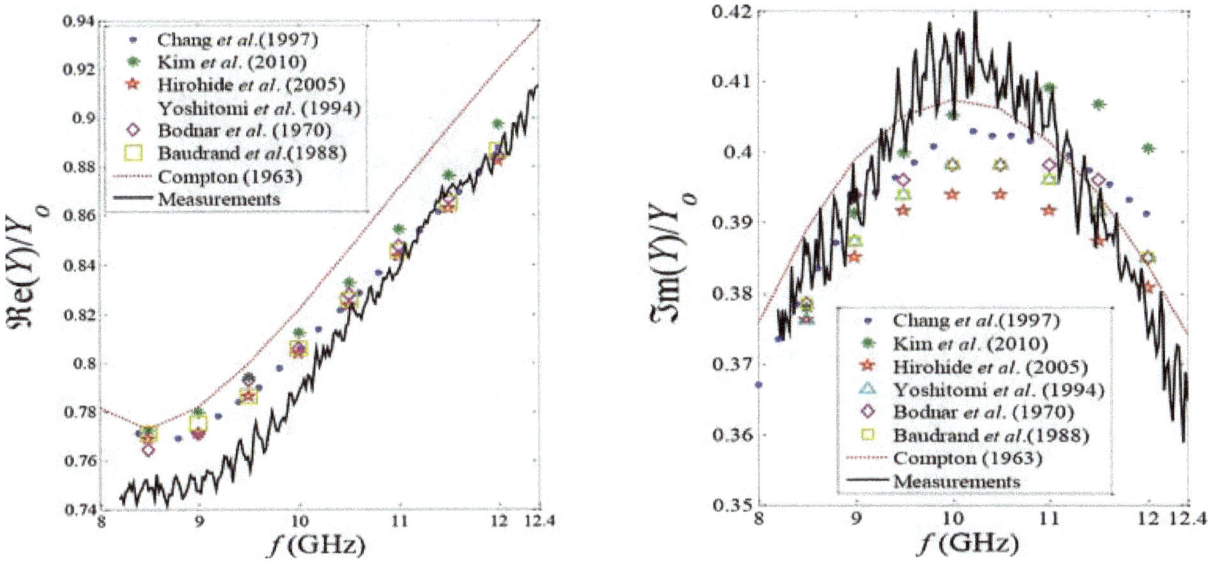

Figure 14. Comparison of real part, Re(Y/Y_o), and imaginary part, Im(Y/Y_o), of the normalized admittance for air.

$$\begin{bmatrix} 1 & 0 & 0 & 0 & S_{12m\_Thru} & 0 & 0 & 0 & 0 & 0 & 0 & 0 \\ 0 & S_{11m\_Thru} & -1 & 0 & 0 & 0 & -S_{12m\_Thru} & 0 & 0 & 0 & 0 & 0 \\ 0 & 0 & 0 & 0 & S_{22m\_Thru} & -1 & 0 & 0 & 0 & 0 & 0 & 0 \\ 0 & S_{21m\_Thru} & 0 & 1 & 0 & 0 & -S_{22m\_Thru} & 0 & 0 & 0 & 0 & 0 \\ 1 & -S_{11m\_Short} & 1 & 0 & 0 & 0 & 0 & 0 & 0 & 0 & 0 & 0 \\ 0 & 0 & 0 & 0 & -S_{12m\_Short} & 0 & -S_{12m\_Short} & 0 & 0 & 0 & 0 & 0 \\ 0 & -S_{21m\_Short} & 0 & 0 & 0 & 0 & 0 & 0 & 0 & 0 & 0 & 0 \\ 0 & 0 & 0 & 1 & -S_{22m\_Short} & 1 & -S_{22m\_Short} & 0 & 0 & 0 & 0 & 0 \\ 1 & 0 & 0 & 0 & e^{-j\beta l}S_{12m\_Line} & 0 & 0 & 0 & 0 & 0 & 0 & 0 \\ 0 & e^{j\beta l}S_{11m\_Line} & e^{j\beta l} & 0 & 0 & 0 & -S_{12m\_Line} & 0 & 0 & 0 & 0 & 0 \\ 0 & 0 & 0 & 0 & e^{-j\beta l}S_{22m\_Line} & -e^{-j\beta l} & 0 & 0 & 0 & 0 & 0 & 0 \\ 0 & e^{j\beta l}S_{21m\_Line} & 0 & 1 & 0 & 0 & -S_{22m\_Line} & 0 & 0 & 0 & 0 & 0 \end{bmatrix} \begin{bmatrix} e_{00} \\ e_{11} \\ \Delta_x \\ ke_{33} \\ ke_{22} \\ k\Delta_y \\ k \\ 0 \\ 0 \\ 0 \\ 0 \\ 0 \end{bmatrix} = \begin{bmatrix} S_{11m\_Thru} \\ 0 \\ S_{21m\_Thru} \\ 0 \\ S_{11m\_Short} \\ 0 \\ S_{21m\_Short} \\ 0 \\ S_{11m\_Line} \\ 0 \\ S_{21m\_Line} \\ 0 \end{bmatrix} \tag{8}$$

where $k = e_{10}/e_{23}$, $\Delta_x = e_{00}e_{11}-e_{10}e_{01}$ and $\Delta_y = e_{22}e_{33}-e_{32}e_{23}$. Once the S_{11m_sample}, S_{21m_sample}, S_{12m_sample} and S_{22m_sample} at plane AA' and BB' for the sample under test are measured, the calibrated reflection coefficient, S_{11a_sample} at plane CC' and transmission coefficient, S_{21a_sample} at plane DD' can be calculated as:

$$S_{11a\_sample} = \frac{\left\{ \begin{array}{l} \left(\dfrac{S_{11m\_sample}-e_{00}}{e_{10}e_{01}}\right)\left[1+\left(\dfrac{S_{22m\_sample}-e_{33}}{e_{23}e_{32}}\right)e_{22}\right] \\ -e_{22}\left(\dfrac{S_{21m\_sample}-S_{21m\_Thru}}{e_{10}e_{32}}\right)\left(\dfrac{S_{12m\_sample}-S_{12m\_Thru}}{e_{23}e_{01}}\right) \end{array} \right\}}{D} \tag{9a}$$

$$S_{21a\_sample} \cong \frac{\left(\dfrac{S_{21m\_sample}-S_{21m\_Thru}}{e_{10}e_{32}}\right)}{D} \tag{9b}$$

The denominator, D in (9a) and (9b) is given as:

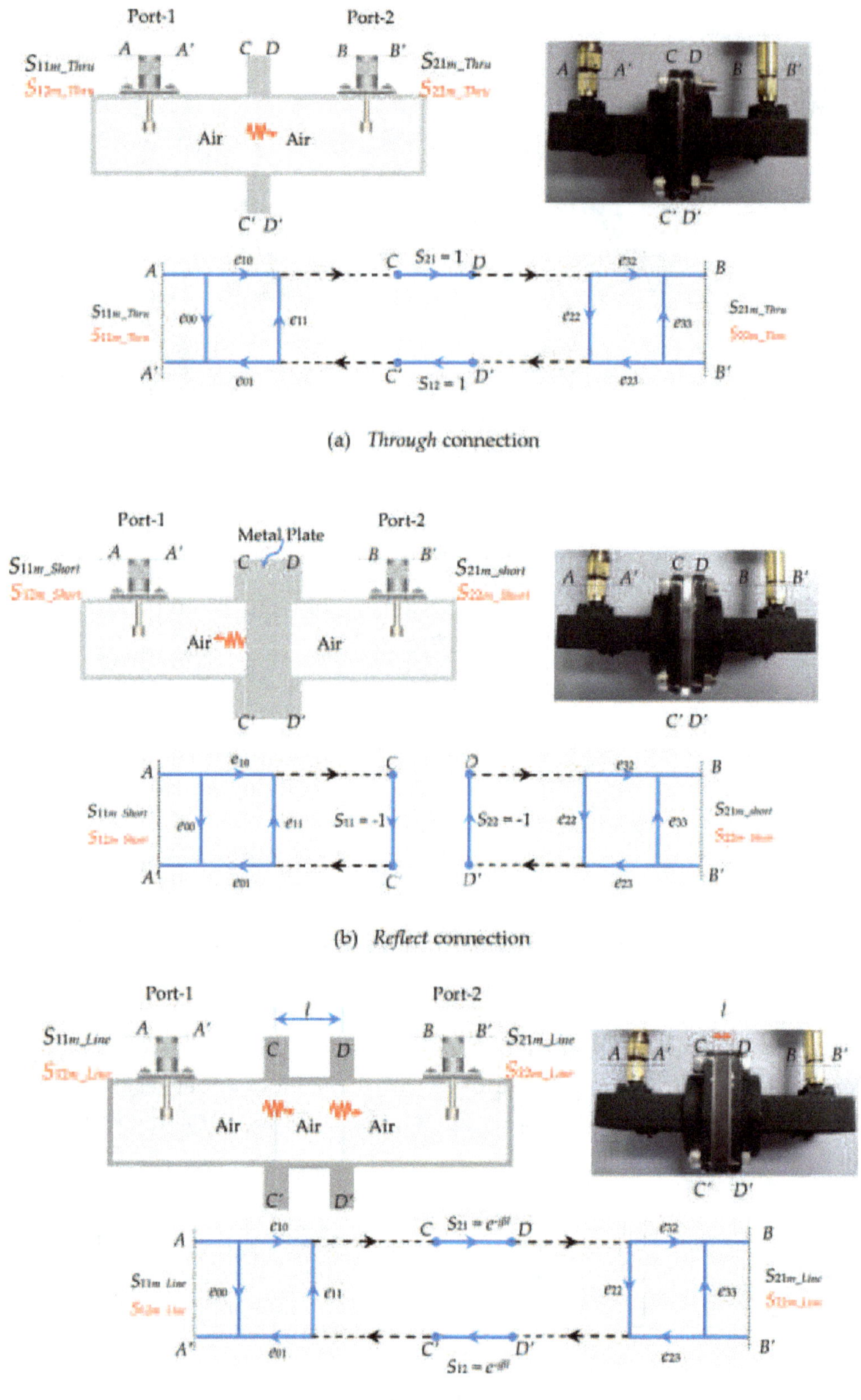

(a) *Through* connection

(b) *Reflect* connection

(c) *Line* connection

Figure 15. through-short-line (TRL) calibration procedures and its network errors. *(a) Through* connection; *(b) Reflect* connection; *(c) Line* connection.

$$D = \left\{ \begin{array}{l} \left[1+\left(\frac{S_{11m\_sample}-e_{00}}{e_{10}e_{01}}\right)e_{11}\right]\left[1+\left(\frac{S_{22m\_sample}-e_{33}}{e_{23}e_{32}}\right)e_{22}\right] \\ -\left(\frac{S_{21m\_sample}-S_{21m\_Thru}}{e_{10}e_{32}}\right)\left(\frac{S_{12m\_sample}-S_{12m\_Thru}}{e_{23}e_{01}}\right)e_{22}e_{11} \end{array} \right\}$$

4. Material parameters extraction

4.1. Reflection measurements (one-port measurements)

There are two methods of determining sample parameters (ε_r, μ_r or σ), which are the direct method and the inverse method. The direct method involves the explicit model to predict the sample under test based on the measured reflection coefficient, S_{11a_sample}, while the inverse method is implemented rigorous integral admittance model to estimate the sample parameters (ε_r, μ_r or σ) using optimization procedures. For coaxial probe measurement cases, the explicit relationship between ε_r and S_{11a_sample} [8] is tabulated in **Table 1**. For rectangular waveguide cases, the measured S_{11a_sample} is transferred to normalized admittance, $\tilde{Y}_{a\_sample}$ through equation: $\tilde{Y}_{a\_sample} = (1 - S_{11a\_sample})/(1 + S_{11a\_sample})$. The predicted value of ε_r is obtained by minimizing the difference between the measured normalized admittance, $\tilde{Y}_{a\_sample}$ and the quasi-static integral model, $\tilde{Y}$ (in **Table 2**) [9, 17] by referring to particular objective function. The procedures of direct method are more straightforward than the inverse method. The detail descriptions of the parameters (Y_o, C and γ_o) and the coefficients (a_1, a_2 and a_3) in Eqs. (10)–(13) can be found in [8, 9, 17].

4.2. Reflection/transmission measurements (two-port measurements)

Conventionally, the complex $\varepsilon_r = \varepsilon_r' - j\varepsilon_r''$ and the $\mu_r = \mu_r' - j\mu_r''$ of the sample filled in the coaxial or rectangular waveguide are obtained by converting the calibrated reflection coefficient, S_{11a_sample} and the transmission coefficient, S_{21a_sample} by using Nicholson-Ross-Weir (NRW) routines [10, 11]. In this section, another alternative method, namely transmission phase shift (TPS) method [24], is reviewed. The TPS method is a calibration-independent and material position-invariant technique, which can reduce the complexity of the de-embedding procedures. The important formulations of the NRW and the TPS methods are tabulated in **Table 3**.

Sample cases	Open-ended coaxial probe
Semi-infinite space sample (**Figure 5a**)	$\varepsilon_r = \left(\frac{Y_o}{j\omega C}\right)\left(\frac{1-S_{11a\_sample}}{1+S_{11a\_sample}}\right)$ (10)
Thin sample backed by metal plate (**Figure 5b**)	$\varepsilon_r = \left(\frac{Y_o}{j\omega C}\right)\left(\frac{1-S_{11a\_sample}}{1+S_{11a\_sample}}\right)(a_1 + a_2 e^{-d/M} + a_3 e^{-2d/M})$ (11)

Table 1. Explicit formulations for open-ended coaxial probe.

Sample cases	Open-ended rectangular waveguide
Semi-infinite space sample (**Figure 5c**)	$$\tilde{Y}=\frac{j8b}{a\gamma_o}\int_0^a\int_0^b\left\{(a-x)\left\{D_1(b-y)\cos\frac{\pi y}{b}+D_2\sin\frac{\pi y}{b}\right\}\times\frac{\exp\left(-jk_1\sqrt{x^2+y^2}\right)}{\sqrt{x^2+y^2}}\right\}dxdy \quad (12)$$ where $D_1=\frac{1}{b^2}\left(\frac{k_1^2}{4\pi}-\frac{\pi}{4b^2}\right)$, $D_2=\frac{1}{\pi b}\left(\frac{k_1^2}{4\pi}+\frac{\pi}{4b^2}\right)$, and $k_1=\frac{2\pi f}{c}\sqrt{\varepsilon_r}$
Thin sample backed by metal plate (**Figure 5d**)	$$\tilde{Y}=\frac{j8b}{a\gamma_o}\int_0^a\int_0^b\chi\frac{\exp\left(-jk_1\sqrt{x^2+y^2}\right)}{\sqrt{x^2+y^2}}dxdy+\frac{j16b}{a\gamma_o}\sum_{n=1}^{\infty}\int_0^a\int_0^b\chi\frac{\exp\left(-jk_1\sqrt{x^2+y^2+4n^2d^2}\right)}{\sqrt{x^2+y^2+4n^2d^2}}dxdy \quad (13)$$ where $\chi=(a-x)\left\{D_1(b-y)\cos\frac{\pi y}{b}+D_2\sin\frac{\pi y}{b}\right\}$

Table 2. Integral admittance formulations for open-ended rectangular waveguide.

	Waveguide factors	Explicit equations
NRW method [10–11]	Coaxial: $\zeta=1$ Waveguide: $\zeta=\sqrt{k_o^2-\left(\frac{\pi}{b}\right)^2}$	$$\varepsilon_r=j\zeta\left(\frac{1-\Gamma}{1+\Gamma}\right)\left(\frac{c}{2\pi fd}\right)\ln\left(\frac{1}{T}\right) \quad (14a)$$
		$$\mu_r=j\frac{1}{\zeta}\left(\frac{1+\Gamma}{1-\Gamma}\right)\left\{\frac{c}{2\pi fd}\ln\left(\frac{1}{T}\right)\right\} \quad (14b)$$
TPS method [24]	Coaxial: $\xi=0$ and $\gamma_o=k_o$	$$\varepsilon'_r=\frac{1}{k_o^2}\left\{\left(\gamma_o+\frac{\phi_{21\_air}-\phi_{21\_sample}}{d}\right)^2+\xi-\alpha^2\right\} \quad (15a)$$
	Waveguide: $\xi=\left(\frac{\pi}{b}\right)^2\gamma_o=\sqrt{k_o^2-\left(\frac{\pi}{b}\right)^2}$	$$\varepsilon''_r=\frac{2\alpha}{k_o^2}\left(\gamma_o+\frac{\phi_{21\_air}-\phi_{21\_sample}}{d}\right) \quad (15b)$$

Table 3. Explicit formulations for reflection/transmission measurements.

where $k_o = 2\pi f/c$ is the propagation constant of free space ($c = 2.99792458\ \mathrm{ms}^{-1}$); b (in meter) are the width of the aperture of the waveguide, respectively; d(in meter) is the thickness of the sample. The expressions for parameters Γ and T in Eqs. (14a, b) can be found in [10, 24]. The ϕ_{21_air} and ϕ_{21_sample} in Eqs. (15a, b) are the measured phase shift of the transmission coefficient in the air (without sample) and the sample, respectively. On the other hand, symbol α (in nepers/meter) is the dielectric attenuation constant for the sample.

You et al. [24] have been mentioned that the uncertainty of the permittivity measurement is high for the low-loss thin sample by using TPS method due to the decreasing of the sensitivity for the transmitted wave through the thin sample, especially for transmitted waves that have longer wavelengths. However, the literature [24] did not discuss how the thickness of the thin sample may affect the uncertainty of measurement using TPS technique in quantitative. From this reasons, the TPS method is reexamined in this section. Various thicknesses of acrylic, FR4 and RT/duroid 5880 substrate samples were placed in the X-band rectangular waveguide and measured for validation. **Figure 16a–c** shows the predicted dielectric constant, ε_r' of the samples using Eq. (15a) at 8.494, 10.006 and 11.497 GHz, respectively. Clearly, the TPS method is capable of providing a stable and accurate measurement for operating frequency in X-band range when the thicknesses of the samples have exceeded 2 cm [25].

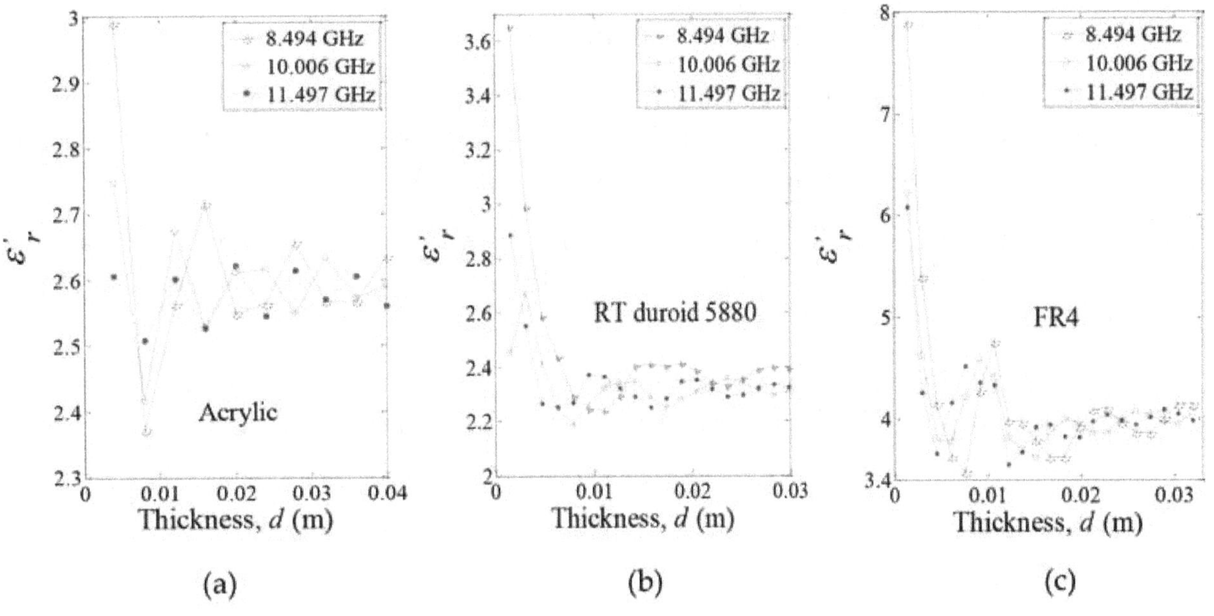

Figure 16. Variations in relative dielectric constant, ε_r with the thickness layer of (a) acrylic, (b) RT/duroid 5880 substrate and (c) FR4, respectively.

5. Conclusion

The brief background of the microwave waveguide techniques for materials characterization is reviewed and summarized. Not only that the measurement methods play an important role, the calibration process is crucial as well. However, most of the literatures have ignored the description of calibration. Measurement without calibration certainly cannot predict the properties of materials accurately. Thus, in this chapter, some of the waveguide calibrations are described in detail.

Author details

Kok Yeow You

Address all correspondence to: kyyou@fke.utm.my

Department of Communication Engineering, Faculty of Electrical Engineering, Universiti Teknologi Malaysia, Johor, Malaysia

References

[1] Hakki B W, Coleman P D: A dielectric resonator method of measuring inductive capacities in the millimeter range. IRE Transactions on Microwave Theory and Techniques. 1960; **8(7)**: 402–410.

[2] Courtney W E: Analysis and evaluation of a method of measuring the complex permittivity and permeability of microwave insulators. IEEE Transactions on Microwave Theory and Techniques. 1970; **18(8)**: 476–485.

[3] Tosaka T, Fujii K, Fukunaga K, Kasamatsu A: Development of complex relative permittivity measurement system based on free-space in 220–330 GHz range. IEEE Transaction on Terahertz Science and Technology. 2015; **5(1)**: 102–109.

[4] Abbas Z, Pollard R D, Kelsall R W: Complex permittivity measurements at Ka-Band using rectangular dielectric waveguide. IEEE Transactions on Instrumentation and Measurement. 2001; **50(5)**: 1334–1342.

[5] Kemptner E, Thurner S: Free space material characterization for microwave frequencies. 6th European Conference on Antennas and Propagation (EUCAP). 2011: 3513–3515.

[6] You K Y, Mun H K, You L L, Jamaliah S, Abbas Z: Small and slim coaxial probe for single rice grain moisture sensing. Sensors. 2013;**13(3)**: 3652–3663.

[7] You K Y, Jamaliah S, Abbas Z: Effects of length and diameter of open-ended coaxial sensor on its reflection coefficient. RadioEngineering. 2012; **21(1)**: 496–503.

[8] You K Y, Then Y L: Simple calibration and dielectric measurement technique for thin material using coaxial probe. IEEE Sensors Journal. 2015; **15(10)**: 5393–5397.

[9] You K Y, Abbas Z, Malek M F A, Cheng E M: Non-destructive dielectric measurements and calibration for thin materials using waveguide-coaxial adaptors. Measurement Science Review. 2014; **14(1)**: 16–24.

[10] Nicolson A M, Ross G F: Measurement of the intrinsic properties of materials by time-domain techniques. IEEE Transactions on Instrumentation and Measurement. 1970; **19(4)**: 377–382.

[11] Weir W B: Automatic measurement of complex dielectric constant and permeability at microwave frequencies. Proceedings of the IEEE. 1974; **62(1)**: 33–36.

[12] You K Y. RF Coaxial Slot Radiators: Modeling, Measurements, and Applications. USA: Artech House; 2015. ISBN: 978-1-60807-822-6.

[13] Kraszewski A, Stuchly M A, Stuchly S S: ANA calibration method for measurements of dielectric properties. IEEE Transactions on Instrumentation and Measurement. 1983; **32 (2)**: 385–386.

[14] Da Silva E F, McPhun M K: Calibration techniques for one-port measurement. Microwave Journal. 1978; **21(6)**: 97–100.

[15] Chang C W, Chen K M, Qian J: Nondestructive determination of electromagnetic parameters of dielectric materials at X-band frequencies using a waveguide probe system. IEEE Transactions on Instrumentation and Measurement. 1997; **46(5)**: 1084–1092.

[16] Kim J H, Enkhbayar B, Bang J H, Ahn B C:New formulas for the reflection coefficient of an open-ended rectangular waveguide radiating into air including the effect of wall thickness or flange. Progress in Electromagnetics Research M. 2010; **12**: 143–153.

[17] Compton R T Jr. The Aperture Admittance of a Rectangular Waveguide Radiating into a Lossy Half-Space. Technical Report, 1691-1, Columbus, Ohio: Ohio State University; 1963.

[18] Ganchev S I, Bakhtiari S, Zoughi R: A novel numerical technique for dielectric measurement of generally lossy dielectrics. IEEE Transactions on Instrumentation and Measurement. 1992; **41(3)**: 361–365.

[19] Yoshitomi K, Sharobim H R: Radiation from a rectangular waveguide with a lossy flange. IEEE Transactions on Antennas and Propagation.1994; **42(10)**: 1398–1403.

[20] Hirohide, Serizawa, Hongo K: Radiation for a flanged rectangular waveguide. IEEE Transactions on Antennas and Propagation.2005; **53(12)**: 3953–3962.

[21] Bodnar D G, Paris D T: New variational principle in electromagnetic. IEEE Transactions on Antennas and Propagation.1970; **18(2)**: 216–223.

[22] Baudrand H, Tao J, Atechian J: Study of radiation properties of open-ended rectangular waveguides. IEEE Transactions on Antennas and Propagation. 1988; **36(8)**: 1071–1077.

[23] Engen G F, Hoer C A: Thru-reflect-line: an improved technique for calibrating the dual six-port automatic network analyzer. IEEE Transactions on Microwave Theory and Techniques. 1979; **27(12)**: 987–993.

[24] You K Y, Lee Y S, Zahid L, Malek M F A, Lee K Y, Cheng E M: Dielectric measurements for low-loss materials using transmission phase-shift method. JurnalTeknologi. 2015; **77 (10)**: 69–77.

[25] You K Y: Effects of sample thickness for dielectric measurements using transmission phase-shift method. International Journal of Advances in Microwave Technology. 2016; **1(3)**: 64–67.

5

Review on Microwave Metamaterial Structures for Near-Field Imaging

Hatem El Matbouly

Abstract

In the past decade, metamaterials have attracted a lot of attention because of their abilities to exhibit unusual electromagnetic properties. These properties are exploited in designing functional components and devices for many potential applications. In this chapter, we review the theory and design of metamaterial structures for microwave near-field imaging/microscopy. The chapter highlights metamaterial microwave components to obtain super-resolution and manipulating subwavelength images. Moreover, a review on surface plasmons manipulation at microwave frequencies is presented as a technique to enhance the transmission for near-field microscopy. The chapter presents as well a survey on the near-field imaging instrumentation and its advances mainly for microwave and THz regimes.

Keywords: metamaterials, microscopy, near-field, microwave, novel electromagnetic structures, subwavelength imaging, surface plasmons

1. Introduction

The technique of microwave near-field microscopy is used to characterize materials by measuring the microwave electromagnetic response of materials on length scales shorter than the wavelength of the incident electromagnetic wave. With this technique one can achieve a spatial resolution at a submicrometer level compared to conventional far-field imaging technique, which has a resolution limitation [1]. This technique may facilitate new measurements on systems that cannot be studied directly with scanned probes, such as buried features or single molecule detection [2] as well as in biomedical applications [3].

New advances in metamaterials research are making unnatural electromagnetic phenomena become realizable [4]. The purpose of this chapter is to focus on using metamaterial and its exciting properties for the enhancement of near-field microscopy and imaging.

This chapter starts with a general overview on the near-field imaging/microscopy and its applications including a description of the microwave near-field imaging system and its components. Metamaterial design technique will be discussed as a step to design novel structures that allow further improvements in spatial resolution.

The chapter concludes with a review on another possibility to enhance microwave near-field microscopy that is the manipulation of surface plasmons at microwave and millimeter-wave frequencies. This can enhance the transmission through subwavelength near-field apertures.

2. Overview on microwave near-field imaging/microscopy

2.1. Backgrounds

The conventional far-field optical imaging system suffers from resolution limitations, this is due to Abbe's theory that restrains the spatial resolution to a diffraction limit [5]. In case of far-field imaging, forming an image is subjected to the Rayleigh criterion, which is the condition to resolve the diffraction pattern of two points separated by a distance of Δy. This condition is given by $\Delta y = 0.61\ \lambda/\mathrm{NA}$, where λ is the wavelength and (NA) is the numerical aperture of the microscope [6]. On the contrary, subwavelength spatial resolution imaging can only be achieved by using near-field technique suggested first by E.A Synge in early 1928 (see the next section for a survey on near-field technique). The near-field results in the excitation of an electric dipole moment within a sample, the idea is to probe the dipole radiation in the near-field region, $\vec{E}_{rad}$ is the radiated electric field and $\vec{B}_{rad}$ is the radiated magnetic field of the dipole radiation which are given by [7]

$$\vec{E}_{\mathrm{rad}} = \frac{1}{4\pi\varepsilon_0}\left\{k^2(\hat{r}\times\vec{p})\times\hat{r}\frac{e^{ikr-\omega t}}{r} + (3r(\hat{r}\cdot\vec{p})-\vec{p})\left(\frac{1}{r^3}-\frac{ik}{r^2}\right)e^{ikr-\omega t}\right\} \quad (1)$$

$$\vec{B}_{\mathrm{rad}} = \frac{1}{4\pi}\left\{k^2(\hat{r}\times\vec{p})\left(\frac{1}{r^3}-\frac{ik}{r^2}\right)e^{ikr-\omega t}\right\} \quad (2)$$

where $\vec{p}$ is the dipole moment, $\vec{k}$ is the wave number and $\vec{r}$ is the distance between the source and the observation point. The first part of Eq. (1) is the propagating far field of the dipole radiation that cannot be used for subwavelength imaging because the diffraction limits the image resolution. However, the second part is the near field of the dipole radiation and there is no diffraction problem associated with using it since it has a nonpropagating nature. In this case the spatial resolution will not be limited by the wavelength if the nonpropagating near field is used [7, 8].

2.2. Description of a microwave near-field imaging microscope

A schematic diagram of the near-field microwave microscope is shown in **Figure 1**. The near-field microscope consists of a microwave source, a transmission line resonator coupled to the microwave generator, a near-field probe connected to the transmission line, a detector to measure the reflected signal from the resonator and a feedback circuit [9].

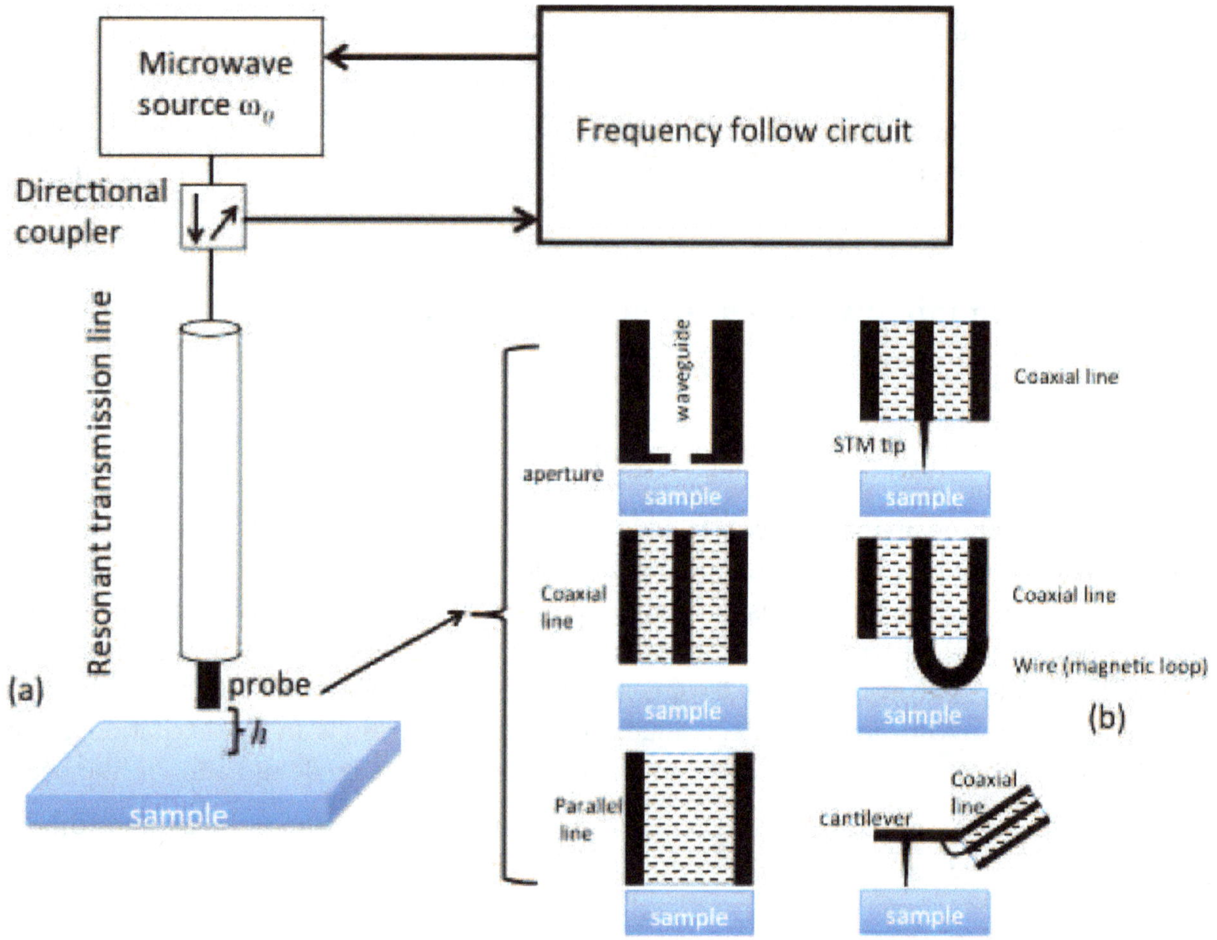

Figure 1. The schematic diagram of the near field microscope (a) the structure of the microscope and (b) different near field probe structures.

The probe is an antenna-like structure operating at the subwavelength, which is used to scan in close proximity of the sample's surface. As shown in **Figure 1(a)**, the probe can take different forms for instance it can be an aperture in an opaque screen, an electric wire formed by a sharpened rod or scanning tunneling microscopy (STM) tip, an atomic force microscopy (AFM) tip, an electrically open flush end of a transmission line, a magnetic loop, or a variety of other geometries [10]. The electromagnetic response of the surface is measured as the probe is scanned over the sample [11]. The probe can be either in contact or at a separation h from the sample surface, which is usually much less than the probe characteristic length-scale D, this ensures a good signal level from the sample surface [11, 12]. In the near zone of an antenna, the structure of electric and magnetic fields is more complicated as their distribution is strongly dependent on the antenna geometry as well as electrodynamics properties of the surrounding region [13–15].

The design of near-field probes can be divided into two main categories, aperture-based probes and apertureless probes [16]. For aperture-based probes, a subwavelength aperture is used to limit the illuminating source, or to collect the local near-field response. It can be a subwavelength hole in a metallic film, a metallic coated pipette, or a tapered optical fiber as illustrated in **Figure 1(b)**.

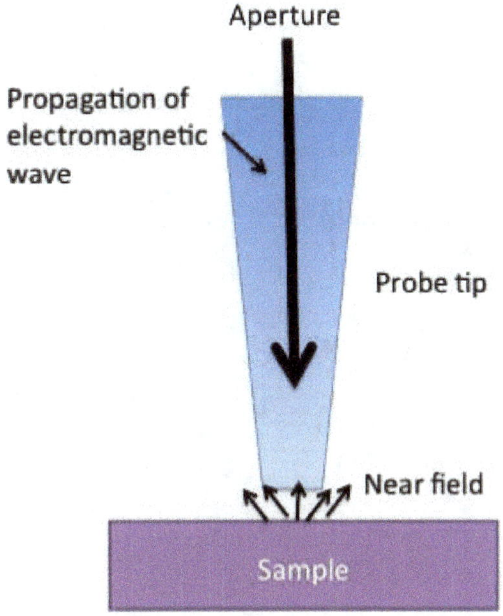

Figure 1 (b). Near-field imaging using aperture type probe.

By positioning the aperture in the close proximity of the sample, the transmitted wave through the aperture only represents the local properties of the sample. Thus, in principle, subwavelength spatial resolution becomes possible by reducing the aperture size and raster scanning the sample surface [17]. The concept was first experimentally demonstrated by Ash and Nicholls [26] (see the next section for a survey on near-field technique), who resolved a $\lambda/60$ grating structure by using λ = 3 cm microwave (10 GHz). In optical frequencies, the spatial resolutions are typically tens of nanometers when using a tapered single mode optical fiber probe. However, the aperture probes suffer from strong signal attenuation. For a hole in a metallic film, the electric field transmission is proportional to $(D/\lambda)^3$, where D is the diameter of the hole [11, 13]. For an optical fiber probe, the image contrast suffers from a strong decrease when the aperture diameter exceeds a critical value. So the signal intensity will be far too small to be detected when the aperture size is further reduced to achieve a better spatial resolution. On the other hand, for the long wavelength of THz radiation, according to the wavelength dependence, the transmission efficiency is extremely low and makes the THz aperture near-field imaging impractical.

An alternative approach is to allow a propagating wave to irradiate some large portion of the sample, but to employ a field-concentrating feature, such as a tip, is to enhance the probe-sample interaction locally. **Figure 2** shows the probing using apertureless near-field probe.

Using a metallic probing tip, the incident wave is scattered by the tip-surface system. Usually, the scattering wave is modulated by a dithering tip and measured in nonspecular directions due to reduced background noise. The wave scattering depends on the local surface properties. By measuring the scattering wave when scanning over a sample, the surface properties can be mapped. In principle, the spatial resolution for an apertureless probe is only limited by the size of the probing tip. Resolution as small as 10 angstrom has been reported with this technique at optical frequencies and a subwavelength factor of 10^{-6} was achieved at the microwave region [18].

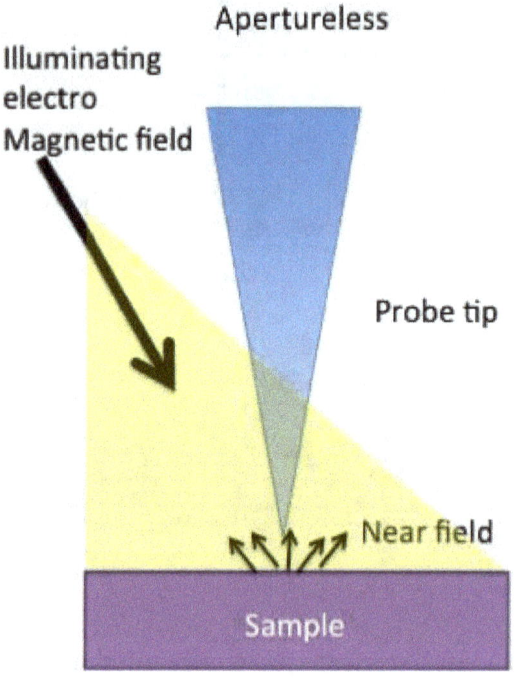

Figure 2. Near-field imaging using apertureless type probe.

3. Survey on near-field imaging from microwave to terahertz frequencies

The idea of near-field microscopy was first demonstrated theoretically by Synge in 1928 [19]. He used an opaque screen with a small subwavelength diameter hole (10 nm in diameter) at 10 nm above the surface of an optically transparent flat sample. The transmitted light is collected by raster scanning the surface. Detailed calculations of the near-field distribution of an aperture were conducted later by Bethe [20] and Bouwkamp [21, 22]. A large fraction of high-spatial frequency evanescent waves is contained in the near field, which can lead to subwavelength spatial resolution [23, 24].

In the 1960s, when the idea of near-field imaging transferred to practice, ferromagnetic resonance microwave microscopy has been developed to characterize ferromagnetic material thin films, for example, Soohoo developed a microwave magnetic probe capable of measuring the spatial variation of the magnetic properties of materials [25] using a cavity made of superconducting material.

A detection system using quasi-optical hemispherical resonator was used at microwave frequencies by Ash and Nicholls [26]. In this experiment, a small aperture (1.5 mm diameter) similar to Synge geometry is used. The aperture is scanned over a sample with a microwave signal at 10 GHz. The sample was harmonically distance modulated at a fixed frequency and the reflected signal was phase-sensitively detected at this modulation frequency to improve sensitivity to sample contrast. This microscope demonstrated contrast sensitivity to metal films on dielectric substrates, as well as bulk dielectrics with dielectric constant differing by only about 10%.

Nonresonant scanned aperture probes were also developed in the 1960s. In the case of nonresonant probe, a transmission line (coaxial or waveguide) delivers a microwave signal to the probe aperture where the fringe electric and magnetic fields from the aperture interact with the sample. Some part of the signal is stored locally in evanescent and near-zone waves, some is absorbed by the sample, some is reflected back up the transmission line and some is scattered away as far-field radiation. By monitoring the scattered or reflected signals as a function of probe height and position, an image of the sample response can be constructed. An example of this type of probe was developed by Bryant and Gunn [27]. They measured the reflected signal from a tapered open-ended coaxial probe with an inner conductor diameter of 1 mm. The probe was used for quantitative measurements of semiconductor resistivities between 0.1 and 100 Ω cm^{-1}.

The development of near-field optical microscopes using optical waveguides beyond cutoff [28, 29] affected the development of new microwave microscopes. The first of these microwave systems operating at 2.5 GHz was developed by Fee et al. to measure the reflected signal from a nonresonant open-ended coaxial probe [30]. A 500-μm in diameter conductor coaxial probe is used and the central conductor was sharpened to a tip of 30 μm in radius. An image of copper grid lines showed a spatial resolution of about 30 μm (or $\lambda/4000$). Other waveguide structure microscopes based on a scanned aperture were developed by Golosovsky et al. [31, 32] and Bae et al. [33–35].

The first THz near-field microscopy was demonstrated in [36]. In this work, a tapered metallic waveguide was used as the aperture to limit the size of the incident ultrashort THz pulses. The probe aperture is similar to the tapered optical fiber tips used in most of scanning near-field optical microscopy. With the aperture diameter $\sim\lambda/3$ (100 μm), a spatial resolution better than $\lambda/4$ was demonstrated for a gold pattern on silicon substrate. A collection mode THz scanning near-field microscopy was developed by Mitrofanov et al. [37–39]. It uses an aperture-type probe to collect the near field instead to filter the illuminating light. He showed improvements in sensitivity and resolution by integrating the probe and a photo-conducting antenna THz detector in the near-field region of the aperture. THz near-field imaging with a dynamic aperture was demonstrated by Wynne et al. and Chen et al. [40, 41]. In his work, the dynamic aperture was formed by a transient photo-carrier layer induced by a THz near-field imaging. In case that the optical pulse arrives earlier than the THz pulse, the THz transmission is modulated by the induced photo-carrier layer, whose size is determined only by the optical beam focus. This approach has the potential to improve spatial resolution and image contrast, while it is limited to semiconductor surface.

Apertureless THz near-field imaging has been also investigated [42, 43]. In their work, van and Planken used a metallic probing tip with THz waves to image down to 18 μm in size. Moreover, a scattering type apertureless THz scanning near-field microscope has been demonstrated by Wang et al. [44]. THz radiation is scattered and modulated by the dithering metallic probing tip in the tip-surface system, the THz signal is detected in nonspecular directions [45].

4. Metamaterial for near-field imaging manipulation

Conventional imaging devices or lenses suffer from diffraction limit which constrain the image resolution. In addition, conventional imaging cannot propagate the evanescent waves which carry the subwavelength information due to the exponential decay exhibited in natural materials. These issues can be overcome using metamaterials. For instance, image focusing, rotation, lateral shift and image magnification are achievable with subwavelength resolutions using a two- or three-dimensional metamaterials structures, usually with negative refraction properties, to achieve resolution beyond the diffraction limit (ideally, infinite resolution). Such behavior is enabled by the capability of double-negative materials to yield negative phase velocity.

4.1. Metamaterial design by electromagnetic field transformation

Electromagnetic metamaterials are periodic structures than can artificially manipulate the electromagnetic properties of media. They can be engineered for the radio-frequency (RF) and microwave range as well as for terahertz and optical frequencies. Metamaterials allow the possibility of attaining mediums with properties, which have not been considered before, such as, negative or close to zero refractive index since they are inaccessible in nature. In addition, they can mix the electric and magnetic response of a material (chirality) [46, 47]. The first step to design a metamaterial medium is using the electromagnetic transformation to get the new medium properties (i.e., transformed permittivity and permeability). Transformation electromagnetics are based on the invariance of Maxwell's equations with respect to coordinate transformations; this enables a user-defined coordinate transformation to be translated into electric and magnetic material parameters. **Figure 3** shows the relations between the different coordinates [48, 49].

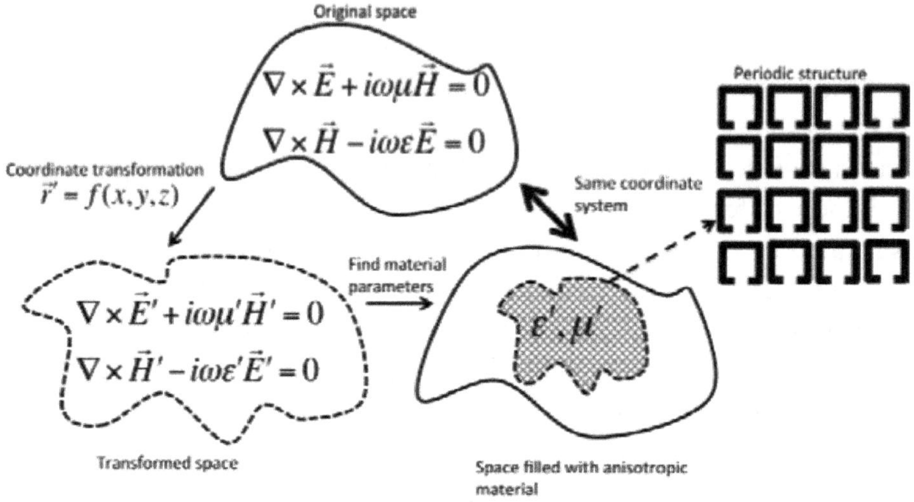

Figure 3. Transformation electromagnetic for metamaterial medium.

Once the transformed permittivity and permeability and their spatial distribution are obtained, the structure of unit cell (or particle) is designed by optimizing its geometric parameters. The most common unit cell structure is the split ring resonator (SRR). The unit cell can be a resonant or nonresonant. In case of the resonant unit cell, the losses are high at the resonance frequency; however, they have a large dynamic rang. The nonresonant unit cell has the advantage of small losses but with low dynamic rang [49].

4.2. Metamaterials medium types

The electromagnetic mediums are categorized as isotropic or anisotropic. In the isotropic electromagnetic medium the permittivity, ε and permeability, μ, of the medium are uniform in all directions of the medium. The dispersion relation in that medium is $k = \omega\sqrt{\mu\varepsilon}$. One simple example of an isotropic electromagnetic medium is the free space. However, the medium is called anisotropic when the permittivity, ε and permeability, μ, of a medium depend upon the directions of field vectors and hence, they are in tensor form. Based on Maxwell equations in an anisotropic medium the dispersion relation can be obtained by solving the matrix equation [49, 50]:

$$det\left[\overline{\overline{k}} \cdot \overline{\overline{\varepsilon}}^{-1} \cdot \overline{\overline{k}} + k^2{}_0\overline{\overline{\mu}}\right] = 0 \tag{3}$$

where $\overline{\overline{\mu}}, \overline{\overline{\varepsilon}}$ and $\overline{\overline{k}}$ are all matrices and $k^2{}_0 = \omega^2\mu_0\varepsilon_0$ (μ_0, ε_0 are the permeability and permittivity of free space, respectively). Metamaterials are composite structures in which the electromagnetic wave propagation depends on the orientation of structured metallic unit cells. Usually, these unit cells are designed to have a certain permittivity and permeability responses by optimizing its geometric parameters. One common example of unit cell structure is the split ring resonator (SRR) (see Section 4.1) [51]. The anisotropic metamaterial is defined with its permittivity and permeability tensor as follows

$$\overline{\overline{\varepsilon}} = \begin{pmatrix} \varepsilon_x & 0 & 0 \\ 0 & \varepsilon_y & 0 \\ 0 & 0 & \varepsilon_z \end{pmatrix} \text{ and } \overline{\overline{\mu}} = \begin{pmatrix} \mu_x & 0 & 0 \\ 0 & \mu_y & 0 \\ 0 & 0 & \mu_z \end{pmatrix} \tag{4}$$

where the diagonal elements of $\overline{\overline{\varepsilon}}$ and $\overline{\overline{\mu}}$ could have positive or negative values. To understand the propagation properties of a wave in such medium a TE plane wave is considered. Using Eqs. (3) and (4), one can get the anisotropic dispersion relation of the medium

$$\frac{k_y^2}{\varepsilon_z\mu_x} + \frac{k_x^2}{\varepsilon_z\mu_y} = \frac{\omega^2}{c^2} \tag{5}$$

where c is the speed of light in free space and k is the wave vector of the propagating wave. For losses material, the plane wave solution in the anisotropic medium can be either a propagating wave or evanescent wave depending on the sign of the wave number k_x and hence, the medium can be classified as in **Table 1** [49].

Medium type	Material parameters	Wave property	Cut off conditions
Cutoff media	$\varepsilon_z\mu_y > 0, \frac{\mu_y}{\mu_x} > 0$	Propagating	$k_y < k_c$
		Evanescent	$k_y \geq k_c$
Anti-cutoff media	$\varepsilon_z\mu_y < 0, \frac{\mu_y}{\mu_x} < 0$	Evanescent	$k_y < k_c$
		Propagating	$k_y \geq k_c$
Never-cutoff media	$\varepsilon_z\mu_y > 0, \frac{\mu_y}{\mu_x} < 0$	Propagating	All real k_y
Always-cutoff media	$\varepsilon_z\mu_y < 0, \frac{\mu_y}{\mu_x} < 0$	Evanescent	No real k_y

Table 1. Anisotropic media for a z-polarized TE plane wave.

4.3. Construction of compensated bilayer of anisotropic metamaterials for imaging

Veselago introduced the bilayer compensation to illustrate the ability to focus on the electromagnetic wave from a point source by a flat slab of composite material with a negative refraction. The proposed structure was a bilayer of vacuum and a left-hand material (with permittivity and permeability equal to –1) where the left-hand material compensates for the propagation effect associated with an equal length of vacuum. This bilayer structure is called Veselago lens (or the perfect lens) [52]. The same idea can be implemented using positive and negative refracting layers of anisotropic metamaterials to implement near-field focusing in the same way as Veselago lens. **Figure 4** shows the bilayer of anisotropic metamaterials in two dimensions.

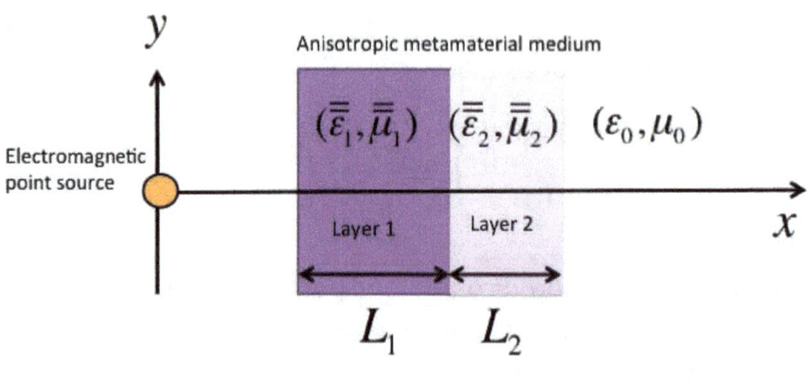

Figure 4. Compensated anisotropic metamaterial medium structure.

Depending on the sign of the permittivity and permeability of the two layers, one can get different material types for the compensated bilayers. For instance, compensated bilayer can be made of combination of positive and negative refracting layers of never-cut off medium (NCM) or anticutoff medium (ACM). **Table 2** shows examples of metamaterials that could make up the compensated bilayer [49].

Bilayer type	Layer	ε_x	ε_y	ε_z	μ_x	μ_y	μ_z	Material
Veselago lens	L1	1	1	1	1	1	1	Free space
	L2	-1	-1	-1	-1	-1	-1	Left hand material
NCM compensated bilayer	L1	-1	1	1	-1	1	1	Positive refractive NCM
	L2	1	-1	-1	1	-1	-1	Negative refractive NCM
ACM compensated bilayer	L1	1	-1	-1	-1	1	1	Positive refractive ACM
	L2	-1	-1	1	1	-1	-1	Negative refractive ACM

Table 2. Compensated bilayer anisotropic metamaterial types.

5. Near-field subwavelength image manipulating through compensated anisotropic metamaterial

5.1. Superlens

As mentioned earlier, the performance limitation of conventional lenses is due to the diffraction limit. In conventional imaging the field emitting from an object is a superposition of plane waves with dispersion relation given by [53]

$$k_z = \sqrt{\frac{\omega^2}{c^2} - (k_x^2 + k_y^2)} \tag{6}$$

In case of energy propagation in the +z-direction, the only real values are the ones allowed for and hence all the components of k_z contains the spectrum of the image which are transmitted and refocused by an ordinary lens. On the other hand, if k_z is imaginary, i.e., $\frac{\omega^2}{c^2} < (k_x^2 + k_y^2)$, the wave is an evanescent wave whose amplitude decays as the wave propagates along the z-axis. This leads to a loss of the high-frequency components of the wave, which contain information about subwavelength features of the object being imaged. The super lens with negative-index materials can overcome this problem. Since in such materials transporting the electromagnetic energy in the +z-direction requires negative wave vector. This amplifies the evanescent wave and all components of the angular spectrum can be transmitted without distortion [53].

Super lensing can be achieved with bilayer of anisotropic metamaterial, i.e., (NCM) and (ACM) bilayers. However, the mechanism of constructing perfect image with subdiffraction resolution in the case of (NCM) and (ACM) bilayers is different than the mechanism of super lens with single-negative index layer (Veselago lens). As discussed in the previous section, the subwavelenght resolution, in case of perfect lens, is achieved thought the amplification of the evanescent wave which by consequence recovers the subwavelength features. In the case of NCM or ACM bilayers lens, the evanescent components of the wave are converted to a propagating mode in the bilayer region and then back to evanescent component at the back surface of the bilayer which creates and image beyond the diffraction limit. A comparison between the evanescent wave propagation in different bilayers is shown in **Figure 5** [49, 54].

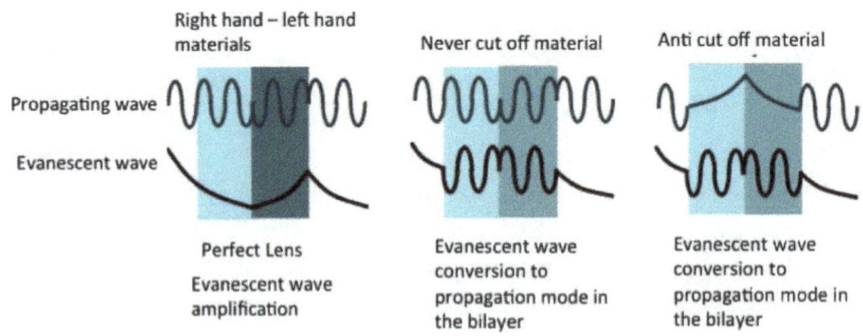

Figure 5. Different evanescent wave propagation in three types of metamaterial bilayer structures.

It is important to mention that the electric or magnetic loss is considered a major problem in the implementation of metamaterials in the imaging structures. The reason is that these losses can limit the resolution of the perfect lens. These losses are due to the imaginary part of the permittivity and permeability tensors. In general, the beamwidth of the electric field is degraded at the back surface due to losses compared to the source plane. The effect of degradation is different depending on the type of the bilayer used for the super lens. In case of the left-hand material lens the electric field degradation is about $\lambda/5$ while for NCM or ACM is $\lambda/10$. This indicates that the NCM and ACM lenses are less sensitive to material parameters losses than the left-hand material lens [49, 55, 56].

Another issue with the bilayer structures is the retardation effect. This effect is due to the difficulty to realize a completely compensated bilayer in practical cases. As with the losses, the retardation effect degrades the electric field beamwidth and the NCM or ACM are less sensitive to that degradation than the left-hand material. This is why anisotropic metamaterial medium is considered advantageous since it can approximate the ideal situation [49, 57].

5.2. Bilayer lens using transmission line metamaterial

The bilayer of anisotropic metamaterial for imaging can be realized using L-C transmission line network. Different types of anisotropic metamaterials such as ACM and NCM can be constructed by L-C loaded microstrip line grids operating at certain frequency bandwidth. The planner microstrips are periodically loaded by serial capacitance in z- or y-direction with or without shunting inductance loaded at the center of the node. The unit cell of the planner transmission line metamaterial is illustrated in **Figure 6** [47, 49, 58].

For z-polarized electromagnetic wave the dispersion relation is dependent on ε_z, μ_x and μ_y as indicated in Eq. (5). Dispersion relation of a transmission line metamaterial can be obtained using periodic transmission line theory [59, 60]. For example, under the effective medium approximation, the dispersion relation of the unit cell, shown in **Figure 6(a)**, is

$$\frac{k^2_y}{2\beta^2(1{-}2\omega_0/\omega\beta d)}+\frac{k^2_x}{2\beta^2}=1 \tag{7}$$

where $\omega_0 = 1/2C_yZ_0$, Z_0, β and d are the characteristic impedance, the propagation constant and the length of the transmission line section.

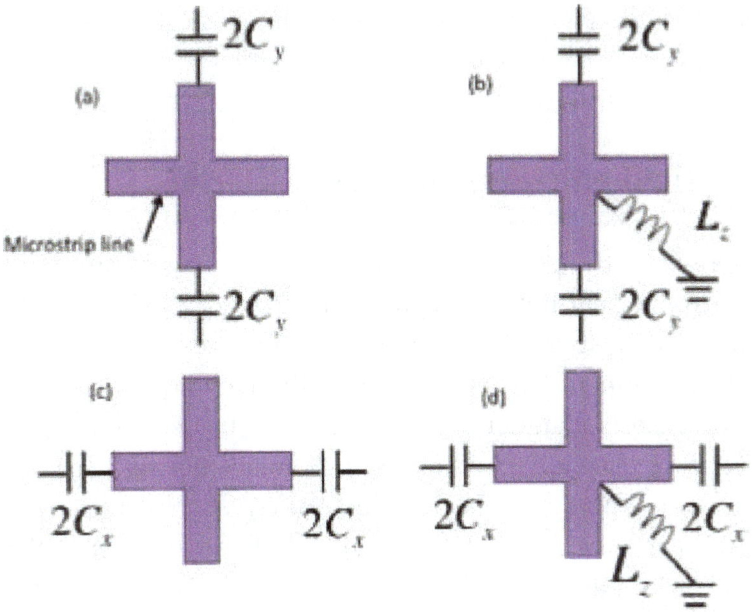

Figure 6. Different unit cells of loaded transmission line grid for implementing different anisotropic metamaterials. (a) positive refraction NCM (b) positive refraction ACM (d) negative refraction ACM (d) negative refraction NCM.

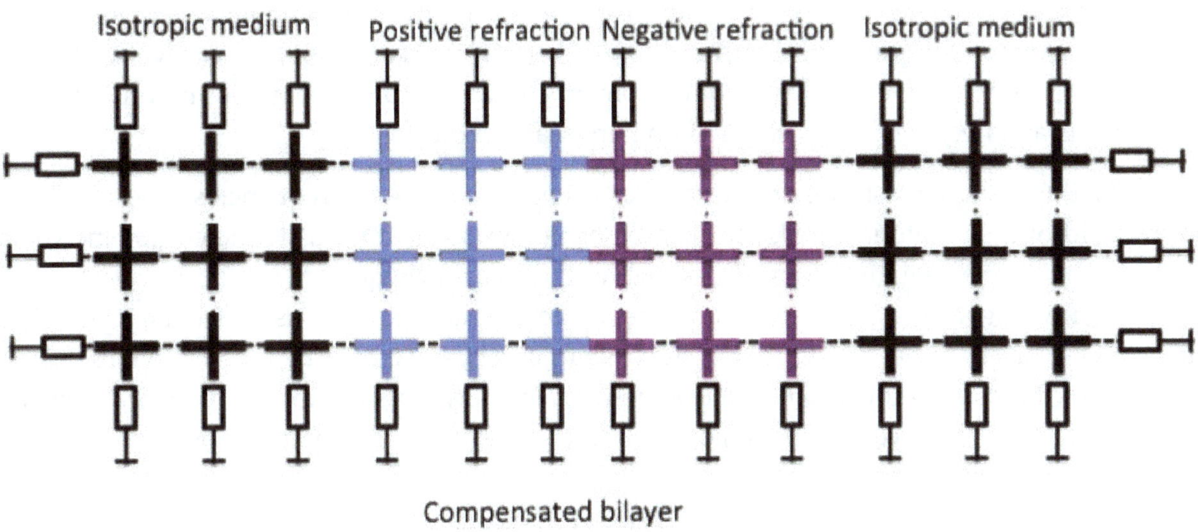

Figure 7. The schematic diagram of a 2D compensated bilayer using loaded L-C metamaterial transmission line.

Using these four unit cells in a network, one can construct a 2D compensated bilayer lens composed of positive NCM/negative NCM or positive ACM/negative ACM. **Figure 7** shows the schematic diagram of the bilayer medium. The loaded capacitances and inductances need to be chosen to satisfy the requirement of the bilayer medium see **Table 2**.

The imaging properties of the compensated bilayer structure are investigated experimentally in the microwave range [61]. The results show that the electromagnetic wave from a point source has been focused on the image plane with beamwidth of 0.1 λ which is below the diffraction limit. This focusing result is well below the left-hand material planer lens, which is 0.21 λ [49, 61]. The NCM bilayer lens has better subwavelength resolution because it is less sensitive to material losses.

6. Plasmonic and field enhancement metamaterials

The structures presented in the previous section such as bilayers and superlens can be integrated in the system of a near-field microwave microscope [62]. This can improve image measurements by improving spatial resolution, this helps in studying surfaces that cannot be imaged directly with a scanned probes. Moreover, the technique of manipulation of surface plasmons for enhancing electromagnetic propagation can be used in imaging system at microwave frequencies. Surface plasmons are known in optical frequency [63]. They can be excited on a metal by an external plane-wave beam. This can enhance the electric field at the surface compared with that in the incident beam and it has been suggested that this effect is an important aspect of surface-enhanced Raman scattering [64]. Surface plasmons have been investigated for extraordinary optical transmission through subwavelength hole arrays [65]. This overcomes the constraint of low transmittivity of subwavelength apertures. Ebbesen et al. observed that the maxima the transmission efficiency could exceed unity (when normalized to the area of the holes), which are orders of magnitude greater than predicted by standard aperture theory [65]. A review that summarizes the basic principles and achievements in the field of plasmonic for subwavelength guiding has been presented [66]. This review chapter discusses as well the potential future developments and applications of nanophotonic devices and circuits and near-field microscopy with nanoscale resolution [67].

Manipulating surface plasmons at microwave and millimeter-wave frequencies is possible due to the field of metamaterials [68]. This can be achieved using composite dielectric where the property of having negative permittivity below the plasma frequency has been investigated [69]. This investigation showed that thin wires structure reduces the plasma frequency having longer wavelength to be diffracted and the system can be described by an effective dielectric constant of the plasma form. This achievement opens new possibilities for microwave structures to produce enhanced transmission through subwavelength apertures as in the case of photonic structures.

7. Conclusion

This chapter introduced a review on metamaterial structures for near-field microscopy. A general overview on the near-field imaging/microscopy instrumentation is first presented followed by a survey on the research done on this topic. The metamaterial design based on transformation electromagnetic as well as a summary on metamaterial media was introduced. The bilayer metamaterial medium is discussed for super-resolution structures that can manipulate subwavelength images and evanescent waves. Finally, the manipulation of surface plasmons is presented as a technique to enhance the transmission for near-field microscopy.

Author details

Hatem El Matbouly

Address all correspondence to: hatem.el-matbouly@lcis.grenoble-inp.fr

LCIS, Grenoble-INP, Grenoble Alpes University, Valence, France

References

[1] Eric B, Jay KT, "Near-field optics: microscopy, spectroscopy and surface modi cation beyond the diraction limit", Science, Vol. 257, No. 5067, pp. 189–195, 1992.

[2] Xiaoliang SX, Robert CD, "Probing single molecule dynamics", Science, New Series, Vol. 265, No. 5170, pp. 361–364, 1994.

[3] Chen H, Ma S, Wu X, Yang W, Zhao T, "Diagnose human colonic tissues by terahertz near-field imaging", Journal of Biomedical Optics, Vol. 20, No. 3, art. no. 036017, 2015.

[4] Yuandan D, Tatsuo I, "Promising future of metamaterials", IEEE Microwave Magazine, March, 2012.

[5] Born M, Wolf E, "Principles of optics", Cambridge University Press, Cambridge, 6th edition, 1998.

[6] http://hyperphysics.phy-astr.gsu.edu/hbase/phyopt/raylei.html

[7] Jackson J, "Classical Electrodynamics", John Wiley, New York, 3rd edition, 1998.

[8] Alexander NR, Nadezhda VY, "Electrodynamics of microwave near-field probing: Applicationto medical diagnostics", Journal of Applied Physics, Vol. 98, arct. no. 114701, 2005.

[9] Steven MA, Steinhauer DE, Feenstra BJ, Vlahacos CP, Wellstood FC, "Near Field Microwave Microscopy of Materials Properties", H. Weinstock and M. Niseno (eds), Microwave Superconductivity, Kluwer Academic Publishers. Printed in the Netherlands, 2001.

[10] https://www.photonics.ethz.ch/fileadmin/user_upload/Courses/NanoOptics/nfoprobes_rev-1.pdf

[11] Steven MA, Vladimir VT, Andrew RS, "Principles of Near-Field Microwave Microscopy", July 17, 2006. http://citeseerx.ist.psu.edu/viewdoc/download?doi=10.1.1.569.1377&rep=rep1&type=pdf

[12] Schelkuno S, "Antennas, Theory and Practice". Wiley and Sons, Inc., New York, 1952.

[13] Osofsky SS, Schwarz SE, "Design and performance of a noncontacting probe for measurements on high-frequency planar circuits", IEEE Transactions on Microwave Theory and Techniques, Vol. 40, No. 8, pp. 1701–1708, 1992.

[14] Jean JB, Rémi C, "Image formation in near-field optics", Progress in Surface Science, Vol. 56, No. 3, pp. 133–237, 1997.

[15] Alexander NR, "Quasistatics and electrodynamics of near-field microwave microscope", Journal of Applied Physics, Vol. 115, arct. no. 084501, 2014.

[16] Björn TR, "High-frequency near-field microscopy", Review of Scientific Instruments, Vol. 73, No. 7, 2002.

[17] Steinhauer DE, Vlahacos CP, Dutta SK, Wellstood FC, Steven MA, "Surface Resistance Imaging with a Scanning Near-Field Microwave Microscope", Applied Physics Letters, Vol. 71, pp. 1736,1997. http://arxiv.org/abs/cond-mat/9712142v1

[18] Lewis G, et al, "Apertureless scanning near-field optical microscopy: A comparison between homodyne and heterodyne approaches", Journal of Optical Society B, Vol. 23, May 2006.

[19] Synge EH, "A suggested method for extending the microscopic resolution into the ultra-microscopic region". Philosophical Magazine, Vol. 6, pp. 356, 1928.

[20] Bethe HA, "Theory of diffraction by small holes", Physical Review, Vol. 66, No. 7–8, pp. 163–182, 1944.

[21] Bouwkamp CJ, "On the freely vibrating circular disk and the diffraction by circular disks and apertures", Physica, Vol. 16, No. 1, pp. 1–16, 1950.

[22] Bouwkamp CJ, "Diffraction Theory", Reports on Progress in Physics, Vol. 17, 1954.

[23] Massey GA, Davis JA, Katnik SM, Omon E, "Subwavelength resolution far-infrared microscopy", Applied Optics, Vol. 24, No. 10, pp. 1498–1501, 1985.

[24] Grober RD, Rutherford T, Harris TD, "Modal approximation for the electromagnetic field of a near-field optical probe", Applied Optics, Vol. 35, No. 19, pp. 3488–3495, 1996.

[25] Soohoo RF, "A microwave magnetic microscope", Journal of Applied Physics, Vol. 33, No. 3, pp. 1276–1277, 1962.

[26] Ash EA, Nicholls G, "Super-resolution aperture scanning microscope", Nature, Vol. 237, No. 5357, pp. 510–512, 1972.

[27] Bryant CA, Gunn JB, "Noncontact technique for the local measurement of semiconductor resistivity", Review of Scientific Instruments, Vol. 36, No. 11, pp. 1614–1617, 1965.

[28] Pohl DW, Denk W, Lanz M, "Optical stethoscopy: Image recording with resolution lambda/20". Applied Physics Letters, Vol. 44, No. 7, pp. 651–653, 1984.

[29] Betzig E, Isaacson M, Lewis A, "Collection mode near-field scanning optical microscopy", Applied Physics Letters, Vol. 51, No. 25, pp. 2088–2090, 1987.

[30] Fee M, Steven C, Hänsch TW, "Scanning electromagnetic transmission line microscope with sub-wavelength resolution", Optics Communications, Vol. 69, No. 3–4, pp. 219–224, 1989.

[31] Golosovsky M, Dan D, "Novel millimeter-wave near-field resistivity microscope", Applied Physics Letters, Vol. 68, No. 11, pp. 1579–1581, 1996.

[32] Copty A, Golosovsky M, Davidov D, Frenkel A, "Localized heating of biological media using a 1-W microwave near-field probe", IEEE Transactions on Microwave Theory and Techniques, Vol. 52, No. 8 II, pp. 1957–1963, 2004.

[33] Bae J, Okamoto T, Fujii T, Mizuno K, Nozokido T, "Experimental demonstration for scanning near-field optical microscopy using a metal micro-slit probe at millimetre wavelengths", Applied Physics Letters, Vol. 71, No. 24, pp. 3581–3583, 1997.

[34] Nozokido T, Iibuchi R, Bae J, Mizuno K, Kudo H, "Millimeter-wave scanning near-field anisotropy microscopy", Review of Scientific Instruments, Vol. 76, No. 3, art. no. 033702, 2005.

[35] Nozokido T, Bae J, Koji M, "Scanning near-field millimeter-wave microscopy using a metal slit as a scanning probe", IEEE Transactions on Microwave Theory and Techniques, Vol. 49, No. 3, pp. 491–499, 2001.

[36] Hunsche S, Koch M, Brener I, Nuss M, "THz near-field imaging", Optical Communications, Vol. 150, No. 22, 1998.

[37] Mitrofanov O, et al. "Collection mode near field imaging with 0.5 THz pulses", IEEE Journal of Selected Topics in Quantum Electronics, Vol. 7, No. 4, pp. 630–640, 2001.

[38] Mitrofanov O, et al. "Near field microscope probe for far infrared time domain measurements", Applied Physics Letters, Vol. 77, pp. 591–593, 2000.

[39] Mitrofanov O et al. "Terahertz near field microscopy based on a collection mode detector", Applied Physics Letters, Vo. 77, pp. 3496–3498, 2000.

[40] Wynne K, Carey J, Zawadzka J, Jaroszynski D, "Tunneling of single cycle terahertz pulses through waveguide", Optical Communications, Vo. 176, pp. 429–435, 2000.

[41] Chen Q, Jiang Z, Xu G, Zhang XC, "Near field terhertz imaging with a dynamic aperture", Optical Letters, Vol. 25, p. 11222, 2000.

[42] van der VN, Planken P, "Electro optic detection of sub wavelength terahertz sport sizes in the near field of a metal tip", Applied Physics Letters, Vol. 81, p. 1558, 2002.

[43] Planken P, van der VN, "Spot size reduction in terahertz apertureless near field imaging", Optical Letters, Vol. 29, p. 2306, 2004.

[44] Wang K, Barkan A, Mittleman B, "Propagation effects in aperture less near field optical antennas", Applied Physics Letters, Vol. 84, p. 305, 2004.

[45] Chen HT, Kersting R, Cho G, "Terahertz imaging with nanometer resolution", Applied Physics Letter, Vol. 83, p. 3009, 2003.

[46] George VE, Michael S, "Transforming Electromagnetics using Metamaterials", IEEE magazine, April, 2012.

[47] Yuandan D, Tatsuo I, "Promising Future of Metamaterials", IEEE magazine, April, 2012.

[48] Nathan BK, David RS, John BP, "Electromagnetic design with transformation optics", Proceedings of the IEEE, Vol. 99, No. 10, 2010.

[49] Tie JC, David R, Smith RL, "Metamaterials Theory, Design and Applications", Springer, 2010.

[50] Eroglu A, "Wave Propagation and Radiation in Gyrotropic and Anisotropic Media", DOI 10.1007/978-1-4419-6024-5_2, Springer Science+Business Media, LLC, 2010.

[51] Guven K, Ozbay E, "Near field imaging in microwave regime using double layer split-ring resonator based metamaterial", Opto-Electronics Review, Vol. 14, No. 3, pp. 213–216, 2006.

[52] Viktor GV, "The electrodynamics of substances with simultaneously negative values of ε and μ", Reviews of Topical Problems, Soviet Physics Uspekhi, 1968.

[53] Penday J, "Manipulating the near field with metamaterials", Optics & Photonics News, Sep. 2004.

[54] Smith DR, Willie JP, Vier DC, Nemat-Nasser SC, Schultz S, "Composite medium with simultaneously negative permeability and permittivity", Physical Review Letters, Vol. 84, No. 18, 2000.

[55] David RS, David S, Jack JM, Pavel K, Patrick R," Partial focusing of radiation by a slab of indefinite media", Applied Physics Letters, Vol. 84, p. 2244, 2004.

[56] Shen JT, Platzman PM, "Near field imaging with negative dielectric constant lenses", Applied Physics Letters, Vol. 80, p. 3286, 2002.

[57] David RS, David S, Marshall R, Sheldon S, Ramakrishna SA, John BP, "Limitations on subdiffraction imaging with a negative refractive index slab", Applied Physics Letters, Vol. 82, p. 1506, 2003.

[58] Lai A, Itoh T, Caloz C, "Composite right/left-handed transmission line metamaterials", IEEE Microwave Magazine, Vol. 5, No. 3, 2004.

[59] David MP, "Microwave Engineering", 4th Edition, Wiley & Sons, 2012.

[60] Yijun F, Xiaohua T, Yan C, Tian J, "Electromagnetic wave propagation in anisotropic metamaterials created by a set of periodic inductor-capacitor circuit networks", Physics Review B, Vol. 72, art. no. 245107, 2005.

[61] Anthony G, George VE, "Overcoming the diffraction limit with a planar left-handed transmission-line lens", Physics Review Letter, Vol. 92, art. no. 117403, 18 March 2004.

[62] Muhammed SB, Omar MR, "Experimental and numerical study of sensitivity improvement in near-field probes using single-negative media", IEEE Transactions on Microwave Theory and Techniques, Vol. 57, No. 12, pp. 3427–3433, 2009.

[63] Stefan AM, Harry AA, "Plasmonics: Localization and guiding of electromagnetic energy in metal/dielectric structures", Journal of Applied Physics, Vol. 98, art. no. 011101,2005.

[64] Andrey KS, Vladimir MS, "Electrodynamics of Metamaterials", World Scientific Publishing Co. Pte. Ltd. 2007.

[65] Ebbesen TW, Lezec HJ, Ghaemi HF, Thio T, Wolff PA, "Extraordinary optical transmission through sub-wavelength hole arrays", Nature, Vol. 391, pp. 667–669, 1998.

[66] Dmitri KG, Sergey IB, "Plasmonics beyond the diffraction limit", Nature Photonics, Vol. 4, pp. 83–91, 2010.

[67] Nicholas F, Hyesog L, Cheng S, Xiang Z, "Sub-diffraction-limited optical imaging with a silver superlens", Science, Vol. 308, p. 22, April 2005.

[68] Megel et al. "Enhanced milimeter wave transmission through quasioptical subwaveleght perforated plate", IEEE Transaction on Microwave and Propagation, Vol. 53, No. 6, pp. 2146–2154, 2006.

[69] Pendry JB, Holden AJ, Robbins DJ, Stewart WJ, "Low frequency plasmons in thin-wire structures", Journal of Physics: Condensed Matter, Vol. 10, pp. 371–3821998.

6

Advanced Filtering Waveguide Components for Microwave Systems

Snežana Stefanovski Pajović, Milka Potrebić and Dejan V. Tošić

Abstract

Microwave filter design is a popular topic in the area of modern microwave engineering. Novel technologies, novel applications and more demanding component miniaturiza-tion are some of the key drivers for the development of novel microwave filters. For the systems operating with high power and low losses, waveguide filters represent sustainable solutions, in spite of their size. Herein, a method for the advanced bandpass and bandstop waveguide filter design is presented. Properly designed printed-circuit inserts, with simple resonators, are used as resonating elements inside a standard rectangular waveguide. In this manner, multi-band bandpass or bandstop waveguide filters are developed. Multiple resonant frequencies can be obtained using single insert, with properly positioned resonators. Filter design is exemplified by numerous three-dimensional electromagnetic models of the considered structures, equivalent micro-wave circuits and fabricated devices. Some of the advantages of the proposed design are simplicity, ease of implementation, possibility of miniaturization and experimental verification. The waveguide filters considered here are designed to operate in the X frequency band, so their application is recognized with the radar and satellite systems. Further improvement of the proposed method is possible, according to the future use of the presented devices.

Keywords: bandpass filter, bandstop filter, multi-band filter, waveguide technology, planar inserts, 3D EM modeling and simulation

1. Introduction

Microwave filters are essential components of every modern communication system operating at microwave frequencies. In general, filters are used to pass the signals at chosen frequencies and to block or weaken the undesired signals. Microwave filter design represents an important topic in the area of modern microwave engineering. The researchers are developing novel microwave filters using various technologies, keeping their focus on the novel applications and more demanding device miniaturization, but also bearing in mind the basic filter requirement —to meet a given specification [1].

This chapter focuses on the waveguide filter design. As high quality-factor devices [2], waveguide filters still represent sustainable solutions for the systems where high power and low losses are required (e.g., satellite systems and radar systems).

Waveguide filters can be implemented in different ways. Various solutions based on the waveguide structure modification, by adding branches, are known for decades and are widely implemented. Furthermore, numerous implementations using substrate integrated waveguide (SIW) technology can be found. Herein, waveguide filters using discontinuities inside the waveguide structure are of interest. These discontinuities can have various shapes and positions [3]. Numerous examples, along with the equivalent circuits and closed-form expressions to calculate the parameters of discontinuities, are given in [4].

In this chapter, waveguide filters using printed-circuit discontinuities are considered. These inserts can be differently positioned inside the waveguide, and some classification can be made accordingly. In the available literature, various solutions can be found for the filters using inserts placed in the E-plane of a rectangular waveguide. Actually, this is a widespread approach, and an example of such design applied to the bandpass filter is presented in [5]. Split-ring resonators (SRRs) are often used as resonating elements for the bandstop filter design. Implementations of bandstop filters using E-plane inserts with SRRs can be found in [6], with a single rejection band, and in [7], with multiple rejection bands. Another approach is based on the use of H-plane inserts. Similarly as for the E-plane insert, various types of resonators can be implemented on the H-plane insert, as well. For the bandpass filters, complementary split-ring resonators (CSRRs) are widely used, as presented in [8, 9], for the filter with a single pass band. Dual-band filter with SRRs is proposed in [10], while compact filter solutions can be found in [11].

The main advantage of the implementations based on the concept of using inserts in the waveguide, compared to the solutions with modified waveguide structure, is simpler design and ease of fabrication.

A method for the advanced bandpass and bandstop waveguide filter design has been developed and reported. It is based on the use of printed-circuit discontinuities, with relatively simple resonators, acting as resonating elements. Multiple resonant frequencies can be obtained using single insert, with properly positioned resonators. The goal is to implement multi-band bandpass or bandstop waveguide filters using printed-circuit inserts, having better characteristics (e.g., bandwidth, return loss, insertion loss and structure size), compared to the ones published in the available literature. The method considers three-dimensional electromagnetic (3D EM) modeling of the structures, generating equivalent microwave circuits, optimization and device miniaturization. Relatively simple design, ease of implementation and experimental verification are important aspects of the proposed design guidelines.

The waveguide filters considered here are designed to operate in the X frequency band (8.2–12.4 GHz), so they can be used as components of radar and satellite systems of various purposes. Proposed design method allows their further improvement in accordance with their potential future use. We have been investigated various planar structures, as well, which might be suitable for miniaturization, see for example [12–15].

The chapter is organized as follows. Section 2 will briefly cover waveguide fundamentals, while Section 3 will provide basic concepts of microwave filter design. Section 4 will elaborate on the method for the waveguide filter design, explaining 3D EM modeling and simulation, equivalent microwave circuits and experimental verification of the considered waveguide filters. Sections 5 and 6 will introduce bandpass and bandstop waveguide filters using the presented method for an advanced waveguide filter design, respectively. Section 7 will focus on the multi-band filter design. Finally, the obtained results and findings will be outlined in conclusion.

2. Brief review of waveguide fundamentals

Hollow waveguides represent structures for transmission of electromagnetic waves, consisting of a single transmission line [16]. Depending on the shape of their cross-section, they can be classified as rectangular or circular. Actually, rectangular waveguide is the most commonly used waveguide component [17]. Hollow metal waveguide is a shielded structure, so there is practically no electromagnetic coupling with the surrounding devices. In spite of their size at lower frequencies, they still represent sustainable solution for the systems operating with high power (e.g., transmitters and radars) and low losses (e.g., satellite systems). Also, they can be used for the implementation of high quality-factor resonators.

Rectangular waveguide is used for the transmission of transverse electric (TE) and transverse magnetic (TM) waves. **Figure 1** depicts the rectangular waveguide for the wave propagating along the *z*-axis. We can assume that the conductor is perfect, while the waveguide is filled by the homogeneous and lossless dielectric. The analysis of the wave propagation assumes solving the wave equation, taking into account the boundary conditions on the waveguide walls.

For the TE wave propagation, with axial electric field $E_z = 0$, it is required to solve the wave equation as a function of H_z, $\Delta_t H_z + K^2 H_z = 0$, or in Cartesian coordinates $\left(\partial^2 H_z / \partial x^2\right) + \left(\partial^2 H_z / \partial y^2\right) + K^2 H_z = 0$, where $K^2 = \gamma^2 + k^2$, $\gamma = j\beta$ is propagation coefficient (β is phase coefficient), $k = \omega\sqrt{\varepsilon\mu}$. Scalar function H_z, given in Cartesian coordinates, $H_z(x, y, z) = H_z(x, y, 0)e^{-\gamma z}$, represents the solution of the given wave equation, and each non-zero field component can be expressed as a function of H_z. By solving the wave equation, the following expression is obtained for H_z: $H_z(x,y,z) = H_0\cos(m\pi x/a)\cos(n\pi y/b)e^{-j\beta z}$, $m = 0,1,2,...,$ $n = 0,1,2,...,$ where H_0 is a complex constant. Starting from this solution, the other field components of the TE$_{mn}$ mode can be derived [17].

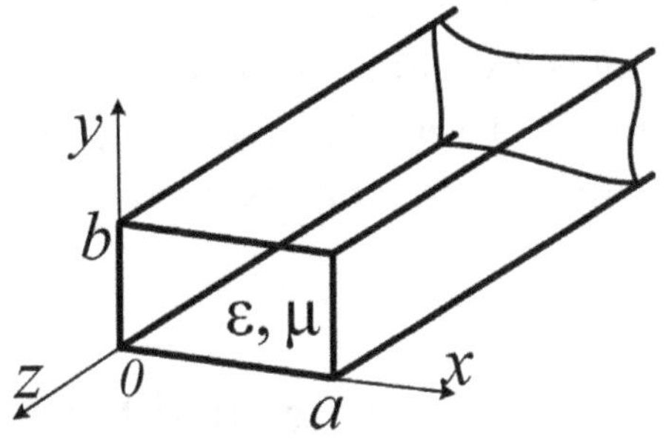

Figure 1. Rectangular waveguide (wave propagation is along *z*-axis).

Similarly, for the TM wave propagation, with axial magnetic field $H_z = 0$, it is required to solve the wave equation as a function of E_z, $\Delta_t E_z + K^2 E_z = 0$, or in Cartesian coordinates $\left(\partial^2 E_z/\partial x^2\right) + \left(\partial^2 E_z/\partial y^2\right) + K^2 E_z = 0$. Scalar function E_z, given in Cartesian coordinates, $E_z(x,y,z) = E_z(x,y,0)e^{-\gamma z}$, represents the solution of the given wave equation, and each non-zero field component can be expressed as a function of E_z. The following expression is obtained for E_z, by solving the wave equation: $E_z(x,y,z) = E_0\sin(m\pi x/a)\sin(n\pi y/b)e^{-j\beta z}$, $m = 1,2,...,$ $n = 1,2,...,$ where E_0 is complex constant. Starting from this solution, the other field components of the TM$_{mn}$ mode can be obtained, as in [17].

For each TE$_{mn}$ or TM$_{mn}$ mode, the frequency from which that mode starts to propagate, called a cut-off frequency, can be defined as: $f_c = (c/2)\sqrt{(m/a)^2 + (n/b)^2}$ $c = 1/\sqrt{\varepsilon\mu}$.

The wavelength of the guided wave in the waveguide is calculated as: $\lambda_g = \lambda_0/\sqrt{1 - \left(f_c/f\right)^2}$,where λ_0 is the wavelength in the vacuum at frequency *f*. The following equations are used to determine the wave impedance for the TE$_{mn}$ and TM$_{mn}$ mode: $Z_{TE} = \sqrt{\mu/\varepsilon}/\sqrt{1 - \left(f_c/f\right)^2}$, $Z_{TM} = \sqrt{\mu/\varepsilon}\sqrt{1 - \left(f_c/f\right)^2}$.

The dominant mode of propagation is the one with the lowest cut-off frequency. In case $a > b$, the dominant mode is TE$_{10}$. The propagation of the dominant mode is of interest, and it will be considered throughout the chapter. Its cut-off frequency is $f_{cTE_{10}} = c/2a = 1/(2a\sqrt{\varepsilon\mu})$, and a set of field equations for this mode is: $E_x(x,y,z) = 0$, $E_y(x,y,z) = -j\omega\mu H_0(a/\pi)\sin(\pi x/a)e^{-j\beta z}$, $H_x(x,y,z) = j\beta H_0(a/\pi)\sin(\pi x/a)e^{-j\beta z}$, $H_y(x,y,z) = 0$, $H_z(x,y,z) = H_0\cos(\pi x/a)e^{-j\beta z}$, where $\beta = \sqrt{\omega^2\varepsilon\mu - (m\pi/a)^2 - (n\pi/b)^2}$ denotes the phase coefficient.

Maximum power in the waveguide is limited by the dielectric breakdown, that is, by the critical field in the waveguide. Furthermore, in case the waveguide is not matched to its ports, so there is a reflected wave beside the incident one, the maximum power will be decreased by the value of the reflected power.

For the waveguide filters considered in this chapter, standard WR-90 rectangular waveguide is used. According to **Figure 1**, dimensions of its cross-section are a = 22.86 mm and b = 10.16 mm. The frequency range for this waveguide is 8.2–12.4 GHz. The cut-off frequency for the dominant mode of propagation TE_{10} is 6.56 GHz. Next, higher mode, TE_{20}, starts propagating from 13.12 GHz, which is in accordance with the observed frequency range for this waveguide.

3. Basic concepts of microwave waveguide filter design

This section briefly covers resonators, as building blocks of the filters, and filter design principles, with the focus on the waveguide filters.

3.1. Resonators

The resonators can be characterized as fundamental passive microwave components and basic elements of the frequency selective circuits, such as filters. They are capable to store frequency-dependent electric and magnetic energy [2]. Under ideal circumstances, resonator is electromagnetically isolated part of the space, without losses, so once excited electromagnetic field can be indefinitely retained [16]. However, in real situations, every resonator loses its energy due to various losses (e.g., losses in conductors and dielectric) and the coupling with the surrounding components, which is usually weak. This means that the energy must be continuously added to the resonator, in order to compensate for the losses and maintain the excited field.

The main parameters which qualify a resonator are its resonant frequency and the quality factor (Q-factor). In general, Q-factor reflects the resonator performance [17].

Resonators can be implemented in different ways: for the frequencies up to 1 GHz as *LC* circuits and for the frequency range 1 MHz–10 GHz as transmission line sections, while for the frequency range 1–100 GHz as waveguide resonators.

Planar resonators can be classified as resonators implemented in planar technologies, such as the microstrip technology. Quarter-wave and half-wave planar resonators are widely used for the filter design, either as short-circuited or opened lossless transmission-line section of electrical length $\theta = \pi/2$ or $\theta = \pi$ (at resonant frequency $\omega_0 = 2\pi f_0$), respectively. Besides the short-circuited quarter-wave resonators (QWRs), we have developed various models of the waveguide filters using SRRs (for bandstop filters) and CSRRs (for bandpass filters). Filters using these resonators can have one or more pass bands or rejection bands [18–24].

3.2. Basic filter design principles

Filters can be defined as frequency selective networks, with two or more ports, used to pass signals at chosen frequencies and weaken other signals (mostly by reflecting them) [16]. Depending on their use, they can be implemented in different ways. However, the filter design procedure is pretty much determined; it consists of several steps which can be classified as follows [1, 16, 25]: specification, approximation, synthesis, simulation model, implementation, study of imperfections and optimization.

Specification represents a set of criteria that a filter should meet [1, 16]. Approximation is a mathematical representation of the frequency response, so the filter specification can be met and the filter itself can be implemented. It is usually given as an attenuation, in dB, for the low-pass filter, and based on that, approximations for the other types of filters can be derived. Some of the most commonly used approximations are Butterworth, Chebyshev, elliptic (Cauer), Bessel and Gaussian function.

Synthesis assumes generating the filter schematic, using ideal elements, which fulfills transfer function given by approximation. Filter simulation model is a filter schematic with real components instead of ideal ones. Filter implementation is actually a laboratory prototype—fabricated device; it is used to experimentally validate the filter response by measurements, in order to evaluate the simulation model and modify it, if necessary. Study of imperfections investigates various parasitic effects caused by the use of real components. Optimization assumes systematic variation of the filter parameters, using some numerical method, in order to meet the specification.

The filter transfer function is the ratio of the Laplace transforms of the output signal and the input signal assuming zero initial conditions, and the filter is treated as a linear time-invariant two-port network. The frequency response is the transfer function on the imaginary axis. Microwave filter response is typically represented by scattering parameters (S-parameters) [16].

Generic representation of the filter as a two-port network is depicted in **Figure 2a**. At the first port, filter is excited by a real generator, with internal resistance, and the second port is terminated by a resistor.

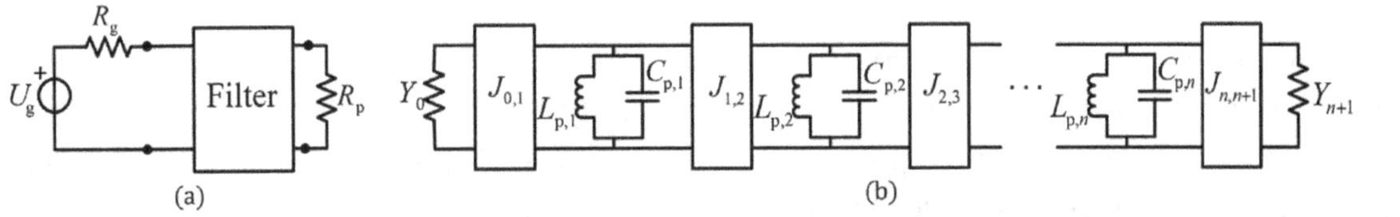

Figure 2. (a) Generic representation of the filter as a two-port network, (b) bandpass filter schematic with parallel LC circuits and admittance inverters.

A filter prototype is a ladder LC network of the low-pass filter with ideal elements. Starting from the prototype, the parameters of the real filter elements can be obtained, by using frequency and impedance transformations [2, 26, 27]. In this manner, bandpass and bandstop filters, consisting of parallel and series LC circuits, can be obtained. Often, it is convenient to have a network with *LC* circuits only in parallel branches or only in series branches. This can be achieved by adding inverters.

Immittance inverter (a common name for the impedance and admittance inverter) is a two-port, frequency independent, network without losses. Impedance inverter can be used to transform parallel admittance into series impedance, and admittance inverter can be used to transform series impedance into parallel admittance. **Figure 2b** depicts the bandpass filter schematic, with admittance inverters and parallel *LC* circuits. The following equations are used to calculate the characteristic inverter admittance (J): $J_{0,1} = \sqrt{Y_0 B C_{p,1}/\Omega_{p,\mathrm{proto}} g_0 g_1}$, $J_{i,i+1} = (B/\Omega_{p,\mathrm{proto}})\sqrt{C_{p,i} C_{p,i+1}/g_i g_{i+1}} (i = 1, \ldots, n-1)$, $J_{n,n+1} = \sqrt{Y_{n+1} B C_{p,n}/\Omega_{p,\mathrm{proto}} g_n g_{n+1}}$, $L_{p,i} = 1/\omega_0^2 C_{p,i}$ $(i = 1, \ldots, n)$. Parameters g_i are those used for the prototype filter, as well as $\Omega_{\mathrm{p,proto}}$, while B denotes the bandwidth. The values of the elements Y_0, Y_{n+1}, C_i can be arbitrarily adopted, and the filter response will be identical to the prototype response if the parameters $J_{i,i+1}$ are calculated using aforementioned equations.

Immittance inverters can be implemented in different ways. In this chapter, inverters as quarter-wave transmission line sections are of interest. To be more precise, since waveguide filters are considered, inverters are implemented as waveguide sections.

4. Method for advanced waveguide filter design using printed-circuit discontinuities

Advanced waveguide filter design method assumes the development of resonating elements employed as printed-circuit inserts, placed inside a rectangular waveguide [28]. The goal is to obtain one or more resonant frequencies using a single resonating insert, which can be achieved by optimal layout of the resonators on the plate. Desired solution might assume uncoupled resonators, in order to be able to tune each resonator independently for each frequency band. In this manner, multi-band filter can be implemented. Herein, SRRs, CSRRs, QWRs and similar types of simple resonators are used on the printed-circuit inserts as resonating elements.

Waveguide filters can be implemented using H-plane or E-plane inserts. For the higher-order multi-band filters with H-plane inserts, it is necessary to properly implement inverters between the resonators, for each center frequency. For the considered filters, waveguide section of length equal to $\lambda_g/4$ (λ_g – guided wavelength in the waveguide) is used as an inverter. Furthermore, for the higher-order bandpass filter, folded H-plane inserts are used in order to meet requirement regarding quarter-wave inverter implementation for each center frequency. For the compact filter design, miniaturized inverters are considered, by introducing properly designed additional plates between H-plane resonating inserts. The other solution for the waveguide filter design is based on the E-plane insert implementation, with properly coupled resonators on the plate. Along with the waveguide filter design, structures for precise positioning of inserts are developed and used for the measurement of the frequency responses.

Based on the presented filter design method, an algorithm has been developed, that assumes-following steps: making 3D EM models, optimizing resonator and filter parameters, investigating frequency responses, (optionally) generating microwave circuits and, finally, experimental verification by measuring the response of the fabricated prototype. In order to complete these steps, software tools supporting full-wave simulations have been used, as well as the tools for microwave circuit simulations. Experimental verification is performed to validate simulation results and verify the filter design method.

Various bandpass and bandstop waveguide filters have been developed using proposed method [28, 29]. The same standard rectangular waveguide (WR-90) is used for each implementation in the chosen frequency band, since there is no modification of the waveguide structure. The dominant mode of propagation (TE_{10}) is of interest. Filters are designed to operate in the X frequency band (8.2–12.4 GHz).

For precise modeling of the waveguide filters, WIPL-D software (http://www.wipl-d.com) is used, as powerful software capable to perform EM simulations of the complex 3D structures. It allows us to work with a large number of elements (nodes, plates, generators, etc.) and variables. In terms of numerical calculations, the software is based on the method of moments (MoM), and theoretical background is elaborated in detail in [30]. Besides the network parameters ($S/Z/Y$), it can efficiently calculate near field, far field (radiation) and current distribution.

WIPL-D software provides the possibility of precise geometrical modeling of the structure, meaning it can be completely defined using various elements and parameters available in the software. Virtually, every structure can be modeled in this manner. For example, structures can be modeled as pure metallic, pure dielectric or composite (metallic and dielectric). Every structure primarily consists of nodes, as essential elements, and they are further used to create wires, plates, junctions and generators. As an excitation, an ideal delta-function generator is used. Graphical interface allows us to visually track the modeling procedure.

WIPL-D software provides the possibility to take into account conductor and dielectric losses. The losses due to the surface roughness and the skin effect in conductors are taken into account, so the conductivity of the metal plates, for the X frequency band, is set to $\sigma = 20$ MS/m. Dielectric losses are given as parameters of the domain used to model the considered dielectric.

There are several features in WIPL-D software which can be used to increase the accuracy of the obtained results (edge manipulation) or simplify the modeling procedure (symbols, grids, imaging, symmetry, etc.). All these features are thoroughly explained in the software documentation (http://www.wipl-d.com).

Besides the modeling of the waveguide filter itself, it is important to properly model its excitation, that is, its ports. Each waveguide port (one port on each waveguide end) can be represented as a short-circuited waveguide section with a monopole, fed by an ideal delta-function generator [28]. This section should be long enough so each evanescent mode ceases toward the port end, thus only desired mode propagates in the considered structure. In order to obtain required filter response, that is, to determine its S-parameters, de-embedding technique is applied, so the influence of the ports on the amplitude response is eliminated. This technique is explained in detail at http://www.wipl-d.com.

In order to generate microwave circuits of the considered filters, the following software tools have been used: WIPL-D Microwave Pro (http://www.wipl-d.com) and NI AWR Microwave Office (http://www.awrcorp.com). The equivalent microwave circuits have been generated using lumped elements (for the resonant circuits) and waveguide sections (for the inverters). The goal is to develop an equivalent circuit as simple as possible, starting from the known equivalent circuits of the corresponding elements, thus allowing for a relatively simple filter optimization.

The printed-circuit inserts are fabricated on the machine MITS Electronics FP-21TP (http://www.mitspcb.com). Experimental verification is performed by measuring the frequency characteristics of the fabricated filters using network analyzer Agilent N5227A (http://www.keysight.com), as shown in **Figure 3**.

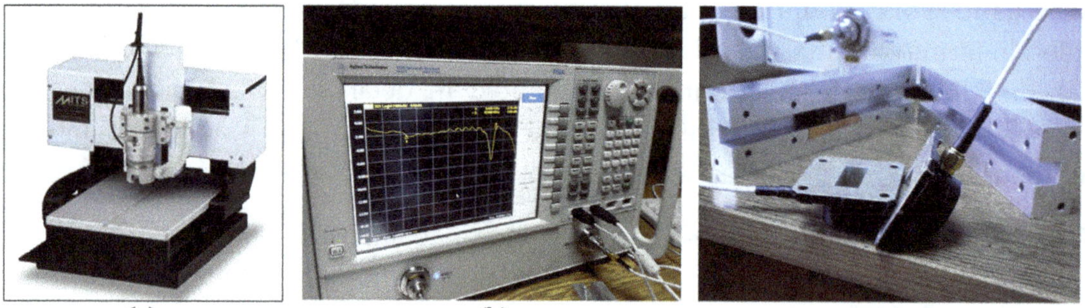

(a) (b) (c)

Figure 3. (a) MITS Electronics FP-21TP machine, (b) Agilent N5227A network analyzer, (c) ports used for the measurements.

5. Bandpass filters

Bandpass waveguide filter design is based on the use of CSRRs. Starting from the models given in [9], various types of waveguide resonators and filters have been developed and proposed. The inserts are placed in the H-plane of the standard rectangular waveguide. The 3D EM models and equivalent microwave circuit of the CSRR on the multi-layer planar insert can be found in [31, 32]. Such CSRR can be further upgraded by adding a central section on the inner side, providing for resonant frequency fine-tuning, as proposed in [32].

Third-order bandpass waveguide filter has been developed using aforementioned printed-circuit inserts with CSRRs [31]. For the printed circuits, RT/Duroid 5880 substrate (http://www.rogerscorp.com) is used, and its parameters are relative permittivity $\varepsilon_r = 2.2$, losses $\tan\delta = 0.0009$, substrate thickness $h = 0.8$ mm, metallization thickness $t = 0.018$ mm. The inserts are placed in the transverse planes of the waveguide (**Figure 4a**). Filter is designed to operate at center frequency of 11.95 GHz and 3-dB bandwidth of 500 MHz, so the parameters of the CSRRs are tuned accordingly. In order to properly design higher-order filter, it is necessary to take into account the inverters between the resonating elements (waveguide sections of length equal to $\lambda_g/4$ at 11.95 GHz). The obtained amplitude response is shown in **Figure 4b**.

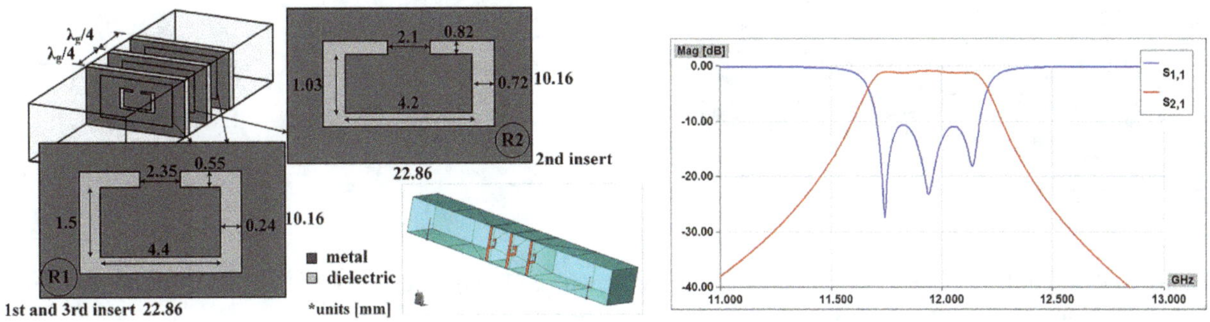

(a) (b)

Figure 4. Third-order bandpass filter using CSRRs on the printed-circuit inserts: (a) 3D model, (b) amplitude response.

6. Bandstop filters

Bandstop filters presented here are implemented using SRRs and QWRs on the multilayer planar inserts. Depending on the filter design, the inserts can be placed in the H-plane or E-plane of the rectangular waveguide, as will be elaborated in this section.

The basic model of the waveguide resonator uses printed-circuit insert with dimensions equal to those of the waveguide cross-section (**Figure 5a**, type 1). SRR is used as a resonator. Such model can be found in the available literature [10, 11], and it is also investigated in detail in [33]. Another planar insert with SRR (type 2), introduced in [34, 35], is implemented as a small dielectric plate (it is significantly smaller than the waveguide cross-section), attached to the top and bottom waveguide walls by thin dielectric strips. The waveguide resonator using this type of resonating insert is also shown in **Figure 5a**. In both cases, inserts are designed using RT/Duroid 5880 substrate (its parameters are given in Section 5). Dimensions of the SRR in both cases are optimized to achieve f_0 = 9 GHz. The amplitude responses of the waveguide resonators using previously described inserts are compared in **Figure 5b**. Although resonant frequencies are slightly moved apart, it is notable that the return loss beyond the stop band has lower values for the type 2 insert. Therefore, better matching in the pass band can be obtained using smaller planar insert. This is an important result for the higher-order filter design.

For this model of the resonator, the equivalent circuit is proposed (**Figure 5c**). The parameters of the *RLC* circuit are calculated using following equations [28]: $R = \left(2\left|S_{11}(j\omega_0)\right|Z_0\right)/\left(1-\left|S_{11}(j\omega_0)\right|\right)$, $L = \left(2B_{3dB}Z_0\left|S_{11}(j\omega_0)\right|\right)/\omega_0^2$, $C = 1/\left(2B_{3dB}Z_0\left|S_{11}(j\omega_0)\right|\right)$, where R represents resistance, L inductance, C capacitance, ω_0 is the resonant frequency in [rad/s], $|S_{11}(j\omega_0)|$ is the amplitude characteristic of S_{11} at resonant frequency, B_{3dB} is a 3-dB bandwidth, Z_0 represents the nominal impedance and it is calculated according to equations given in Section 2.

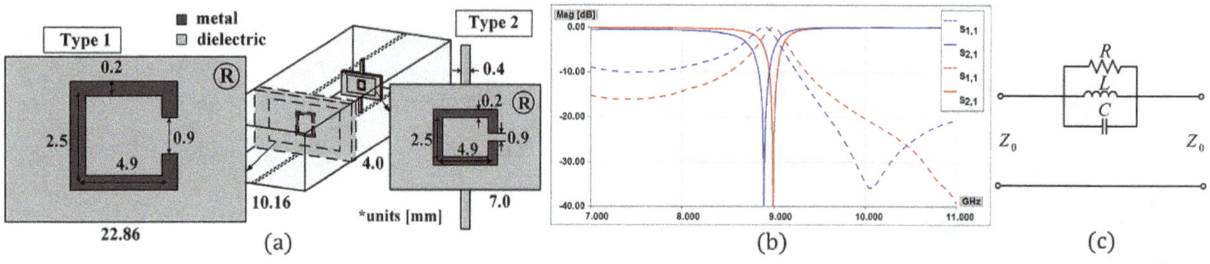

Figure 5. Waveguide resonator using printed-circuit insert with SRR: (a) type 1 and 2, (b) comparison of amplitude responses: blue—type 1 insert, red—type 2 insert, (c) equivalent circuit.

Besides SRRs, properly designed QWRs can be also used for the bandstop filter design. They are short-circuited to top or bottom waveguide wall, depending on the implementation. The insert with the QWRs can be placed in the H-plane [36] or E-plane [37] of the rectangular waveguide.

6.1. Third-order H-plane bandstop filter using printed-circuit insert with SRR

The third-order bandstop filter using aforementioned type 2 planar inserts is shown in **Figure 6a**. Small dielectric plates are centrally positioned in the waveguide. The filter is designed for the center frequency of 9 GHz and the 3-dB bandwidth of 320 MHz, so dimensions of the SRRs are optimized accordingly. The obtained amplitude response is given in **Figure 6b**.

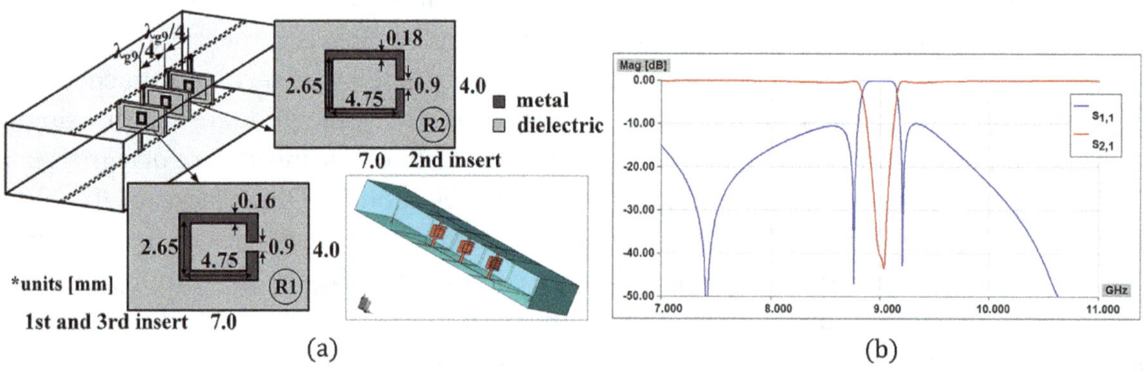

Figure 6. Third-order bandstop filter using printed-circuit insert with SRR (type 2): (a) 3D model, (b) amplitude response.

6.2. E-plane bandstop waveguide filter using QWRs

Herein, second-order bandstop filter using E-plane insert with QWRs is considered. Filter specification requires the center frequency of 10 GHz and 3-dB bandwidth of 680 MHz. Two QWRs, with identical dimensions, are used to achieve this, and they are positioned on the insert as shown in **Figure 7a**. For the planar insert, copper clad PTFE/woven glass laminate (TLX-8) (http://www.taconic-add.com) is used as a substrate, and its parameters are $\varepsilon_r = 2.55$, $\tan\delta = 0.0019$, $h = 1.143$ mm, $t = 0.018$ mm.

Figure 7. Bandstop filter using E-plane insert with QWRs: (a) 3D model, (b) fabricated insert, (c) comparison of amplitude responses: blue—simulation results, red—measurement results.

The amplitude response is analyzed for various values of the distance between the QWRs. It has been shown that this variation primarily influences the bandwidth; to be more precise, by increasing the distance, the bandwidth becomes narrower.

For the considered filter model, the coupling between two QWRs is investigated in terms of their mutual distance, as proposed in [26, 38]. By increasing the distance, the coupling gets weaker. Based on the obtained results for the 3-dB bandwidth and the coupling coefficient k, it can be concluded that these results are in accordance with the statements in the available literature [26], that is, B_{3dB} and k change in the same manner.

In order to validate the proposed design, the planar insert for the filter containing QWRs is fabricated (**Figure 7b**), and the obtained response is experimentally verified using the WR-90 waveguide. **Figure 7c** shows comparison of the simulation and measurement results, and their good matching confirms filter design.

Proposed filter can be further upgraded to achieve bandwidth fine-tuning, by adding a coupling element, realized as a SRR and positioned between the QWRs on the insert (**Figure 8a**). The positions of the QWRs and their mutual distance are the same as in the original model without coupling element. The amplitude response is shown in **Figure 8b**. The obtained results show that, by adding the coupling element between the QWRs, the bandwidth becomes narrower for 50 MHz (which is nearly 7.37% compared to the reference value of 678 MHz for the second-order filter).

In order to investigate the possibility for bandwidth fine-tuning, using the coupling element, its printed line width is varied, while the distance between the QWRs remains the same. In this case, the center frequency practically does not change, while the bandwidth is slightly changed. This is in accordance with the purpose of the coupling element to fine-tune the bandwidth, without modifying the filter response.

Similarly as for the filter with two QWRs, the coupling between the resonators is also analyzed for this case, for the chosen distance between the resonators and different values of the printed line width of the coupling element. The obtained results and conclusions regarding the proposed filter design can be summarized as follows. By adding the coupling element between the QWRs: (1) the bandwidth is decreased, (2) the coupling is weakened, (3) the bandwidth can be tuned (decreased) without increasing the distance between the QWRs, thus providing for the device miniaturization.

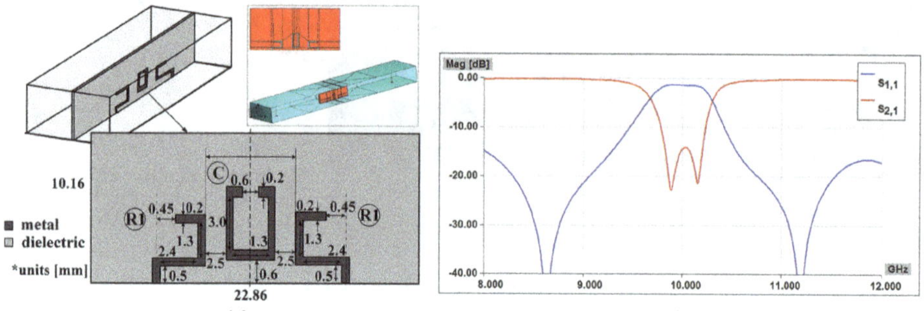

(a) (b)

Figure 8. Waveguide filter using E-plane insert with QWRs and coupling element: (a) 3D model, (b) amplitude response.

7. Multi-band waveguide filter design

Multi-band filter design represents a significant research field, particularly for the modern communication systems. One of the advantages of the presented method for the waveguide filter design is that it can be used to relatively easy develop multi-band bandpass or bandstop filters. Namely, multiple resonant frequencies can be obtained using single insert with properly designed and positioned resonators. It is recommended to have uncoupled resonators on the insert, since in that case, each frequency band can be independently controlled by tuning only particular resonator [28]. Waveguide resonator with multiple resonant frequencies can be obtained in this manner. It can be designed using H-plane insert with multiple CSRRs or similar resonating elements for the bandpass characteristic, as in [32, 39–41], or with SRRs for the bandstop characteristic, as in [33, 34]. H-plane insert with QWRs, providing multiple resonant frequencies and having bandstop characteristic, is introduced in [36]. Herein, compact dual-band bandpass filter with CSRRs and dual-band bandstop filter are elaborated in detail. They both use H-plane inserts.

7.1. Compact dual-band bandpass filter with H-plane inserts

Design starts from the resonating insert with two CSRRs in order to obtain two resonant frequencies, 9 GHz and 11 GHz. Flat metal insert with two CSRRs can be placed in the H-plane of the rectangular waveguide, as explained in [40, 41].

Since it is necessary to properly implement inverters for each center frequency, for higher-order filters, folded insert is introduced as a suitable solution [40, 41]. The second-order filter using two identical folded metal inserts is shown in **Figure 9**. Dimensions of the CSRRs are the same for both inserts and optimized to obtain f_{01} = 9 GHz, $B_{3dB\text{-}1}$ = 450 MHz, f_{02} = 11 GHz, $B_{3dB\text{-}2}$ = 650

MHz. The length of the plate connecting parts of the folded insert with CSRRs is $l_{pl} = (\lambda_{g9GHz} - \lambda_{g11GHz})/8$. In this manner, quarter-wave inverters are implemented for both center frequencies, as shown in **Figure 9a**.

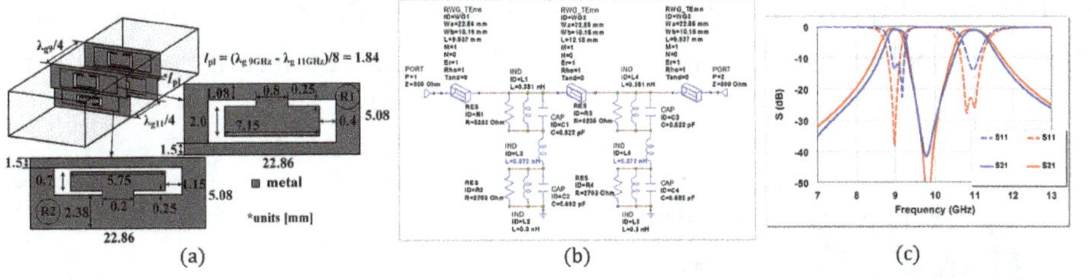

Figure 9. Second-order bandpass filter using folded metal inserts with CSRRs: (a) 3D model, (b) equivalent microwave circuit, (c) comparison of amplitude responses for the 3D EM model (blue) and equivalent circuit (red).

The equivalent microwave circuit is generated for the proposed second-order filter (**Figure 9b**). Two separate networks are used for each insert, and the waveguide section is added between them as a quarter-wave inverter. The short plate providing the fold is represented by an inductor. The following equations are derived [28] and used to calculate the parameters of the *RLC* circuits: $R = \left(Z_0\left|S_{21}(\mathrm{j}\omega_0)\right|\right)/\left(2\left(1-\left|S_{21}(\mathrm{j}\omega_0)\right|\right)\right)$, $L = \left(B_{3dB}Z_0\left|S_{21}(\mathrm{j}\omega_0)\right|\right)/\left(2\omega_0^2\right)$, $C = 2/\left(B_{3dB}Z_0\left|S_{21}(\mathrm{j}\omega_0)\right|\right)$, where R represents resistance, L inductance, C capacitance, ω_0 is the resonant frequency in [rad/s], $|S_{21}(\mathrm{j}\omega_0)|$ is the amplitude characteristic of S_{21} at resonant frequency, B_{3dB} is a 3-dB bandwidth, Z_0 represents the nominal impedance and it is calculated according to equations given in Section 2. The inductances of the additional inductors between *RLC* circuits, for each insert, are tuned. The port impedance is set to Z_0 = 500 Ω, which is the wave impedance at 10 GHz (the frequency between the targeted ones, 9 GHz and 11 GHz). The amplitude responses obtained for the 3D EM model and circuit are compared in **Figure 9c**, and their good mutual agreement validates proposed equivalent representation.

The next step is to develop compact solution, based on the use of miniaturized inverters. Precisely, quarter-wave inverter between the resonating inserts, for each center frequency, is replaced by a shorter waveguide section and additional, properly designed, metal insert. In this manner, the length of the device decreases, by shortening the inverters, without degrading filter response. The realization details, including 3D EM models and the equivalent microwave circuit, of the filter with inverters of length equal to $\lambda_g/8$ for each center frequency, can be found in [40, 41]. The inverters are halved compared to the original model, and additional metal plate for miniaturization (with properly designed slots) is centrally positioned between the resonating inserts, to preserve the original filter response.

In order to simplify fabrication and experimental verification, presented filter can be further modified. Therefore, a compact filter with flat inserts is proposed. This means that the inserts with CSRRs, as well as the additional insert for miniaturization, are flat. However, the inverters for both center frequencies are not miniaturized in the same manner. The distance between the CSRRs with f_{01} = 9 GHz is set to $\lambda_{g9GHz}/8$, while the normalized length of the inverter between the CSRRs with $f_{02} = 11$ GHz is equal to $0.18\lambda_{g11GHz}$ (**Figure 10a**). Dimensions of the slots on the additional metal insert are tuned to preserve the original filter response, without miniaturization. Comparison of amplitude responses of the second-order dual-band filter before and after applying inverter miniaturization, with the same and different normalized lengths of the inverters, is shown in **Figure 10b**. Since the obtained results match well, compact filter with flat inserts is confirmed as a solution more suitable for experimental verification.

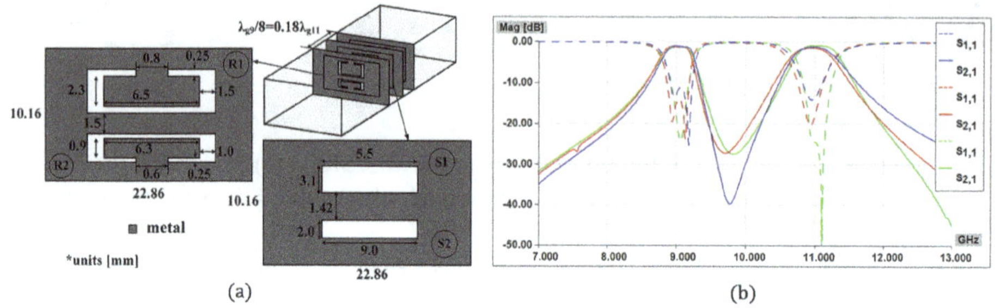

Figure 10. (a) Compact bandpass filter with unequal inverter miniaturization, (b) comparison of amplitude responses of the filter without inverter miniaturization (blue), with equal (red) and unequal (green) inverter miniaturization.

In order to confirm the proposed compact filter solution, the amplitude response of the second-order dual-band filter with flat metal inserts and unequal inverter miniaturization is experimentally verified. Resonating inserts and additional plate for miniaturization are fabricated using metal foil of thickness 100 μm. However, it is necessary to have stable inserts inside the waveguide to successfully measure the amplitude characteristics. Therefore, we propose fixtures for supporting metal inserts. They are fabricated using FR-4 epoxy substrate ($\varepsilon_r = 4.5$, $h = 1.0$ mm, $\tan\delta = 0.02$). Structures for precise positioning of inserts are depicted in **Figure 11a**, and the detailed description of their design can be found in [41]. The photographs of the fabricated inserts and the structures for precise positioning are given in **Figure 11b**. Comparison of the simulated and measured amplitude responses is shown in **Figure 11c**, confirming the compact filter solution by a good agreement of the obtained results.

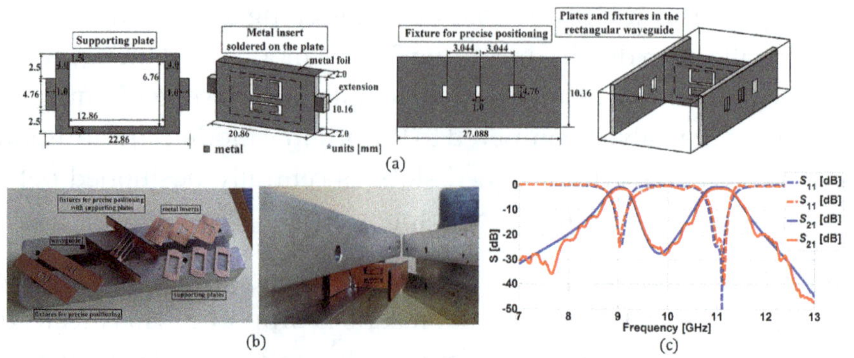

Figure 11. Experimental verification: (a) structures for precise positioning of inserts, (b) photographs of the fabricated components, (c) comparison of simulated (blue) and measured (red) amplitude responses for the compact filter.

7.2. Third-order dual-band bandstop filter using planar insert with SRR

Herein, third-order bandstop filters using printed-circuit inserts with SRRs, introduced in Section 6, are considered. In this manner, the use of such inserts for the higher-order multi-band filters is exemplified. First, third-order bandstop filters with f_0 = 9 GHz and f_0 = 11 GHz, each having a single rejection band, are designed. Based on that, by combining elements of these single-band filters, third-order dual-band filter is designed with f_{01} = 9 GHz and f_{02} = 11 GHz, and each rejection band has the same 3-dB bandwidth of 335 MHz [35].

For the bandstop filter, it is convenient to have *LC* circuits, representing resonators, connected in series; thus, the inverters are used and the parameters of the resonators are determined as explained in [26]. Microwave circuit of this filter is shown in **Figure 12a**.

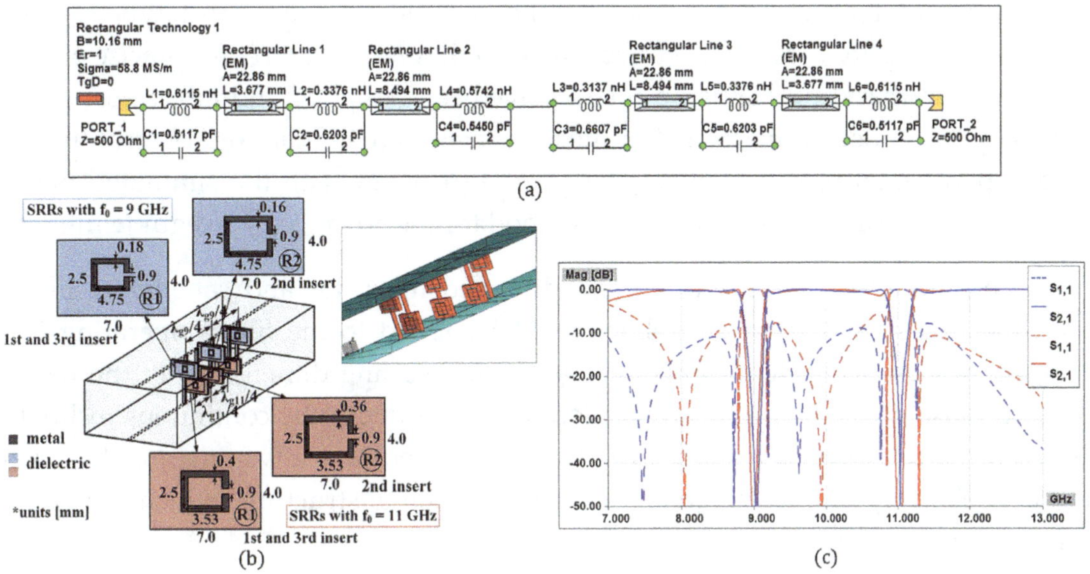

Figure 12. Third-order dual-band bandstop filter: (a) microwave circuit, (b) 3D model, (c) comparison of amplitude responses (blue—3D EM model, red—microwave circuit).

The 3D model of the filter is shown in **Figure 12b**. The inserts are optimally positioned so the inverters can be properly realized between the corresponding resonators, as explained in details in [35]. Comparison of the amplitude responses of the microwave circuit and 3D EM model is given in **Figure 12c**, showing good matching of the obtained results and confirming the proposed filter design.

The experimental measurements for the filter using the presented inserts cannot be easily performed without finding a way to have stable inserts in the waveguide. Therefore, a structure for precise positioning, in a form of ladder fixtures, is introduced in [42]. Since we use dielectric inserts, the fixtures are also designed using dielectric substrate although they can be made as multi-layer planar structures. These fixtures are attached to the top and bottom waveguide walls, and the inserts are attached to them. As shown in [42], the use of the fixtures does not degrade the filter response, so they can be used for precise positioning of inserts in the measurement procedure.

8. Conclusion

Starting from the microwave filter design principles, a method for the advanced waveguide filter design has been developed. It is based on the use of printed-circuit discontinuities as resonating elements. Relatively simple resonators, such as SRRs, CSRRs and QWRs, have been used for the printed-circuit insert design. In spite of their simple forms, these types of resonators provide a lot of possibilities to tune the amplitude response. Since multiple resonant frequencies can be obtained using single insert, these inserts can be employed for the multi-band filter design. Therefore, various implementations of novel waveguide filters have been made using H-plane or E-plane inserts. For bandpass filters, metal or multilayer planar inserts with CSRRs have been used. Also, a compact filter solution has been proposed. Furthermore, bandstop filters have been designed using planar inserts with SRRs and QWRs. Besides 3D EM simulations and microwave circuits of the considered filters, prototypes have been fabricated so the amplitude responses can be measured. Also, structures for precise positioning of inserts, for bandpass and bandstop filters, have been designed and fabricated, since they are necessary to smoothly perform the measurements on the laboratory prototypes. The obtained experimental results have shown good agreement with the simulated ones, thus confirming the application of the proposed method for the advanced waveguide filter design.

Although considered waveguide filters are designed for radar and satellite systems, it is expected that the same design guidelines can be applied to the filters operating in other frequency bands by using the proper waveguide and scaling dimensions of the resonating components. Also, it can be further upgraded to keep up with the continuous technological development (e.g., tunable filter design). The most important advantage of the proposed method is a relatively simple design and implementation of structures which can be used in modern communication systems.

Acknowledgements

This work was supported by the Ministry of Education, Science and Technological Development of the Republic of Serbia under Grant TR32005.

Author details

Snežana Stefanovski Pajović*, Milka Potrebić* and Dejan V. Tošić

*Address all correspondence to: stefanovskisnezana@gmail.com and milka_potrebic@etf.rs

School of Electrical Engineering, University of Belgrade, Belgrade, Serbia

References

[1] Lutovac MD, Tošić DV, Evans BV. Filter Design for Signal Processing using MATLAB and Mathematica. Upper Saddle River, NJ: Prentice Hall; 2001 (Translated in Chinese. Beijing, P. R. China: Publishing House of Electronics Industry, PHEI; 2004).

[2] Cameron RJ, Kudsia CM, Mansour RR. Microwave Filters for Communication Systems: Fundamentals, Design, and Applications. New Jersey: John Wiley & Sons; 2007.

[3] Hunter I. Theory and Design of Microwave Filters. London: The Institution of Engineering and Technology; 2006.

[4] Marcuvitz N. Waveguide Handbook. Stevenage, UK: Peter Peregrinus Ltd., on behalf of the Institution of Electrical Engineers; 1986.

[5] Budimir D, Glubokov O, Potrebic M. Waveguide filters using T-shaped resonators. Electronics Letters. 2011;27(1):38–40. doi:10.1049/el.2010.2958

[6] Jitha B, Nimisha CS, Aanandan CK, Mohanan P, Vasudevan K. SRR loaded waveguide band rejection filter with adjustable bandwidth. Microwave and Optical Technology Letters. 2006;48(7):1427–1429. doi:10.1002/mop.21641

[7] Kehn MNM, Quevedo-Teruel O, Rajo-Iglesias E. Split-ring resonator loaded waveguides with multiple stopbands. Electronics Letters. 2008;44(12):714–716. doi: 10.1049/el:20081108

[8] Ortiz N, Baena JD, Beruete M, Falcone F, Laso MAG, Lopetegi T, Marques R, Martin F, Garcia-Garcia J, Sorolla M. Complementary split-ring resonator for compact waveguide filter design. Microwave and Optical Technology Letters. 2005;46(1):88–92. doi: 10.1002/mop.20909

[9] Bahrami H, Hakkak M, Pirhadi A. Analysis and design of highly compact bandpass waveguide filter utilizing complementary split ring resonators (CSRR). Progress in Electromagnetics Research. 2008;80:107–122. doi:10.2528/PIER07111203

[10] Fallahzadeh S, Bahrami H, Tayarani M. A novel dual-band bandstop waveguide filter using split ring resonators. Progress in Electromagnetics Research. 2009;12:133–139. doi:10.2528/PIERL09103103

[11] Fallahzadeh S, Bahrami H, Tayarani M. Very compact bandstop waveguide filters using split-ring resonators and perturbed quarter-wave transformers. Electromagnetics. 2010;30:482–490. doi:10.1080/02726343.2010.483942

[12] Miljanović D, Potrebić M, Tošić DV. Design of microwave multibandpass filters with quasilumped resonators. Mathematical Problems in Engineering. 2015;2015(Article ID 647302):1–14. doi:10.1155/2015/647302

[13] Miljanović D, Potrebić M, Tošić DV, Stamenković Z. Design of miniaturized bandpass filters using quasi-lumped multilayer resonators. Journal of Circuits, Systems and Computers. 2014;23(6):1450083. doi:10.1142/S0218126614500832

[14] Potrebić M, Tošić DV. A novel design of a compact multilayer resonator using double-sided microstrip. Optoelectronics and Advanced Materials-Rapid Communications. 2012;6(3–4):441–445. [online] http://oam-rc.inoe.ro/index.php?option= magazine&op=view&idu=1851&catid=71

[15] Potrebić M, Tošić DV. Selective bandpass filter with concentrated impulse response. Microwave and Optical Technology Letters. 2008;50(11):2772–2777. doi:10.1002/mop.23785

[16] Pozar DM. Microwave Engineering. New York: John Wiley; 2012.

[17] Collin RE. Foundations for Microwave Engineering. New York: McGraw-Hill; 1992 (Reprinted by John Wiley—IEEE Press; 2001).

[18] Bonache J, Martin F, Falcone F, Baena JD, Lopetegi T, Garcia-Garcia J, Laso MAG, Gil I, Marcotegui A, Marques R, Sorolla M. Application of complementary split-ring resonators to the design of compact narrow band-pass structures in microstrip technology. Microwave and Optical Technology Letters. 2005;46(5):508–512. doi:10.1002/mop.21031

[19] Baena JD, Bonache J, Martin F, Marques Sillero R, Falcone F, Lopetegi T, Laso MAG, Garcia-Garcia J, Gil I, Portillo MF, Sorolla M. Equivalent-circuit models for split-ring resonators and complementary split-ring resonators coupled to planar transmission lines. IEEE Transactions on Microwave Theory and Techniques. 2005;53(4):1451–1461. doi:10.1109/TMTT.2005.845211

[20] Bonache J, Gil I, Garcia-Garcia J, Martin F. Complementary split rings resonators (CSRRs): towards the miniaturization of microwave device design. Journal of Computational Electronics. 2006;5(2–3):193–197. doi:10.1007/s10825-006-8843-0

[21] Kim SI, Jang MY, Kee CS, Park I, Lim H. Characteristics of microwave filters based on microstrip photonic bandgap ring structures. Current Applied Physics. 2005;5(6):619–624. doi:10.1016/j.cap.2004.08.008

[22] Fan JW, Liang CH, Dai XW. Design of cross-coupled dual-band filter with equal-length split-ring resonators. Progress in Electromagnetics Research. 2007;75:285–293. doi:10.2528/PIER07060904

[23] Li D, Xie YJ, Wang P, Yang R. Applications of split-ring resonances on multi-band frequency selective surfaces. Journal of Electromagnetic Waves and Applications. 2007;21(11):1551–1563. doi:10.1163/156939307782000271

[24] Wu GL, Mu W, Dai XW, Jiao YC. Design of novel dual-band bandpass filter with microstrip meander-loop resonator and CSRR DGS. Progress in Electromagnetics Research. 2008;78:17–24. doi:10.2528/PIER07090301

[25] Matthaei GL, Young L, Jones EMT. Microwave Filters, Impedance Matching Networks, and Coupling Structures. New York: McGraw-Hill; 1964 (Reprinted by Norwood, MA: Artech House; 1980).

[26] Hong JS. Microstrip Filters for RF/Microwave Applications. New Jersey: John Wiley & Sons; 2011.

[27] Jarry P, Beneat J. Advanced Design Techniques and Realizations of Microwave and RF Filters. New Jersey: John Wiley & Sons; 2008.

[28] Stefanovski SLJ. Microwave waveguide filters using printed-circuit discontinuities [Ph.D. dissertation, advisor: prof. dr Milka Potrebić]. Belgrade: School of Electrical Engineering, University of Belgrade; 2015.

[29] Stefanovski Pajović S, Potrebić M, Tošić D. Microwave bandpass and bandstop waveguide filters using printed-circuit discontinuities. In: Proceedings of the 23rd Telecommunications Forum (TELFOR 2015); N/A, 24–26 November 2015; Belgrade, Serbia. pp.520–527.

[30] Kolundžija BM, Djordjević AR. Electromagnetic Modeling of Composite Metallic and Dielectric Structures. Norwood, MA: Artech House; 2002.

[31] Potrebić MM, Tošić DV, Cvetković ZŽ, Radosavljević N. WIPL-D modeling and results for waveguide filters with printed-circuit inserts. In: Proceedings of the 28th International Conference on Microelectronics N/A, (MIEL 2012); 13–16 May 2012; Niš, Serbia. pp. 309–312.

[32] Stefanovski S, Potrebić M, Tošić D. Design and analysis of bandpass waveguide filters using novel complementary split ring resonators. In: Proceedings of the 11th International Conference on Telecommunications in Modern Satellite, Cable and Broadcasting Services (TELSIKS 2013); N/A, 16–19 October 2013; Niš, Serbia. pp. 257–260.

[33] Stefanovski S, Potrebić M, Tošić D, Cvetković Z. Design and analysis of bandstop waveguide filters using split ring resonators. In: Proceedings of the 11th International Conference on Applied Electromagnetics (PES 2013); N/A, 1–4 September 2013; Niš, Serbia. pp.135–136.

[34] Stefanovski S, Potrebić M, Tošić D. Novel realization of bandstop waveguide filters. Technics. N/A., 2013; pp.69–76 (special edition). [online] http://www.sits.org.rs/include/data/ docs0485.pdf

[35] Stefanovski S, Potrebić M, Tošić D. A novel design of dual-band bandstop waveguide filter using split ring resonators. Journal of Optoelectronics and Advanced Materials. 2014;16(3–4):486–493. [online] http://joam.inoe.ro/index.php?option=magazine&op=view&idu=3461&catid=83.

[36] Stefanovski S, Potrebić M, Tošić D, Cvetković Z. Bandstop waveguide filters with two or three rejection bands. In: Proceedings of the 29th Conference on Microelectronics (MIEL 2014); N/A, 12–15 May 2014; Belgrade, Serbia. pp. 435–438.

[37] Stefanovski S, Potrebić M, Tošić D. A novel design of E-plane bandstop waveguide filter using quarter-wave resonators. Optoelectronics and Advanced Materials—Rapid

Communications. 2015;9(1–2):87–93. [online] http://oam-rc.inoe.ro/index.php? option=magazine&op=view&idu=2487&catid=88

[38] Villarroya RL. E-plane Parallel Coupled Resonators for Waveguide Bandpass Filter Applications [Ph.D. dissertation]. Edinburgh, Scotland, UK: Heriot-Watt University; 2012.

[39] Stefanovski S, Mirković Đ, Potrebić M, Tošić D. Novel design of H-plane bandpass waveguide filters using complementary split ring resonators. In: Proceedings of Progress in Electromagnetics Research Symposium (PIERS 2014); N/A, 25–28 August 2014; Guangzhou, China. pp. 1963–1968.

[40] Stefanovski S, Potrebić M, Tošić D, Stamenković Z. A novel compact dual-band bandpass waveguide filter. In: Proceedings of IEEE 18th International Symposium on Design and Diagnostics of Electronic Circuits and Systems (DDECS 2015); N/A, April 22–24 2015; Belgrade, Serbia, p. 51–56.

[41] Stefanovski S, Potrebić M, Tošić D, Stamenković Z. Compact dual-band bandpass waveguide filter with H-plane inserts. Journal of Circuits, Systems, and Computers. 2016;25(3):1640015 (18 pages). doi:10.1142/S0218126616400156

[42] Stefanovski S, Potrebić M, Tošić D. Structure for precise positioning of inserts in waveguide filters. In: Proceedings of the 21st Telecommunications Forum (TELFOR 2013); N/A, 26–28 November 2013; Belgrade, Serbia, pp. 689–692.

7

Nonlinear Channel Equalization Approach for Microwave Communication Systems

Modar Shbat, Francisco C. Ordaz-Salazar and
Javier S. González-Salas

Abstract

The theoretical principles of intersymbol interference (ISI) and channel equalization in wireless communication systems are addressed. Several conventional and well-known equalization techniques are discussed and compared such as zero forcing (ZF) and maximum likelihood (ML). The main section in this chapter is devoted to an abstract concept of equalization approach, namely, dual channel equalization (DCE). The proposed approach is flexible and can be employed and integrated with other linear and nonlinear equalization approaches. Closed expressions for the achieved signal-to-noise ratio (SNR) and bit error rate (BER) in the case of ZF-DCE and ML-DCE are derived. According to the obtained outcomes, the DCE demonstrates promising improvements in the equalization performance (BER reduction) in comparison with the conventional techniques.

Keywords: channel equalization, intersymbol interference (ISI), zero forcing (ZF) equalization, maximum likelihood (ML) equalization, bit error rate (BER), dual channel equalization (DCE)

1. Introduction

All types of microwave wireless communication systems under different digital modulation schemes and antennas configuration suffer from the channel effects and related problems such as attenuation, signal amplitude and phase distortions, time-varying fading (Doppler shift), multipath fading, and intersymbol interference (ISI). A common and well-known wireless channel modeling method is based on the representation of the channel as a band-limited digital filter, i.e., linear time-invariant (LTI) filter with specific transfer function (impulse response). Thus, to alleviate and reduce the channel effects, especially for multipath fading and ISI, it is possible to design a digital filter with transfer function that is inverse to the transfer function of the associated wireless channel. This digital filter is called the equalizer

[1, Chapter 10]. Additionally, the employment of multiple antennas can also help to mitigate the multipath fading consequences (transmit/receive diversity) and increase the data rate (spatial multiplexing).

In practice, the wireless transmission system sends a sequence of messages (one-shot transmission) where these successive transmissions should not interfere even if they are closely spaced to increase the data rate. This interference between the successive transmissions is referred as intersymbol interference (ISI) that is able to complicate and reduce the detection performance [2, Chapter 4]. The simple symbol-by-symbol (SBS) detector (optimal in the case of additive white Gaussian noise AWGN channel) cannot be the maximum likelihood estimator for a sequence of message under ISI problem. A receiver for succession messages detection is shown in **Figure 1**, where the matched filter outputs are processed by the receiver and SBS detector to generate the estimate $\hat{X}_k$ of the input symbol X_k at time k.

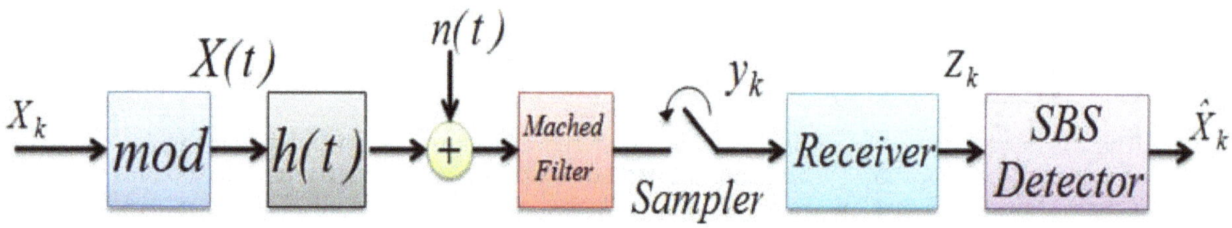

Figure 1. The band-limited channel with receiver and SBS detector.

An equalizer or equalization method can be essentially embedded in the contents of the receiver block presented in **Figure 1**. Different equalization techniques lead to different receiver structures that are not always optimal for detection, but rather are widely implemented as suboptimal cost-effective solutions that alleviate the ISI. Any equalization approach converts the band-limited channel with ISI into memory less appearing channel (AWGN-like) at the receiver output. The wireless channel equalization forms a major challenge in current and future communication networks.

In Section 2 of this chapter, the ISI between successive transmissions is modeled to prove that the distortion caused by this overlapping is unacceptable and some corrective actions should be applied. Some targeted and desired wireless channel responses that exhibit no ISI are discussed in Section 3 with the corresponding Nyquist criterion for the ISI-free channels. In fact, the signal-to-noise ratio (SNR) parameter used to quantify the receiver performance can be consistently considered for equalization techniques evaluation as well. Several conventional and well-known equalization approaches are presented in Section 4 such as zero-forcing equalizer (ZFE), minimum mean square error (MMSE) equalizer, and decision feedback equalizer (DFE). In the last section (Section 5), the proposed equalization approach and its performance are discussed and compared to other equalization techniques under the same initial conditions.

2. Successive message transmission and ISI

The frequency or wireless channel reuse is a common technique to transmit several succession messages separated by the symbol period (T units in time where $1/T$ is the symbol rate). For sending one of M possible messages every T time units, the data rate of the communication system is defined as:

$$R = \frac{log_2(M)}{T}. \quad (1)$$

Increasing the data rate R can be achieved by decreasing T which reduces or narrows the time between message transmissions and as a consequence increases ISI on the band-limited channel. The transmitted signal of K successive transmissions (conveying one of M^K possible messages) is given by

$$x(t) = \sum_{k=0}^{K-1} x_k(t-kT). \quad (2)$$

The detection of M^K messages approach has a complexity that grows exponentially with the block message length K (especially when $K \to \infty$). The SBS detection can be considered as an alternative suboptimal solution that detects independently each of the successive K messages. The ISI problem is the main limitation of the SBS detection approach when the performance degradation increases as T decreases (or R increases). The ISI can be analyzed by rewriting Eq. (2) using the following form:

$$x(t) = \sum_{k=0}^{K-1} \sum_{n=1}^{N} x_{kn}\varphi_n(t-kT), \quad (3)$$

where the original transmissions $x_k(t)$ are decomposed using a standard orthogonal basis set $\{\varphi_n(t)\}$. For instance, in the case of quadrature amplitude modulation (QAM), the baseband basis:

$$\varphi(t) = \frac{1}{\sqrt{T}} sinc\left(\frac{t}{T}\right) \quad (4)$$

produces orthogonal functions for all integer-multiple of T time translations and the successive transmissions sampled at time instants kT are ISI free. Owing to the channel filtering alterations, the filtered basis functions at the channel output are no longer orthogonal and the ISI is introduced.

The band-limited noise-free channel output used for successive transmission of data symbol can be presented as

$$x_p(t) = \sum_{k=0}^{K-1} \sum_{n=1}^{N} x_{kn}\varphi_n(t-kT) * h(t) = \sum_{k=0}^{K-1} \sum_{n=1}^{N} x_{kn}p_n(t-kT), \quad (5)$$

where $p_n(t) = \varphi_n(t)*h(t)$ and when $h(t) \neq \delta(t)$, the functions $p_n(t-kT)$ do not form orthogonal basis (see **Figure 2**). The equalization techniques try to convert the functions $p_n(t-kT)$ to an orthogonal set of functions, thus the SBS detection approach can be applied at the equalized channel output.

The Nyquist criterion [3, Chapter 1] can specify the conditions for ISI-free channel on which SBS detector is optimal. This criterion helps to construct band-limited functions to reduce the

ISI negative effects. Another way to explain the ISI problem is to relate it with the channel frequency response. When consecutive symbols are transmitted using linear modulation over a wireless channel, the frequency response (impulse response) of the channel makes a transmitted symbol to be spread in the time domain. Thus, the ISI is generated because the previously transmitted and currently received symbols are overlapped. The Nyquist theorem could relate the time domain conditions to its equivalent frequency domain ones. Simply, considering the channel impulse response $h(t)$ and the symbol period $h(t)$, the condition of ISI-free response can be expressed as:

$$h(nT) = \begin{cases} 1; & n = 0 \\ 0; & n \neq 0 \end{cases} \text{for all integers } n. \tag{6}$$

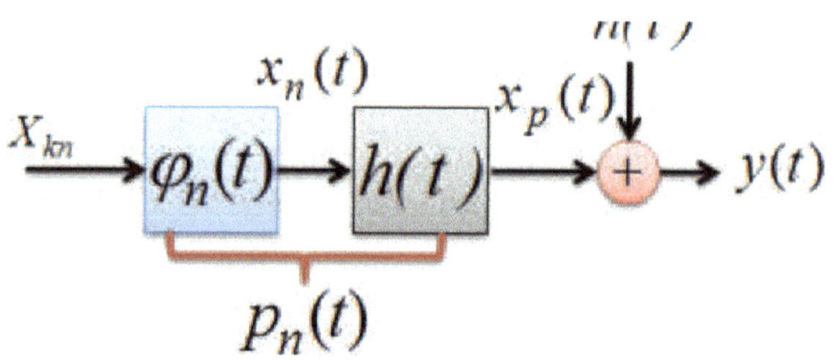

Figure 2. The band-limited channel equivalent impulse response representation.

The last condition can be represented in another form as follows (the Nyquist ISI criterion):

$$\frac{1}{T}\sum_{k=-\infty}^{+\infty} H\left(f-\frac{k}{T}\right) = 1;\ \forall f, \tag{7}$$

where $H(f)$ is the channel frequency response (Fourier transform of $h(t)$). Hence, the $\sin c$ pulse shape allows eliminating the ISI at sampling instants and any filter with excess bandwidth and odd-symmetry around Nyquist frequency can satisfy and meet the ISI-free requirements such as raised cosine filter (RCF).The raised cosine filter with bandwidth equal to $(\frac{1}{2T})(1 + \alpha)$ can be presented using the following form [3]:

$$P_{RC}(t) = \frac{\sin\left(\frac{\pi t}{T}\right)}{\frac{\pi t}{T}} \frac{\cos\left(\frac{\alpha \pi t}{T}\right)}{1-\left(\frac{2\alpha t}{T}\right)^2};\ 0 \leq \alpha \leq 1, \tag{8}$$

where α is the roll-off factor. The RCF impulse and frequency responses are shown in **Figure 3** for different roll-off factor α values.

The eye diagram is a widely used convenient method to observe the effects of ISI and noise introduced by the channel where the quality of the received signal (the ability to correctly recover the symbols and timing) can be illustrated (oscilloscope presentation). The interpretation of the eye diagram gives important information such as:

- sensitivity to timing error or jitter (smaller is better);
- wasted power;
- amount of distortion at sampling instants;
- amount of noise tolerance (larger is better);
- best time for sampling;
- the matching degree between the transmitter and receiver filters;
- the presence of ISI; and
- measurement of eye opening is performed to estimate the achievable BER.

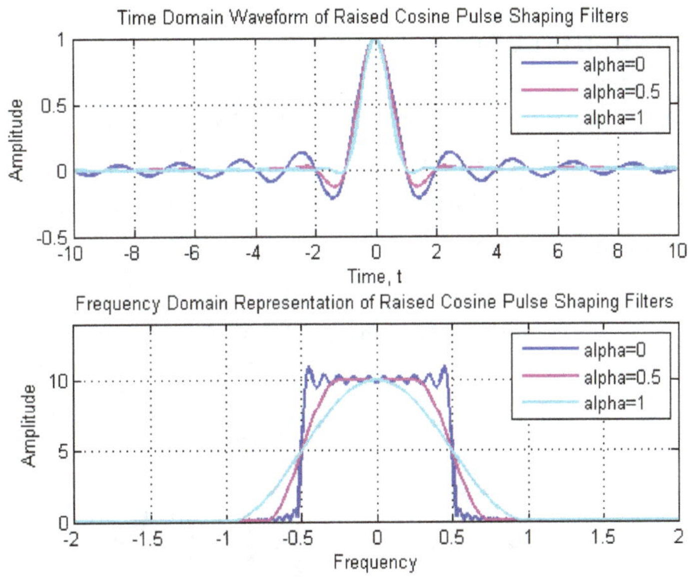

Figure 3. RCF impulse and frequency responses.

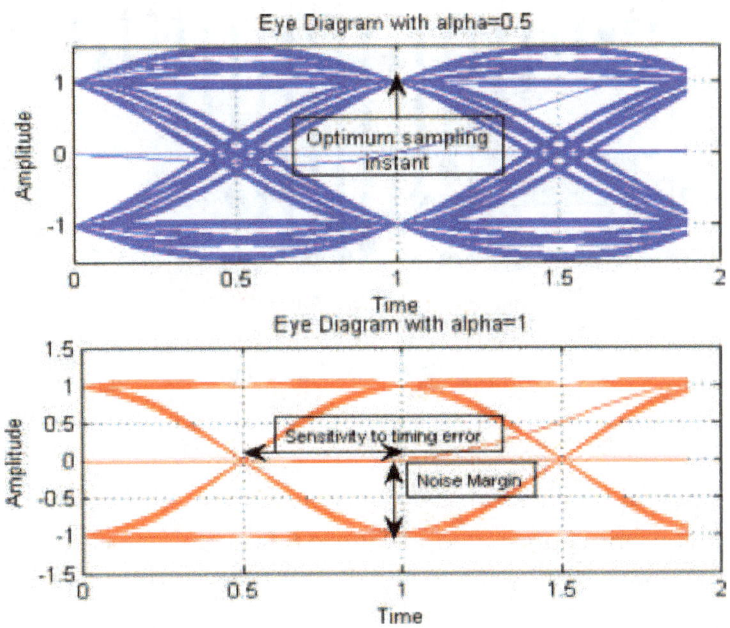

Figure 4. Raised cosine eye diagram for BPSK-modulated symbols.

The eye diagram using raised cosine filtering is presented in **Figure 4** for binary phase shift keying (BPSK)-modulated symbols at two roll-off factor values $\alpha = 0.5; 1$. In general, the ideal sampling instant is at the point where the vertical eye opening is maximum. Under the presence of ISI, the vertical eye opening reduces which leads to higher probability of error (BER). More sensitivity to timing error is presented by smaller horizontal eye opening. From **Figure 4**, it can be observed that the horizontal eye opening is smaller at roll-off factor $\alpha = 0.5$ in comparison with the case when $\alpha = 1$ owing to that the tails of the RCF are stripped down faster. It is important to mention here that higher values of roll-off factor require more transmission bandwidth. Smaller values of roll-off factor α lead to larger errors if the best sampling time (the center of the eye) is not achieved.

3. The equalization main concept

As mentioned before, the equalization technique is the design of a digital filter with inverse or counter transfer function in accordance with the transfer function of the associated wireless channel. This concept is shown in **Figure 5** where the frequency responses of the wireless channel and the equalizer are compared.

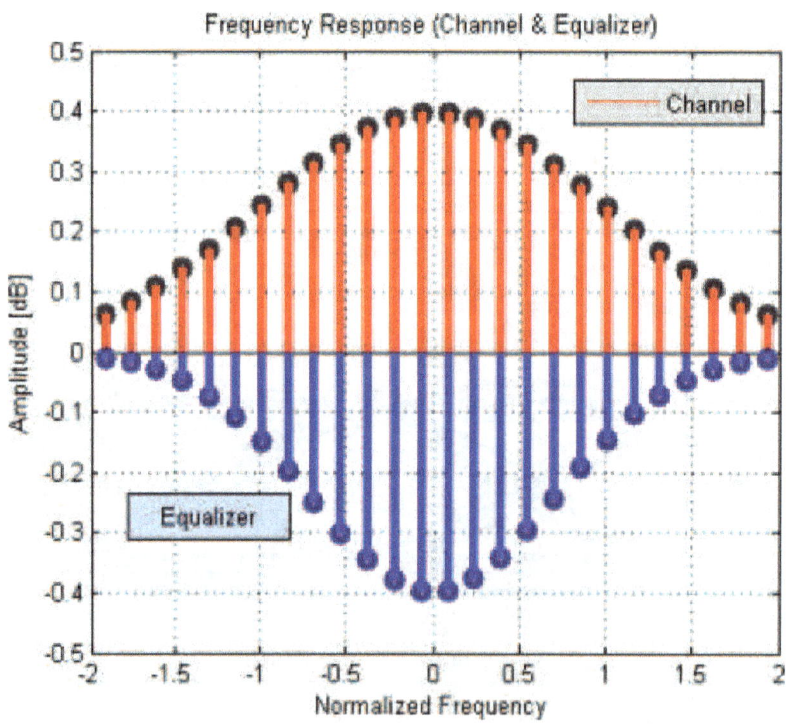

Figure 5. The wireless channel and equalizer frequency responses.

The equalizer design concept is presented in **Figure 6** (a simple block diagram for the channel effect and the equalizer transfer function). The discrete time model that should be considered under the equalization technique designing process is presented using the following form:

$$\hat{C}_k = \sum_{i=0}^{L-1} E_i u_k, \tag{9}$$

where $\hat{C}_k$ is the data sequence at the equalizer output, L is the number of symbols, E_i represents the coefficients of the causal impulse response of the equalizer $H_{EQ}(f)$, and u_k is the input of the equalizer given by

$$u_k = \sum_{j=0}^{L-1} h_j C_k + n_k, \tag{10}$$

where C_k is the transmitted data sequence, the coefficients h_j represents the equivalent causal impulse response (transfer function) of the wireless channel, and n_k is the discrete time additive noise. The channel transfer function $H_{ch}(f)$ can be presented as:

$$H_{ch}(f) = |H_{ch}(f)| \exp\left[j\theta_{ch}(f)\right], \tag{11}$$

where $|H_{ch}(f)|$ is the amplitude response, and $\theta_{ch}(f)$ is the phase response. The bit error rate (BER) reduction is achieved when the value of the error given by

$$e_k = C_k - \hat{C}_k \tag{12}$$

is reduced using equalization.

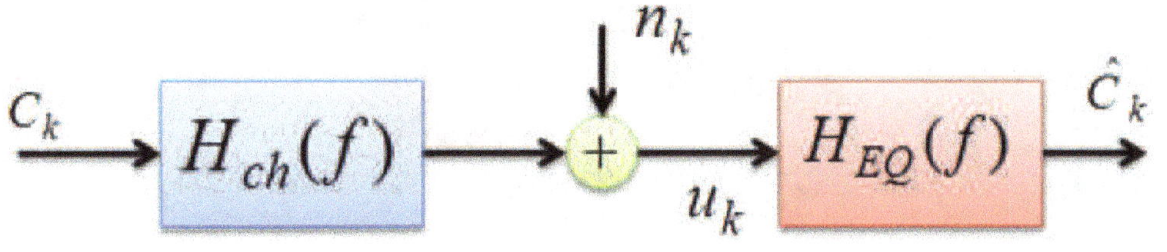

Figure 6. The equalizer main design model.

It is useful at this point to demonstrate the effects of the equalization on the detection performance or quality at the receiver side by presenting the difference between the received symbols with and without applying equalization. This comparison is given in **Figure 7** for the quadrature phase shift keying (QPSK)-modulated symbols. When the equalization is not applied, the uncertainty on the received signal constellation is very high (caused by the channel effect) and as a consequence the detection quality is low (detection performance degradation). Applying the equalization technique, the uncertainty level is lower and the detection performance is improved.

As a last word in this section, the ISI can produce a bias in the SNR value at the receiver. The error e_k presented by Eq. (12) is identical to the additive noise n_k in the case of unbiased detection rule. However, in the case of biased receiver, the probability density function (PDF) of the error e_k is a scaled and shifted version of the noise n_k PDF (the mean value depends on the actual data symbol C_k). The last condition can be formulated as follows:

$$E[\hat{C}_k|C_k] = C_k, \text{Unbiased Receiver} \tag{13}$$

For a given unbiased receiver [4] for a detection decision rule on a general signal constellation C_k, the maximum unconstrained SNR corresponding to the same receiver under any biased decision rule is given by

$$SNR_b = SNR_u + 1, \tag{14}$$

where SNR_b and SNR_u are the SNR for biased and unbiased receivers, respectively. The previous apparent discrepancy is more understandable recalling that the error rate is a monotone function of the SNR only. Thus, from Eq. (14), the negative effect of ISI on the SNR at the receiver is observed.

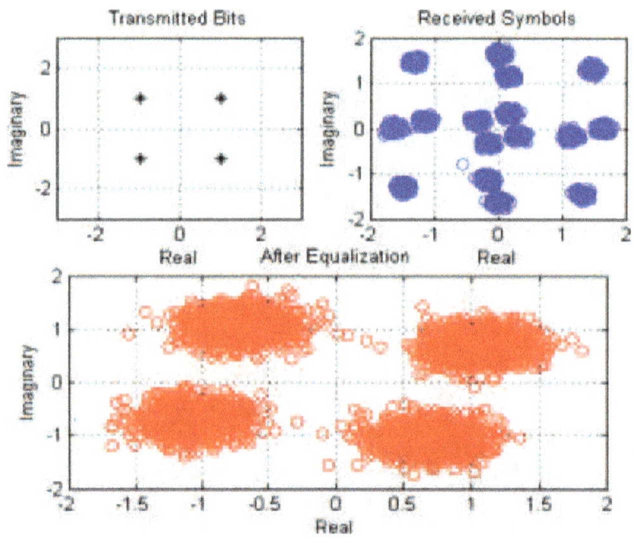

Figure 7. QPSK received symbols with and without equalization.

4. Equalization techniques

In general, the channel equalization techniques are classified to linear and nonlinear algorithms or to blind (without training sequence) and nonblind based on the degree of knowledge. The earliest mentions of digital equalization techniques are made under different design criterions, for example, zero forcing (ZF) equalization [5], minimum mean square error (MMSE) equalization [6], maximum likelihood (ML) equalization [7], decision feedback equalization (DFE) [8], and maximum *a posteriori* (MAP) equalization [9]. In this chapter, a new nonlinear equalization approach is proposed based on the dual channel equalization (DCE) idea with the purpose to improve the equalization performance, namely, the BER reduction, in comparison with the other widely used equalizers under multiple input multiple output (MIMO) wireless channel. The simulation results demonstrate considerable and promising performance improvements applying the suggested equalization approach in comparison with the conventional techniques. In order to help the reader to acquire the basics of equalization concepts, a few well-known and conventional equalization techniques will be reviewed and discussed in the next subsections.

4.1. System model

The complex baseband MIMO wireless channel model is considered with number of transmit antennas equal to N and number of receive antennas equal to M.

The received signal at the input of the receiver (the I/O relation of the MIMO channel) can be presented in the following form:

$$\mathbf{y} = \mathbf{H}\mathbf{x} + \mathbf{z}, \tag{15}$$

where $\mathbf{y}{\in}C^{M1}$ is the vector form of the received signal, $\mathbf{H}{\in}C^{MN}$ is the matrix form of the Rayleigh fading channel with independent and identically distributed (i.i.d) coefficients obeying the circularly symmetric complex Gaussian distribution at zero mean and variance $\sigma_h^2 = 1$, denoted as $h_{ij}\tilde{}CN(0,1)$ for $1{\le}i{\le}M, 1{\le}j{\le}N$, $\mathbf{x}{\in}C^{N1}$ is the vector form of the transmitted signal, and $\mathbf{z}$ is the circularly symmetric complex white Gaussian noise with zero mean and variance σ_n^2, i.e., $\mathbf{z}\tilde{}CN(0,\sigma_n^2\mathbf{I})$. It is assumed that the N data substreams have uniform power which means $\mathbf{x}{\in}C^{N1}$ has a covariance matrix given by

$$E[\mathbf{x}\mathbf{x}^*] = \sigma_x^2\mathbf{I}_N, \tag{16}$$

where $E[.]$ represents the mathematical expectation, $(.)^*$ is the conjugate transpose, and $\mathbf{I}_N$ is NN identity matrix. The signal-to-noise ratio (SNR) at the receiver input is expressed as:

$$SNR = \frac{\sigma_x^2}{\sigma_n^2}. \tag{17}$$

The channel matrix $\mathbf{H}{\in}C^{MN}$ is equalized or inverted by the weight matrix (the equalizer matrix) $\mathbf{W}{\in}C^{MN}$. Thus, the obtained signal $\hat{\mathbf{x}}$ at the equalizer output is defined as follows:

$$\hat{\mathbf{x}} = \mathbf{W}\mathbf{y}. \tag{18}$$

4.2. ZF equalization

ZF is a linear equalization that applies the inverse of the wireless channel frequency response to alleviate the channel effects on the received signal and restore the transmitted symbols. The name is assigned based on the reduction of ISI down to zero in the noise-free channel case (forcing the residual ISI to zero). The ZF equalizer can be designed using finite of infinite impulse response filters (FIR or IIR filters).

Employing ZF equalization technique with matrix $\mathbf{W}_{\mathrm{ZF}}$, the equalizer directly applies the inverse of the channel response to the received signal $\mathbf{y}$. Thus, the ZF equalizer satisfies the following condition:

$$\mathbf{W}_{\mathrm{ZF}}\mathbf{H} = \mathbf{I}_{\mathrm{MN}}. \tag{19}$$

The general form of the ZF equalizer matrix $\mathbf{W}_{\mathrm{ZF}}$ is defined as:

$$\mathbf{W}_{\mathrm{ZF}} = (\mathbf{H}^H\mathbf{H})^{-1}\mathbf{H}^H, \tag{20}$$

where $(.)^H$ represents Hermitian transpose operation. The BER in the case of ZF equalization technique and BPSK modulation can be defined using the following form [10]:

$$BER_{\mathrm{ZF}} = \left[\frac{1}{2}\left(1-\sqrt{\frac{SNR}{1+SNR}}\right)\right]^{M-N+1} \sum_{m=0}^{M-N+1} \left(\begin{matrix} M-N+m \\ m \end{matrix}\right)\left(\frac{1+\sqrt{SNR/1+SNR}}{2}\right), \tag{21}$$

where m is the index of the received signal substream. The last equation is derived considering the Neyman-Pearson (NP) detection criteria. Thus, the SNR of the obtained M decoupled substreams is defined as [10]:

$$SNR_{\mathrm{ZF},m} = \frac{SNR}{[(\mathbf{H}^H\mathbf{H})^{-1}]_{mm}};\ 1 \le m \le M, \tag{22}$$

where $[.]_{mm}$ corresponds to the mth diagonal element. The denominator of the achieved partial SNR given in Eq. (22) can be presented in terms of the mth column $\mathbf{h}_m$ of $\mathbf{H}$ as follows:

$$[(\mathbf{H}^H\mathbf{H})^{-1}]_{mm} = \frac{1}{\mathbf{h}_m^H\mathbf{h}_m - \mathbf{h}_m^H\mathbf{H}_m(\mathbf{H}_m^H\mathbf{H}_m)^{-1}\mathbf{H}_m^H\mathbf{h}_m}. \tag{23}$$

The ZF equalizer tries to null and cancel out all the interfering terms that are sometimes accompanied with noise amplification, for this reason, ZF is not optimal under very noisy channels. In the case of FIR filter use (to deal with noncausal components a decision delay is applied), a complete elimination of ISI problem is not possible owing to the finite filter length. Alternative criterion called peak-distortion criterion can be applied to minimize the maximum possible signal distortion due to ISI at the equalizer output.

4.3. MMSE equalization

The main objective of MMSE equalizer is to minimize the variance of the error signal e_k in Eq. (12). The MMSE equalizer ensures the trade-off between residual ISI and noise enhancement (reduce the total noise power). Under conditions more close to practice, MMSE equalizer can achieve lower BER compared to ZF equalizer at low-to-moderate SNRs. Thus, the MMSE equalization is applied to minimize the value of the mean given by

$$E[(\mathbf{W}_{\mathrm{MMSE}}\mathbf{y}-\mathbf{x})(\mathbf{W}_{\mathrm{MMSE}}\mathbf{y}-\mathbf{x})^H], \tag{24}$$

where $\mathbf{W}_{\mathrm{MMSE}}$ is the MMSE equalizer matrix that is presented by the following expression [11]:

$$\mathbf{W}_{\mathrm{MMSE}} = \left[\mathbf{H}^H\mathbf{H} + \frac{1}{SNR}\mathbf{I}\right]^{-1}\mathbf{H}^H. \tag{25}$$

The MMSE matrix $\mathbf{W}_{\mathrm{MMSE}}$ is multiplied by the received signal vector $\mathbf{y}$ to obtain M decoupled substreams with SNR equal to [11]:

$$SNR_{\text{MMSE},m} = \frac{SNR}{\left[\left(\mathbf{H}^H\mathbf{H} + \frac{1}{SNR}\mathbf{I}\right)^{-1}\right]_{mm}} - 1; 1 \le m \le M, \tag{26}$$

The BER for this equalizer considering the NP criterion and BPSK modulation can be approximated using the following form [11]:

$$BER_{\text{MMSE}} = E[Q\sqrt{2SNR_{\text{MMSE},m}}], \tag{27}$$

where $Q(x) = \frac{1}{\sqrt{2\pi}}\int_x^\infty \exp(-t^2/2)dt$ is the standard Gaussian Q-function. The output SNRs of the M decoupled substreams using ZF and MMSE equalizers are related as follows:

$$SNR_{\text{MMSE},m} = SNR_{\text{ZF},m} + \delta_{\text{SNR}.m}; 1 \le m \le M, \tag{28}$$

where $\delta_{\text{SNR},m}$ is nondecreasing function of SNR ($SNR_{\text{ZF},m}$ and $\delta_{\text{SNR},m}$ are statistically independent). Moreover, the ratio of the output SNR gains for these two equalizers for any full rank channel realization goes to unity (in dB) [11]:

$$10\log_{10}\left(\frac{SNR_{\text{MMSE},m}}{SNR_{\text{ZF},m}}\right) = 10\log_{10}\left(1 + \frac{\delta_{\text{SNR},m}}{SNR_{\text{ZF},m}}\right) \to 0;\ as\ \text{SNR} \to \infty \tag{29}$$

Again the MMSE equalizer can be implemented with FIR and IIR filters and in both cases the error signal e_k in Eq. (12) depends on the estimated symbols $\hat{C}_k$. According to the orthogonality principle of MMSE optimization, the error e_k and the input of the MMSE equalizer must be orthogonal. The achieved SNR can be defined based on the error variance σ_e^2. For instance, in the case of MMSE-IIR equalizer, the SNR is given by

$$SNR_{\text{MMSE, IIR}} = \frac{1 - \sigma_e^2}{\sigma_e^2} \tag{30}$$

which is more general form of SNR in comparison with the form in Eq. (26).

4.4. ML equalization

The ML equalization technique tests all the possible data symbols and chooses the one that has the maximum probability of correctness at the output (optimal in the sense of minimizing the probability of error $P_e = P(\mathbf{x} \neq \hat{\mathbf{x}})$, where $\hat{\mathbf{x}}$ is the estimated or chosen signal at the equalizer output). The Euclidian distance between the received signal vector and the products of all possible transmitted signal vectors is calculated, and the signal with minimum distance is considered. The ML estimation of $\mathbf{x}$ takes the following form [12]:

$$\hat{\mathbf{x}}_{\text{ML}} = \arg\min_{\hat{\mathbf{x}} \in \text{x}} \underbrace{\|\mathbf{y} - \mathbf{H}\hat{\mathbf{x}}\|^2}_{\text{J}} \tag{31}$$

As follows, the ML-based equalizer selects the data sequence $\hat{\mathbf{x}}$ that yields the smallest distance $\text{J}_{\min}$ between the received signal vector $\mathbf{y}$ and the estimated or hypothesized message $\mathbf{H}\hat{\mathbf{x}}$. For NP-based receiver, the obtained SNR at ML equalizer output is related to $\text{J}_{\min}$ and given by

$$SNR_{\text{ML}} = \frac{\mathbf{J}_{\min}}{4\sigma_n^2}. \tag{32}$$

The achieved BER of ML equalizer is based on $\mathbf{J}_{\min}$ as well as defined as [12]:

$$BER_{\text{ML}} = Q\left(\sqrt{\frac{\mathbf{J}_{\min}}{4\sigma_n^2}}\right). \tag{33}$$

In **Figure 8**, a comparison between the ZF, MMSE, and ML equalizer performances in terms of BER as a function of energy per bit E_b to the noise power spectral density N_0 ratio $(\frac{E_b}{N_0})$ is presented in the case of BPSK modulation and for MIMO antenna configuration with $N = M = 2$ and Rayleigh fading channel. As shown in **Figure 8**, the ML equalizer has the best performance and ZF equalizer has the worst one.

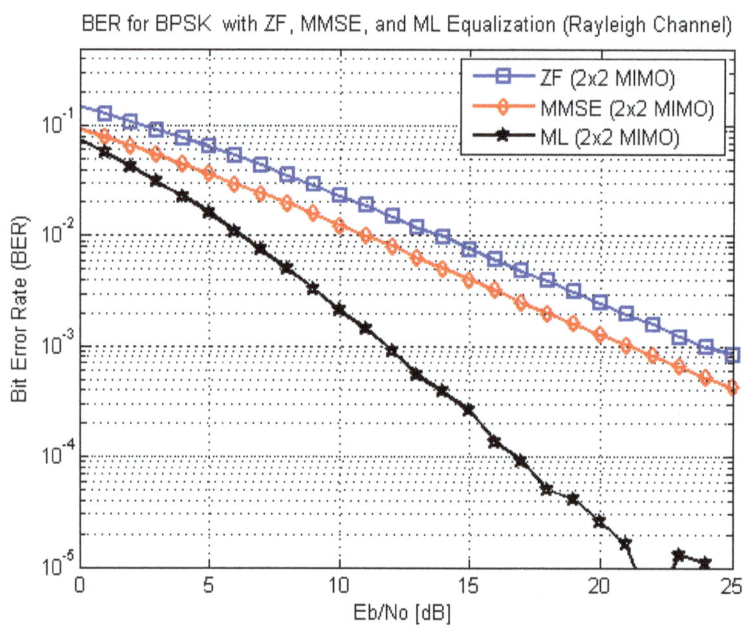

Figure 8. ZF, MMSE, and ML equalizer performances comparison.

4.5. Other equalization techniques

The main linear equalization drawback is that the equalizer filter enhances the noise at the output (increases the noise variance), and additionally, the noise is colored (especially for severely distorted channels). Employing noise prediction technique helps to avoid the described problem. The last short discussion leads to the basic idea about decision feedback equalization (DFE). The DFE structure consists of two filters: a feed forward-filter whose input is the channel output signal and a feedback-filter that feeds back the previous decisions for noise prediction process. This can be achieved by combining linear equalization technique like ZF or MMSE with noise variance prediction (ZF-DFE and MMSE-DFE). A simple block diagram for DFE is presented in **Figure 9**.

The maximum *a posteriori* (MAP) equalizer [13] estimates the transmitted symbol $x[n]$ at discrete-time index n that maximize the *a posteriori* probability $P(x[n] = x|\mathbf{y})$ as follows:

$$\hat{x}[n] = \arg\max P(x[n] = x|\mathbf{y}). \tag{34}$$

The MAP equalizer can be used as SBS detector with maximum-likelihood sequence estimation (MLSE) for transmission over rapidly time-varying (TV) wireless channel as shown in Ref. [13].

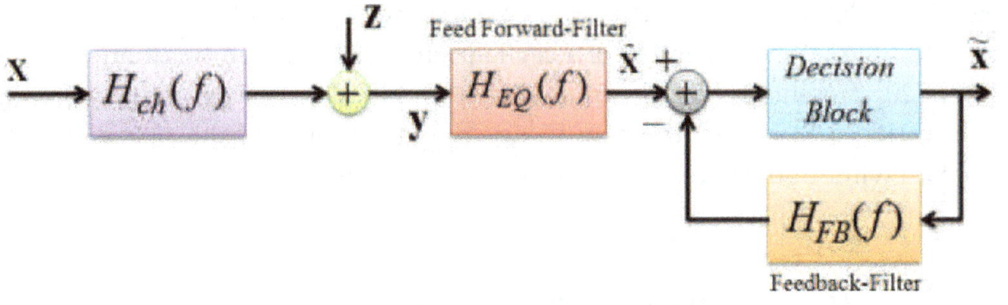

Figure 9. Basic DFE structure.

When some signal properties are used for the determination of the instantaneous error which updates the adaptive filter coefficients or weights, the fractionally spaced equalization (FSE) is an effective approach under the absence of the training sequences (blind equalization) [14]. The FSE receives number of input samples equal to N_{FSE} before it produces one output sample. The adaptive filter weights are updated employing a special algorithm such as constant modulus algorithm (CMA) which uses the constant modularity as the desired signal property. Thus, if the output rate is $1/T$, the input sample rate is N_{FSE}/T. Thus, the tap spacing of the FSE is a fraction of the baud spacing or the transmitted period. The FSE can be modeled as a parallel combination of several baud spaced equalizers known as multichannel model of FSE where the oversampling factor defines the tap spacing as follows:

$$\text{Tap Spacing} = \frac{T}{\text{Oversampling Factor}}. \tag{35}$$

In Ref. [15], a brief discussion about the recent equalization requirements and approaches is presented along with some important related references.

In optical communication field, a novel and efficient multiplier-less finite impulse response filter (FIR) based on chromatic dispersion equalization (CDE) is proposed for coherent receivers [16]. An iterative receiver is designed [17] for joint phase noise estimation, equalization, and decoding in a coded communication system with combined belief propagation, mean field, and expectation propagation (BP-MF-EP). In the frequency domain-based equalization, many important contributions are made recently, for example, in Ref. [18] and for single carrier frequency domain equalization (SC-FDE) in broadband wireless communication systems, a robust design under imperfect channel knowledge is considered based on a statistical channel estimation model where the equalization coefficients are defined under mean square error minimization criterion.

Another work deals with frequency domain equalization for faster-than-Nyquist (FTN) signaling is presented in Ref. [19] for doubly selective channels (DSCs) based on low complexity receivers with variational methods implemented in order to handle the interference of frequency domain symbols instead of using MMSE equalizer that involves high complexity in DSCs.

The frequency domain equalization is also proposed to be employed for broadband power line communications (PLC) as in Ref. [20]. PLC performance can benefit from frequency domain equalization techniques in the context of a cyclic-prefix single carrier modulation schemes. The study in Ref. [20] presents an equalization algorithm based on the properties of complementary sequences (CSs) to reduce the complexity by performing all the operations in the frequency domain without the necessity of noise variance estimator.

5. The new nonlinear equalization approach

5.1. Channel equalization with DCE

The proposed new nonlinear equalization approach applying dual channel equalization (DCE) is presented in this section. The DCE idea can be implemented and coupled with any standard channel equalizers such as ZF or ML equalizers. **Figure 10** shows a simple flow chart for the DCE idea presented as a coupling process between two digital filters. These filters can have the same or different transfer functions and for simplicity of analysis, the similarity case is considered. The topology in **Figure 10** is flexible and able to work with various equalization techniques. The squaring device block $(.)^2$ and the transfer function of the equalization filter $H_{EQ}(f)$ are the sources of the nonlinearity in this approach. The band-stop filter $H_{EQ}(f)$ main function is to filter out the received signal in order to obtain the reference colored noise ψ at the output of the second equalization filter. In fact, the squaring device and the equalization filter form a familiar preliminary design for square-law demodulation or square-law envelop detector.

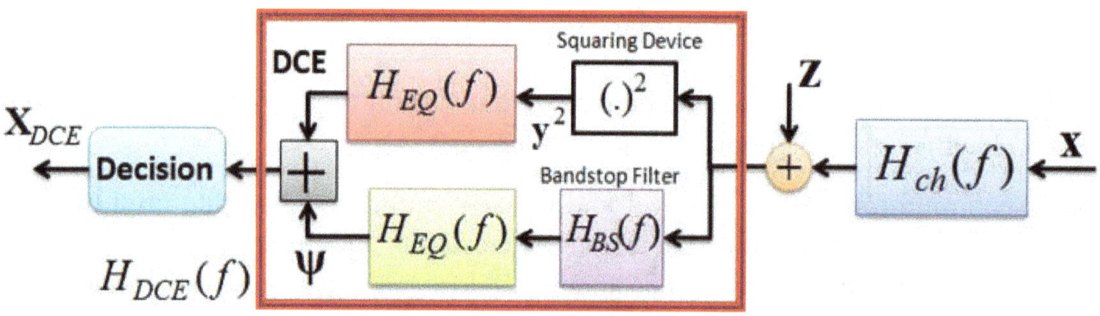

Figure 10. The DCE flow chart.

5.2. ZF equalization with DCE

The system model presented in Section 4.1 is valid in the analysis of ZF-DCE. The signal matrix X_{DCE} at the output can be presented using the following form:

$$\mathbf{X}_{\mathrm{DCE}} = \mathbf{W}_{\mathrm{DCE\text{-}ZF}}\mathbf{y}^2 + \psi = \hat{\mathbf{X}}^2 + \psi, \tag{36}$$

where $\mathbf{W}_{\mathrm{DCE\text{-}ZF}}$ is the DCE matrix that takes the following form:

$$\mathbf{W}_{\mathrm{DCE\text{-}ZF}} = (\mathbf{H}^H\mathbf{H})^{-2}(\mathbf{H}^H)^2. \tag{37}$$

Taking into consideration Eq. (37), the given form in Eq. (36) can be rewritten as:

$$\mathbf{X}_{\mathrm{DCE}} = \hat{\mathbf{X}}^2 + \psi = \hat{\mathbf{X}}^2 + (\mathbf{H}^H\mathbf{H})^{-2}(\mathbf{H}^H)^2\mathbf{Z}_{\mathrm{BS}}, \tag{38}$$

where $\mathbf{Z}_{\mathrm{BS}}$ is the matrix form of the noise at the stopband filter. To derive the closed expressions for the achieved SNR and BER under the use of ZF-DCE, a complete description for the noise component ψ forming at the output is required (the error performance is strongly related to the power of ψ). Using the standard complex matrix decomposition (singular value decomposition SVD), the complex channel matrix ψ can be decomposed as follows:

$$\begin{cases} \mathbf{H} = \mathbf{UTV}^H \\ \mathbf{H}^H\mathbf{H} = \mathbf{UTT}^H\mathbf{U}^H = \mathbf{U}\Lambda\mathbf{U}^H, \end{cases} \tag{39}$$

where $\mathbf{U}$ is MM complex unitary matrix that contains the $\mathbf{HH}^H$ eigenvectors ($\mathbf{U}^H\mathbf{U} = \mathbf{I}_M$, $\mathbf{U}^H = \mathbf{U}^{-1}$), $\mathbf{V}$ is NN complex unitary matrix that contains the $\mathbf{H}^H\mathbf{H}$ eigenvectors, $\mathbf{T} = diag(\lambda_1, ..., \lambda_N)$ is MN real diagonal matrix of the positive square roots of the corresponding eigenvalues, and Λ is MM diagonal matrix. The diagonal elements of $\mathbf{T}$ and Λ are defined by the following form:

$$\lambda_i = \begin{cases} \mathrm{d}_\mathrm{i}^2, \mathrm{i} = 1, 2, ..., \min(N, M) \\ 0, \mathrm{i} = \min(N, M) + 1, ..., \mathrm{M} \end{cases} \tag{40}$$

where the squared singular values $\{d_i^2\}$ are the channel matrix $\mathbf{H}$ eigenvalues. The squared Euclidian norm $\|.\|$ of the channel matrix $\mathbf{H}$ is given by

$$\|\mathbf{H}\|^2 = tr(\Lambda) = \sum_{i=1}^{\min(N,M)} \lambda_i = \sum_{i=1}^{\min(N,M)} d_i^2, \tag{41}$$

where $tr(.)$ is the matrix trace. Based on Eqs. (39) and (41), the power of the noise ψ at the ZF-DCE output can be presented using the expression:

$$\begin{aligned} \|\psi\|^2 &= \|(\mathbf{H}^H\mathbf{H})^{-2}(\mathbf{H}^H)^2\mathbf{Z}_{\mathrm{BS}}\|^2 = \|(\mathbf{VT}^2\mathbf{V}^H)^{-2}(\mathbf{VTU}^H)^2\mathbf{Z}_{\mathrm{BS}}\|^2 \\ &= \|\mathbf{V}^{-2}\mathbf{T}^{-4}(\mathbf{V}^H)^{-2}\mathbf{V}^2\mathbf{T}^2(\mathbf{U}^H)^2\mathbf{Z}_{\mathrm{BS}}\|^2 = \|\mathbf{V}^{-2}\mathbf{T}^{-2}(\mathbf{U}^H)^2\mathbf{Z}_{\mathrm{BS}}\|^2 \\ &= \|(\mathbf{V}^{-1}\mathbf{T}^{-1}\mathbf{U}^H)^2\mathbf{Z}_{\mathrm{BS}}\|^2. \end{aligned} \tag{42}$$

The mathematical expectation of the noise power given by Eq. (42) is defined as follows:

$$E\{\|\psi\|^2\} = E\{\|(\mathbf{V}^{-1}\mathbf{T}^{-1}\mathbf{U}^H)^2\mathbf{Z}_{\mathrm{BS}}\|^2\} = E\{\|(\mathbf{VTU})^{-2}\mathbf{Z}_{\mathrm{BS}}\|^2\} \tag{43}$$

The unitary matrix lemma states that the multiplication with a unitary matrix will not change the vector norm. Thus, owing to the fact that is $\mathbf{V}$ a unitary matrix, the presented form in Eq. (43) can be simplified:

$$
\begin{aligned}
E\{\|\psi\|^2\} &= E\{\|(\mathbf{VTU})^{-2}\mathbf{Z}_{\mathrm{BS}}\|^2\} = E\{\|\mathbf{T}^{-2}\mathbf{U}^{-2}\mathbf{Z}_{\mathrm{BS}}\|^2\} \\
&= E\{tr[(\mathbf{T}^{-2}\mathbf{U}^{-2})\mathbf{Z}_{\mathrm{BS}}\mathbf{Z}_{\mathrm{BS}}^H(\mathbf{U}^2\mathbf{T}^{-2})]\} = tr[(\mathbf{T}^{-2}\mathbf{U}^{-2})E\{\mathbf{Z}_{\mathrm{BS}}\mathbf{Z}_{\mathrm{BS}}^H\}(\mathbf{U}^2\mathbf{T}^{-2})] \\
&= tr[\sigma_z^2(\mathbf{T}^{-2}\mathbf{U}^{-2})(\mathbf{U}^2\mathbf{T}^{-2})] = \sigma_z^2 tr[(\mathbf{T}^{-2}\mathbf{U}^{-2})(\mathbf{U}^2\mathbf{T}^{-2})] = \sigma_z^2 tr[\mathbf{T}^{-4}] \\
&= \sigma_n^2 \sum_{i=1}^{N} \lambda_i^{-2}.
\end{aligned} \tag{44}
$$

where σ_z^2 is the variance of the noise $\mathbf{Z}_{\mathrm{BS}}$ at the bandstop filter output. The last form is obtained considering that $\sigma_z^2 = \sigma_n^2$ (in the frequency band of interest, the noise power spectral density is constant). The SNR at the ZF-DCE output can be defined based on Eq. (44) as follows:

$$
SNR_{\text{ZF-DCE}} = \frac{\sigma_x^2}{\sigma_n^2 \sum_{i=1}^{N} \lambda_i^{-2}}. \tag{45}
$$

The general form used to determine the BER in the case of ZF equalization technique under BPSK modulation can be defined as [10]:

$$
BER_{\mathrm{ZF}} = \int_0^{\infty} Q(\sqrt{2SNR_{\mathrm{ZF}}x})\frac{1}{(M\text{-}N)!}x^{M\text{-}N}e^{-x}dx \tag{46}
$$

The BER form in Eq. (21) [10] is the solution of the last integral in Eq. (46) and together with Eq. (45) can present the final BER closed expression of ZF-DCE as follows:

$$
BER_{\mathrm{ZF}} = \left[\frac{1}{2}\left(1-\sqrt{\frac{\sigma_x^2}{\sigma_x^2+\sigma_n^2\sum_{i=1}^{N}\lambda_i^{-2}}}\right)\right]^{M\text{-}N+1} \sum_{m=0}^{M\text{-}N+1}\binom{M\text{-}N+m}{m}\left(\frac{1}{2}+\frac{1}{2}\sqrt{\frac{\sigma_x^2}{\sigma_x^2+\sigma_n^2\sum_{i=1}^{N}\lambda_i^{-2}}}\right). \tag{47}
$$

As noticed in Eqs. (45) and (47), the derived forms are based on λ_i that also considered as the ith eigenvalue of the sample covariance matrix $\mathbf{R_y}$ given by

$$
\mathbf{R_y} = \frac{1}{N_s}\mathbf{y}\mathbf{y}^H. \tag{48}
$$

where N_s is the number of samples received by each antenna. The direct relation between λ_i and $\mathbf{y}$ can be illustrated using the following form:

$$
\frac{1}{2\sigma_n^2}\sum_{k=0}^{N_s-1} y(k)^H y(k) = \frac{N_s}{2\sigma_n^2}\sum_{i=1}^{M}\lambda_i \tag{49}
$$

5.3. ML equalization with DCE

For this case, the DCE flow chart presented in **Figure 10** can be considered for ML-DCE design excluding the squaring device. The corresponding equalizer or channel matrix takes the following format:

$$\mathbf{W}_{\text{ML-DCE}} = \mathbf{H} = (\hat{\mathbf{x}}^H\hat{\mathbf{x}})^{-1}\hat{\mathbf{x}}^H\mathbf{y} \tag{50}$$

Based on Eq. (31), the ML-DCE estimation is defined as:

$$\hat{\mathbf{x}}_{\text{ML-DCE}} = \arg\min_{\hat{\mathbf{x}}\in\mathbf{x}} \| \mathbf{y}-\hat{\mathbf{x}}(\hat{\mathbf{x}}^H\hat{\mathbf{x}})^{-1}\hat{\mathbf{x}}^H\mathbf{y}\|^2 \tag{51}$$

The presented form in Eq. (51) can be simplified by the following operations:

$$\begin{aligned}
\hat{\mathbf{x}}_{\text{ML-DCE}} &= \arg\min_{\hat{\mathbf{x}}\in\mathbf{x}} [\mathbf{y}-\hat{\mathbf{x}}(\hat{\mathbf{x}}^H\hat{\mathbf{x}})^{-1}\hat{\mathbf{x}}^H\mathbf{y}]^H[\mathbf{y}-\hat{\mathbf{x}}(\hat{\mathbf{x}}^H\hat{\mathbf{x}})^{-1}\hat{\mathbf{x}}^H\mathbf{y}] \\
&= \arg\min_{\hat{\mathbf{x}}\in\mathbf{x}}[\mathbf{y}^H-\mathbf{y}^H\hat{\mathbf{x}}(\hat{\mathbf{x}}^H\hat{\mathbf{x}})^{-1}\hat{\mathbf{x}}^H][\mathbf{y}-\hat{\mathbf{x}}(\hat{\mathbf{x}}^H\hat{\mathbf{x}})^{-1}\hat{\mathbf{x}}^H\mathbf{y}] \\
&= \arg\min_{\hat{\mathbf{x}}\in\mathbf{x}}[\mathbf{y}^H\mathbf{y}-\mathbf{y}^H\hat{\mathbf{x}}(\hat{\mathbf{x}}^H\hat{\mathbf{x}})^{-1}\hat{\mathbf{x}}^H\mathbf{y}-\mathbf{y}^H\hat{\mathbf{x}}(\hat{\mathbf{x}}^H\hat{\mathbf{x}})^{-1}\hat{\mathbf{x}}^H\mathbf{y}+\mathbf{y}^H\hat{\mathbf{x}}(\hat{\mathbf{x}}^H\hat{\mathbf{x}})^{-1}\hat{\mathbf{x}}^H\hat{\mathbf{x}}(\hat{\mathbf{x}}^H\hat{\mathbf{x}})^{-1}\hat{\mathbf{x}}^H\mathbf{y}] \\
&= \arg\min_{\hat{\mathbf{x}}\in\mathbf{x}}[\mathbf{y}^H\mathbf{y}-2\mathbf{y}^H\hat{\mathbf{x}}(\hat{\mathbf{x}}^H\hat{\mathbf{x}})^{-1}\hat{\mathbf{x}}^H\mathbf{y}+\mathbf{y}^H\hat{\mathbf{x}}(\hat{\mathbf{x}}^H\hat{\mathbf{x}})^{-1}\hat{\mathbf{x}}^H\mathbf{y}] \\
&= \arg\min_{\hat{\mathbf{x}}\in\mathbf{x}} \underbrace{[\mathbf{y}^H\mathbf{y}-\mathbf{y}^H\hat{\mathbf{x}}(\hat{\mathbf{x}}^H\hat{\mathbf{x}})^{-1}\hat{\mathbf{x}}^H\mathbf{y}]}_{\mathrm{J_{DCE}}}.
\end{aligned} \tag{52}$$

It is obvious that to minimize the form in Eq. (52), the term $\left(\mathbf{y}^H\hat{\mathbf{x}}(\hat{\mathbf{x}}^H\hat{\mathbf{x}})^{-1}\hat{\mathbf{x}}^H\mathbf{y}\right)$ should be maximized as follows:

$$\begin{aligned}
\mathbf{J}_{\text{ML}} &= \arg\max_{\hat{\mathbf{x}}\in\mathbf{x}} \left(\mathbf{y}^H\hat{\mathbf{x}}(\hat{\mathbf{x}}^H\hat{\mathbf{x}})^{-1}\hat{\mathbf{x}}^H\mathbf{y}\right) = \arg\max_{\hat{\mathbf{x}}\in\mathbf{x}}[\mathbf{y}^H\hat{\mathbf{x}}(\hat{\mathbf{x}}^H\hat{\mathbf{x}})^{-1}\hat{\mathbf{x}}^H\hat{\mathbf{x}}(\hat{\mathbf{x}}^H\hat{\mathbf{x}})^{-1}\hat{\mathbf{x}}^H\mathbf{y}] \\
&= \arg\max_{\hat{\mathbf{x}}\in\mathbf{x}}[\left(\hat{\mathbf{x}}(\hat{\mathbf{x}}^H\hat{\mathbf{x}})^{-1}\hat{\mathbf{x}}^H\mathbf{y}\right)^H\left(\hat{\mathbf{x}}(\hat{\mathbf{x}}^H\hat{\mathbf{x}})^{-1}\hat{\mathbf{x}}^H\mathbf{y}\right)] = \arg\max_{\hat{\mathbf{x}}\in\mathbf{x}} \|\hat{\mathbf{x}}(\hat{\mathbf{x}}^H\hat{\mathbf{x}})^{-1}\hat{\mathbf{x}}^H\mathbf{y}\|^2.
\end{aligned} \tag{53}$$

The related SNR at the ML-DCE output is determined using $\mathbf{J}_{\text{min,DCE}}$ as in the presented form Eq. (32) and the covariance matrix of the reference noise $\mathbf{Z}_{\text{BS}}$ forming at the bandstop filter given by

$$C_{Z_{\text{BS}}}=E[Z_{\text{BS}}^H Z_{\text{BS}}]=\sigma_z^2 I=\sigma_n^2 I \tag{54}$$

Defining the achieved SNR for ML-DCE, it is possible to calculate the BER using the same form in Eq. (**33**). Taking into account the sample covariance matrix $\mathbf{R}_y$ in Eq. (**48**), Eq. (**52**) can be represented as:

$$\hat{\mathbf{x}}_{ML\text{-}DCE} = \arg\min_{\hat{\mathbf{x}}\in\mathbf{x}}[N_s\mathbf{R}_\mathbf{y}-\mathbf{y}^H\hat{\mathbf{x}}(\hat{\mathbf{x}}^H\hat{\mathbf{x}})^{-1}\hat{\mathbf{x}}^H\mathbf{y}] \tag{55}$$

The use of two equalization filters and the bandstop filter (the main DCE idea) helps us to construct new equalization matrices and convert linear equalizers such as ZF equalizer to nonlinear one. Additionally, the reference noise forming at the second equalizer output contributes in the definitions of achieved BER and SNR and can be employed to estimate the noise variance (power) at the equalizer input. In this section, two equalizers are introduced, namely,

ZF-DCE and ML-DCE. The DCE concept can be implemented with other equalization techniques like MMSE and DFE as well. The proposed DCE performance is evaluated in the following subsection.

5.4. Simulation process and results

In this section, *NM* MIMO channel is considered where the number of transmit antennas N= 2 and the number of receive antennas M= 2 under Rayleigh fading channel and for BPSK modulation. The overall simulation process for this system employing the conventional and DCE types of equalization techniques are presented in **Figure 11**. The evolution and comparison criterion is based on the BER (probability of error) as a function of energy per bit E_b to the noise power spectral density N_0 ratio ($\frac{E_b}{N_0}$). The relation between $\frac{E_b}{N_0}$ and the SNR can be simply presented as:

$$SNR = \frac{E_b}{N_0}\frac{D_R}{BW}, \tag{56}$$

where D_R is the data rate and BW is the bandwidth.

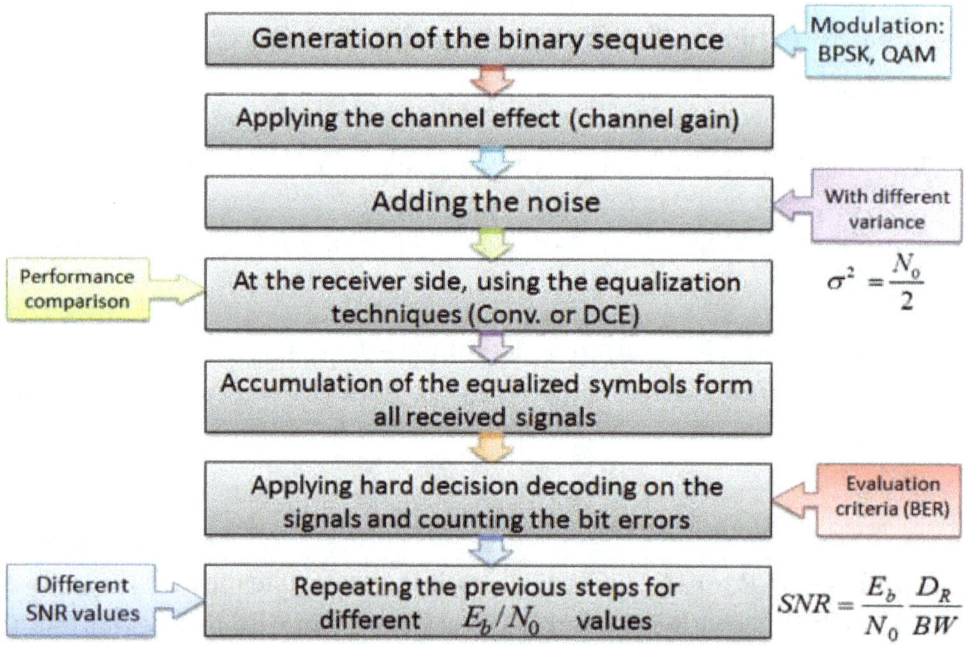

Figure 11. The simulation and evaluation processes.

In **Figure 12**, a comparison between the conventional ZF equalization technique and the modified ZF using dual channel equalization (DCE) is presented. The ZF-DCE approach demonstrates better performance (lower BER) comparing with the conventional ZF at the same $\frac{E_b}{N_0}$ (or SNR) range. The two equalizers have the same performance under relatively lower SNR ($\frac{E_b}{N_0}$≤5[dB]).

The ML-DCE and the conventional ML equalizer performances are compared in **Figure 13** under the same initial conditions but for different SNR ($\frac{E_b}{N_0}$) range with the purpose of demonstrating the effective and efficient SNR values for DCE employment. Once again the ML-DCE outperforms the conventional ML in performance and for lower SNR compared with the case of ZF equalization where the two equalizers present asymptotic performance for SNR less than 0 dB or $\frac{E_b}{N_0}\leq 0[dB]$ (roughly speaking, the ML-DCE performance is slightly better at $\frac{E_b}{N_0}\leq 0[dB]$).

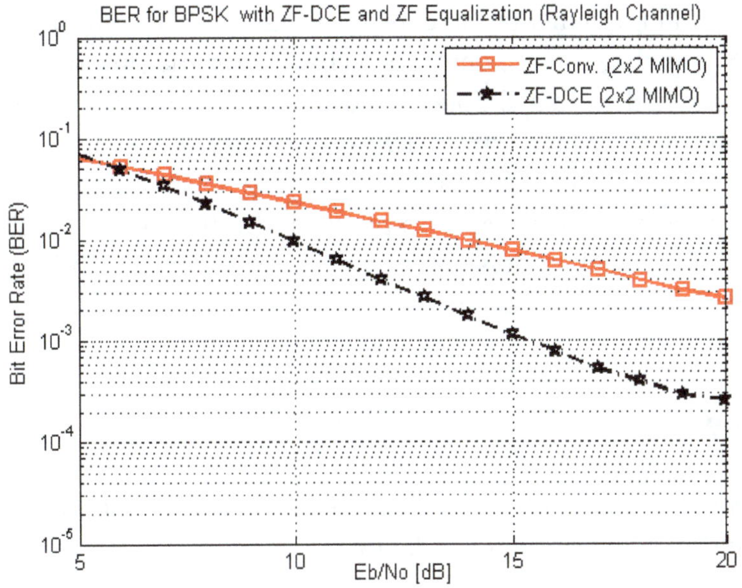

Figure 12. The comparison between conventional ZF and the ZF-DCE.

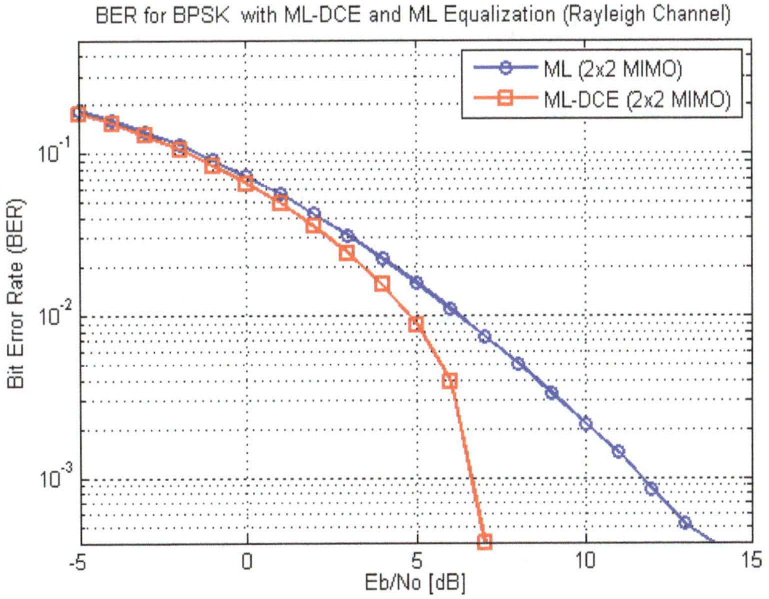

Figure 13. The comparison between conventional ML and the ML-DCE.

6. Conclusion remarks

The discussion of this chapter can be ended by making several remarks. The new equalization approach (DCE) shows promising outcomes in terms of improving the equalization performance by reducing the BER (sequence error probability).The obtained results can be simply generalized for other equalization techniques and for frequency domain equalization as well. Additionally, it is easy to prove that the DCE symbol error performance can be better than that of the conventional equalizers. The type of filters used to design the DCE equalizer is not mentioned or discussed but both FIR and IIR filters are possible and reasonable candidates.

Although the design and implementation complexity issues (the computational cost to design the equalizer and to equalize the channel, respectively) for the proposed above equalization structure are not discussed in this chapter, an ostensible comments can be made. The DCE design relies on other equalization approaches and does not exhibit significant overhead complexity. Thus, the complexity increases employing DCE but not with overwhelming degree. Finally, a complete analysis based on practical conditions and for time-invariant (TIV) and time-varying (TV) channel models under specific scenarios is required before addressing the feasibility of the presented DCE approach.

Acknowledgements

This work is performed by Radio Physics Research Group (RPRG) at Polytechnic University of San Luis Potosi (UPSLP), San Luis Potosi, Mexico.

Author details

Modar Shbat[1*], Francisco C. Ordaz-Salazar[1] and Javier S. González-Salas[2]

*Address all correspondence to: modar.shbat@upslp.edu.mx

1 Telematics Engineering Department, Polytechnic University of San Luis Potosí, San Luis Potosí, México

2 Mathematical Department, Polytechnic University of San Luis Potosi, San Luis Potosí, México

References

[1] Proakis J, Salehi M: Digital Communications. 5th ed: Chapter 10. McGraw-Hill, 2008.

[2] Wong T F, Lok T M: Theory of Digital Communication: Chapter 4. University of Florida, 2004.

[3] Orfanidis S J: Introduction to Signal Processing. Rutgers University, New Jersey, 2010.

[4] Rainfield Y: Unbiased MMSE vs. biased MMSE equalizers. Tamkang Journal of Science and Engineering, 2009; Vol.12, No.1: 45 –56.

[5] Klein A, Kaleh G K, Baier P W: Zero forcing and minimum mean-square-error equalization for multiuser detection in code-division multiple-access channels. IEEE Transaction on Vehicular Technology, 2002; Vol. 45, No. 2: 276–287.

[6] Gersho A: Adaptive equalization in highly dispersive channels for data transmission. Bell System Technical Journal. 1969; Vol. 48: 55–70.

[7] Bottomley G E: Channel Equalization for Wireless Communications: From Concepts to Detailed Mathematics: Chapters 5, 6. Wiley-IEEE Press, 2011.

[8] George D, Bowen R, Storey J: An adaptive decision feedback equalizer. IEEE Transactions on Communication Technology, 1971; Vol. 19, No. 3: 281–293.

[9] Bjerke B A, Proakis J G: Equalization and decoding for multiple-input multiple-output wireless channels. EURASIP Journal on Applied Signal Processing. 2002; No. 3: 249–266.

[10] Ding Y, et al.: Minimum BER block precoders for zero-forcing equalization. IEEE Transactions on Signal Process. 2003; Vol. 51, No. 9: 2410–2423.

[11] Jiang Y, Varanasi M K, Jian L: Performance analysis of ZF and MMSE equalizers for MIMO systems: An in-depth study of the high SNR regime. IEEE Transactions on Information Theory, 2011; Vol. 57, No. 4: 2008–2026.

[12] Forney G: Maximum likelihood sequence estimation of digital sequences in the presence of intersymbol interference. IEEE Transactions on Information Theory. 1972; Vol. 18, No. 3: 363–378.

[13] Barhumi I, Moonen M: MLSE and MAP equalization for transmission over doubly selective channels. IEEE Transactions on Vehicular Technology. 2009; Vol. 58, No. 8: 4120–4128.

[14] Gitlin R D, Weinstein S B: Fractionally-spaced equalization: an improved digital transversal equalizer. The Bell System Technical Journal. 1981; Vol. 60, No. 2: 275–296.

[15] Ma X, Davidson T, Gershman A, Swami A, Tepedelenlioglu C: Advanced equalization techniques for wireless communications. EURASIP Journal on Advances in Signal Processing. 2010; Vol. 2010, Article ID 623540, 2 pages, doi:10.1155/2010/623540.

[16] Martins C, Guiomar F, Amado S, Ferreira R, Ziaie S, Shahpari A, Teixeira A, Pinto A: Distributive FIR-based chromatic dispersion equalization for coherent receivers. Journal of Lightwave Technology, 2016; Vol. PP, No. 99: 1–1. [CE2] doi:10.1109/JLT.2016.2604741.

[17] Wang W, Wang Z, Zhang C, Guo Q, Sun P, Wang X: A BP–MF–EP based iterative receiver for joint phase noise estimation, equalization, and decoding. IEEE Signal Processing Letters. 2016; Vol. 23, No. 10: 1349–1353.

[18] Zhu Y, Zhe P, Zhou H, Huang D: Robust single carrier frequency domain equalization with imperfect channel knowledge. IEEE Transactions on Wireless Communications. 2016; Vol. 15, No. 9: 6091–6103.

[19] Yuan W, Wu N, Wang H, Kuang J: Variational inference-based frequency-domain equalization for faster-than-Nyquist signaling in doubly selective channels. IEEE Signal Processing Letters.2016; Vol. 23, No. 9: 1270–1274.

[20] Moya S, Hadad M, Funes M, Danato P, Carrica D: Broadband PLC-channel equalisation in the frequency domain based on complementary sequences.IET Communications. 2016; Vol. 10, No. 13: 1605–1613.

8

Microwave Active Filter Design

Vincenzo Stornelli, Leonardo Pantoli and
Giorgio Leuzzi

Abstract

A simplified method for the project and design of microwave active filters is presented here. The presented design is based on the use of an active inductor that emulates an inductor behavior by implementing a passive variable phase- and amplitude-compen-sating network and amplifiers, forming a gyrator-C architecture. This method can be applied with success for the design of bandpass filters with very high performances in terms of integration and application from a few hundreds of MHz to tens of GHs with filter high dynamic range and frequency tuning capability.

Keywords: filters, active filter, active inductor, high dynamic range, gyrator

1. Introduction

Nowadays advancements in communications systems, in conjunction with the great increasing request for miniaturization and size reduction of wireless systems and devices, have led to an increase of performances and space reduction of modern communications systems and devices. Several RF building blocks have consequently been successfully implemented as integrated circuits (ICs) with different technologies. In this scenario, analog tunable filters working, in general, at RF and microwave frequencies still remain the most difficult part to be integrated in a single chip [1–18] due to requirements in input and output impedance matching, stability, and dynamic range. In fact, most commercial wireless receivers only use off-chip filters that are typically implemented with discrete components. This scenario leads to the occupation of a large area at the front-end. Even though the integrated version of filters has been presented in several papers, it must be noticed that in monolithic integrated microwave circuits (MMIC) most of the semiconductor area is occupied by passive elements, while active devices play a marginal role in area occupation. It is in fact generally known that passive inductors are largely responsible for the area occupation in IC. Classical spiral inductors, in fact, require large amounts of substrate area or air bridges and present a very limited bandwidth, a general high series resistance, and crosstalk problems. For this reason, it can really be important to replace spiral inductors with active circuits behaving like inductors, but characterized by a very reduced semiconductor area occupation. It is well known that an inductor can be simulated by a gyrator loaded by a capacitance; in the literature and in the commercial market, a lot of solutions have been proposed that use the gyrator principle to simulate an inductive behavior at microwave frequency range [15, 16, 18–32]. Analog filters using Active Inductors (AIs) are good candidates for high-frequency operation; therefore, considerable interest has been shown in their use in active filters. In this perspective, in particular, band-pass filters designed by means of active inductor, usually suffer from a very low dynamic range due to the combination of a high noise level introduced by the AI and the relatively low compression power level of the AI itself. For all these reasons, there are only a few commercial solutions available on the market based on active inductors. Moreover, complex arrangements with many transistors tend to degrade the stability, preventing their use in practical applications. All these considerations show how the availability of simple, stable AIs with relatively high-power handling is therefore a critical issue, for their use as a replacement for spiral inductors in integrated filters especially when high-order filters with tunable characteristics are required. In the following paragraph, we will see how filters using active inductors have good potential to operate at high frequency, with high dynamic range, maintaining a constant quality factor (Q) in tunable applications when coupled with varactors diode, so considerable amount of interest has been shown in their use. A very simple approach for the design of tunable, high-quality factor, stable active inductors by means of an active inductor is presented here.

2. The active inductor design

Generally, active filters suffer from a very low dynamic range (see **Figure 1**) [33–53] due to the combination of a high noise level introduced by the AI and the relatively low compression power level of the AI itself. An interesting way to design active filters, in order to overcome this problem, is through the use of AI. The first high-Q AIs were reported over two decades ago, and the use of these inductors for the implementation of inductor-capacitor (L-C) resonator-type active filters was presented both in several papers and conferences.

However, as already mentioned, high-frequency active bandpass filters designed by means of AI usually suffer from a low dynamic range due to the combination of high noise level introduced by the AI and of low compression point of the AI itself, which limits the maximum handled power. Several kinds of AI have been proposed in the past [1–52], especially grounded inductors that are usually based on gyrator scheme loaded with a capacitor (**Figure 2**). With reference to **Figure 2**, the input voltage through the inverting amplifier (that can be based on a common-source stage) drives the loaded capacitor, producing a 90° delayed voltage; this in turn drives the output current through a noninverting transconductance, typically a source follower.

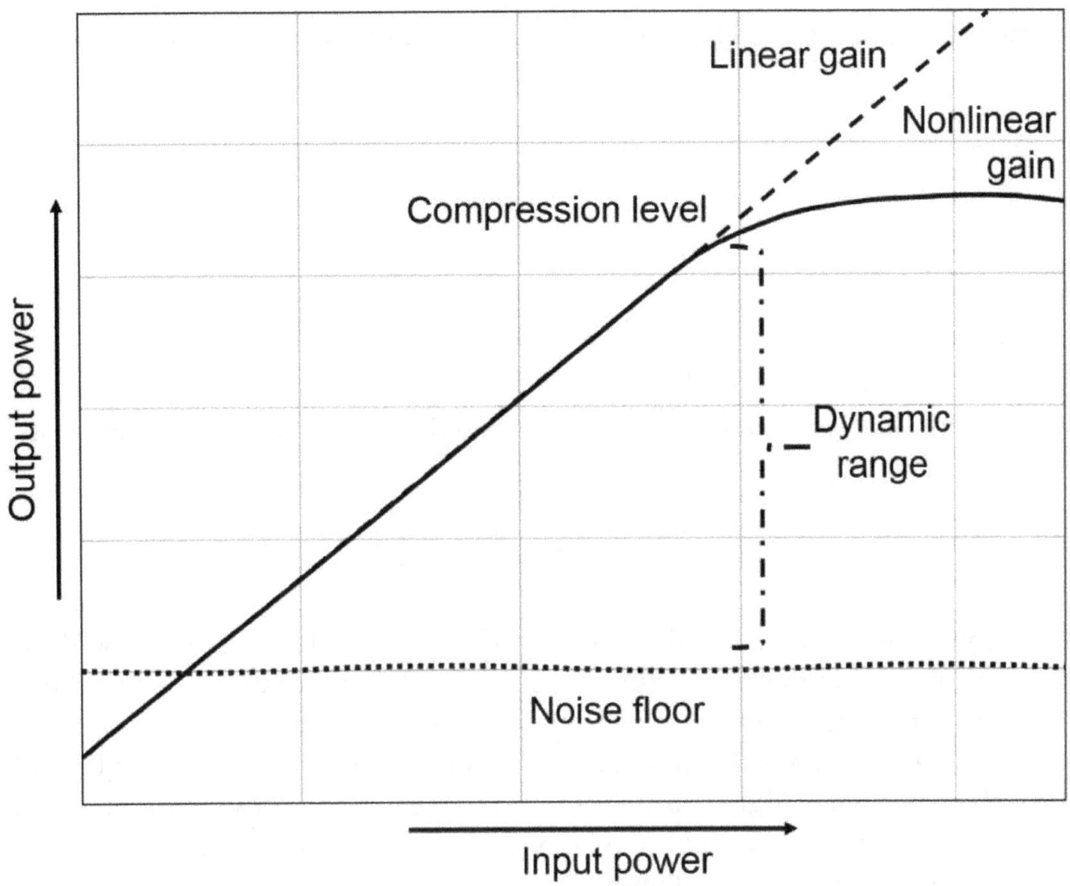

Figure 1. Dynamic range for an active filter.

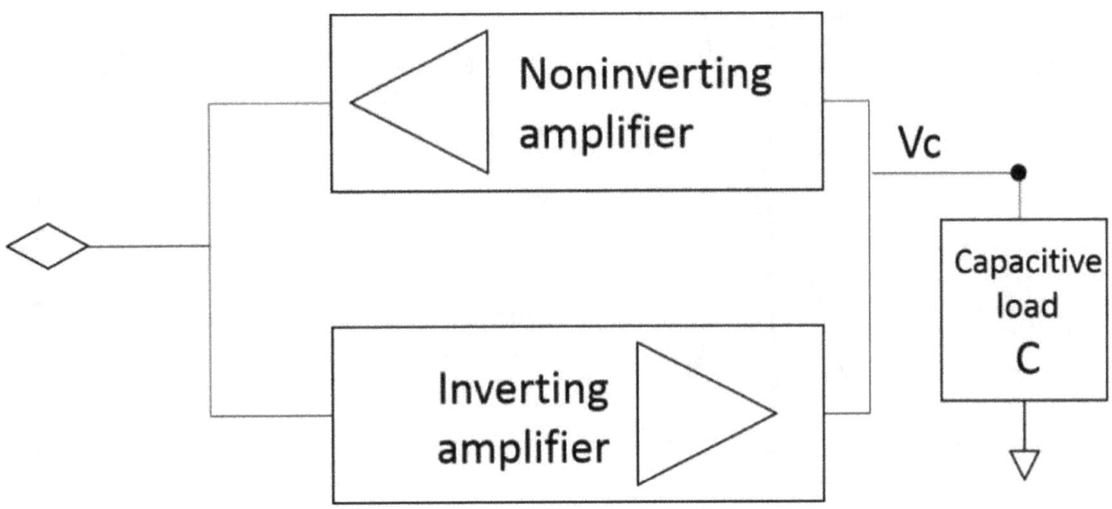

Figure 2. A gyrator-based active inductor scheme.

From **Figure 2** circuit, if we assume ideal amplifiers, the following equations can be derived [tcas]:

$$V_C = -\frac{g_{m1}}{j\omega C} \cdot V_{in} \tag{1}$$

$$I_{in} = -g_{m2} \cdot V_C = \frac{g_{m1} \cdot g_{m2}}{j\omega C} \cdot V_{in} = \frac{1}{j\omega L_{eq}} \cdot V_{in} \tag{2}$$

Finally, the equivalent inductance of the gyrator-based active inductor can be expressed as:

$$L_{eq} = \frac{C}{g_{m1} g_{m2}} \tag{3}$$

Eq. (3), of course, is valid only if the two amplifiers are ideal and zero phase delays and zero losses are present in the architecture. It is important to notice that if nonidealities in the amplifiers are considered, two unwanted effects appear: (1) a parasitic resistance (positive or negative) arises in series to the equivalent inductance; (2) the bandwidth of the inductive impedance becomes finite. **Figure 3a** shows that a phase relation between input less than 90° gives positive resistance, while a phase relation in excess of 90° causes a negative resistance: the AI can now be considered as in **Figure 3b**

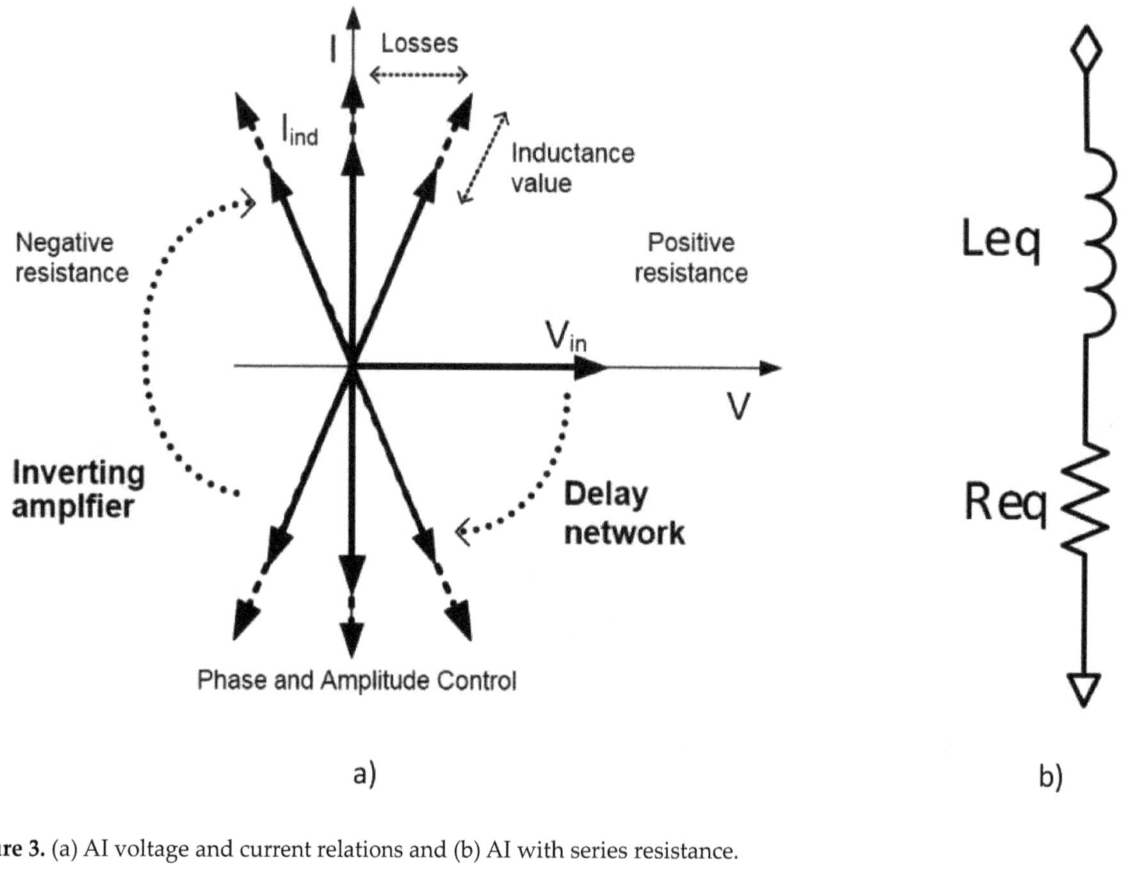

Figure 3. (a) AI voltage and current relations and (b) AI with series resistance.

If we were able to control both these phase relations, without acting on the bias condition of the amplifiers, the linearity of the active component of the gyrator will not be affected. This condition can be reached by the insertion of a suitable variable passive compensation network [35] (**Figure 4**).

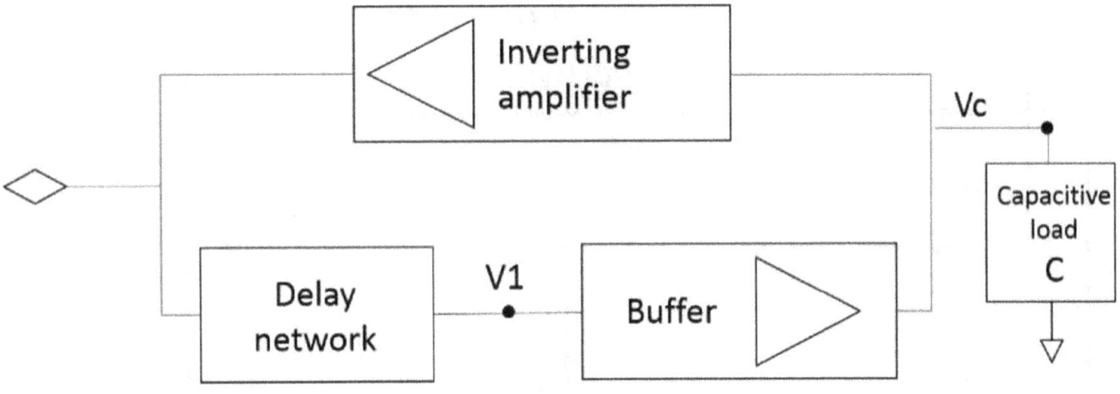

Figure 4. Block scheme of the active inductor with high dynamic range.

It has already been demonstrated in reference [35] that

$$R_{eq} = \frac{1}{g_m A_{var}} \cos(\varphi_{var}) + \frac{\omega R_{par} C_{tot}}{g_m A_{var}} \sin(\varphi_{var})$$
$$L_{eq} = -\frac{1}{g_m A_{var}} \sin(\varphi_{var}) + \frac{\omega R_{par} C_{tot}}{g_m A_{var}} \cos(\varphi_{var}) \tag{4}$$

From Eq. (4), it can be noticed that the phase of the compensating network must be such that the two addends cancel in the expression of the equivalent resistance, in the band of interest. This leads to a complex scenario. In order to simplify the design, a single active block AI and a compensation network can be also adopted as shown in **Figure 5**. In **Figure 6**, an equivalent circuit of the single active block AI [38] is shown. In this case, a simplified model of the capacitance gyrator with a single active block is shown.

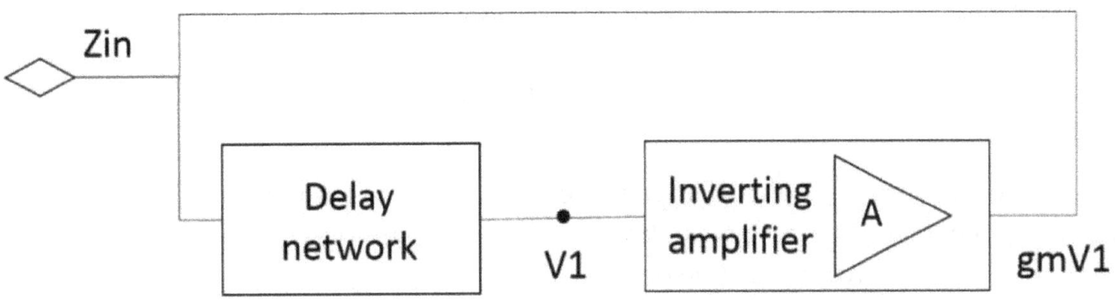

Figure 5. Block scheme of the proposed single active block active inductor.

An exact set of values of the resistor-capacitor (R-C) network gives a correct phase delay between input and output (voltage and current), minimizing the real part of the equivalent impedance. In the ideal case [38], the circuit input impedance is given by the following relation:

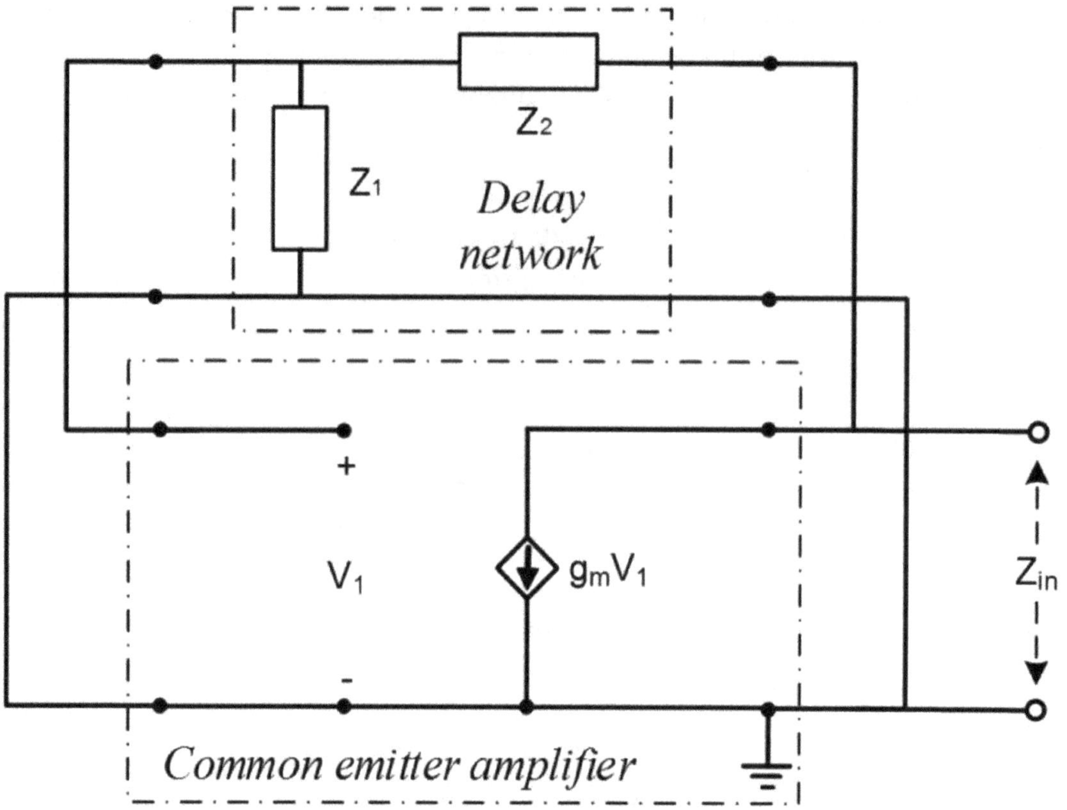

Figure 6. Simplified schematic model of the single-transistor AI.

$$Z_{in} = \frac{1 + j\omega RC}{g_m + j\omega C} \tag{5}$$

Eq. (5) can be also written in a more simple way as real and imaginary parts as in the following:

$$\mathrm{Re}\{Z_{in}\} = \frac{g_m + \omega^2 RC^2}{g_m^2 + \omega^2 C^2} \tag{6}$$

$$\mathrm{Im}\{Z_{in}\} = \frac{\omega g_m RC - \omega C}{g_m^2 + \omega^2 C^2} \tag{7}$$

Now it can be seen that the input impedance of the network is not always inductive [38]. By analyzing the imaginary part, we notice that the imaginary part of the AI impedance has a zero in $\omega = 0$, and it is always positive only if:

$$g_m R > 1 \tag{8}$$

The last is a very important condition if we want to obtain an inductive behavior. By considering also the amplifier block parasitic, we obtain the real and imaginary part of the input impedance of the single transistor (common emitter stage as the transconductance amplifier) gyrator as shown in **Figures 7** and **8**.

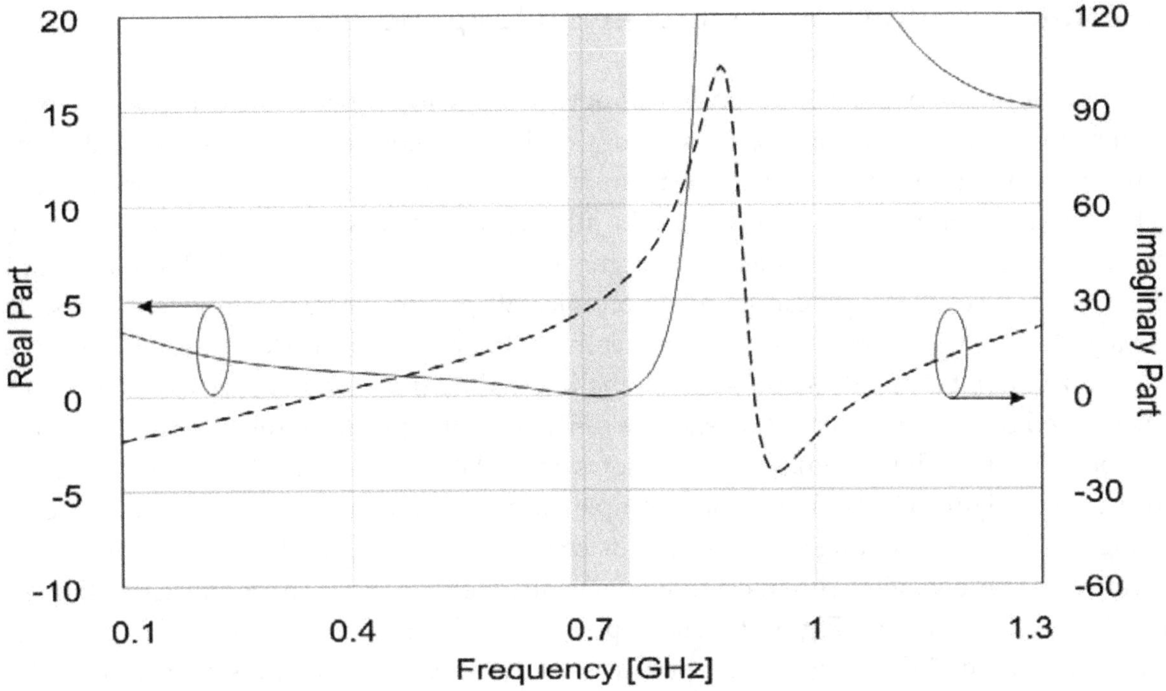

Figure 7. Real and imaginary part of the input impedance of the single transistor gyrator.

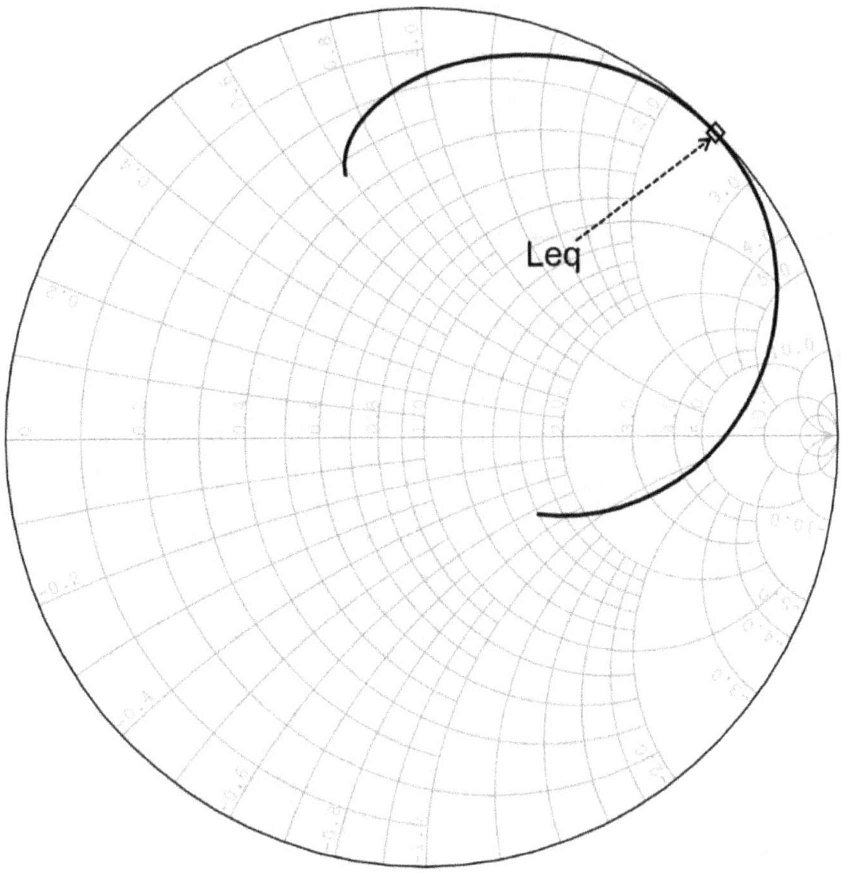

Figure 8. Complex input impedance of the AI.

3. Tunable microwave active filter example application

Thanks to the aforementioned approach, it is now possible to design an AI using, for example, medium-power bipolar transistor. The topology of the compensating network can be designed by means of an optimizer numerical software and the exact values of the network elements can be found by optimization, in order to take into account microstrip discontinuities and lines [38]. The ideal AI application is, of course, in monolithic integrated circuits, but as an example, a preliminary discrete design can be made on microstrip with discrete components. First, the AI design has to be performed, and consequently, a first-order (or greater) passband filter can be optimized (see **Figure 9**). As mentioned, the internal AI transconductance amplifier can be implemented by a bipolar transistor (e.g., the BFP420 bipolar junction transistor by Infineon in common-emitter configuration). **Figure 10** shows a simulated standalone AI real and imaginary part, while in **Figure 11**, the Smith chart complex input impedance is shown. In this case, the topology of the compensating network, and its components value, has been chosen in order to have an inductive behavior with minimum nonnegative series resistance around 2500 MHz. The standalone AI Q is shown in **Figure 12**. The implemented filter response in terms of S-parameters is shown in **Figure 13** showing the design success of microwave active filters by means of AI.

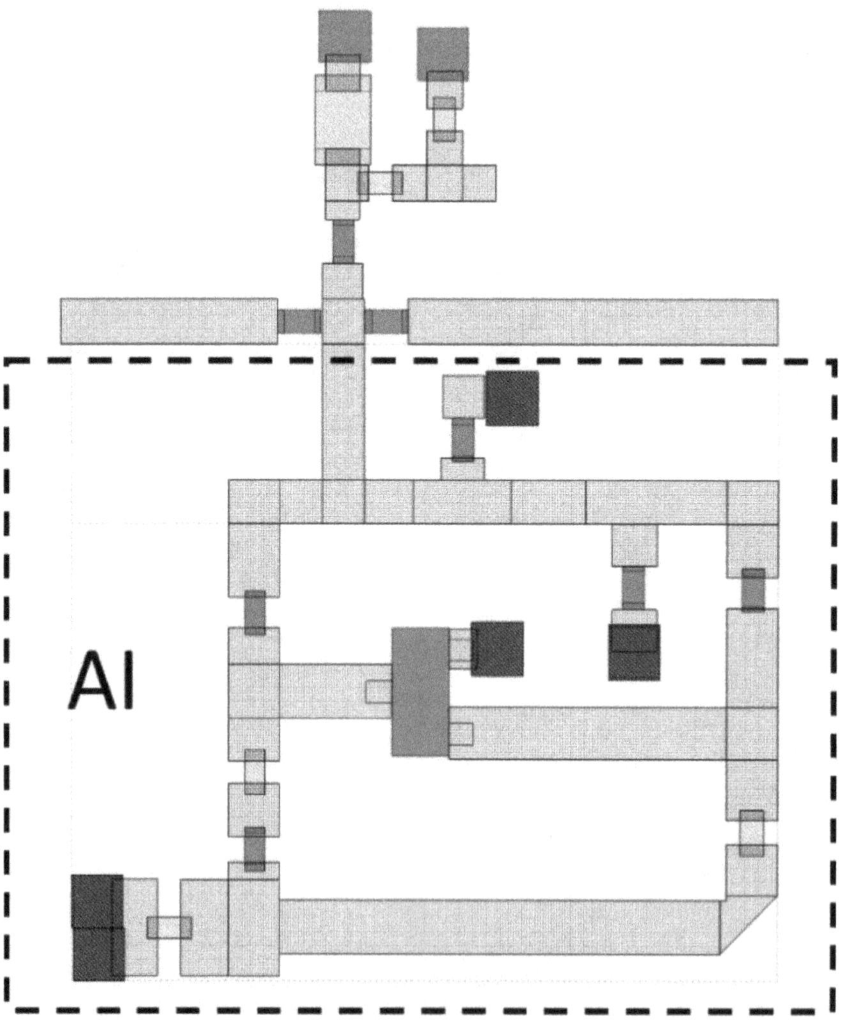

Figure 9. AI-based LC filter.

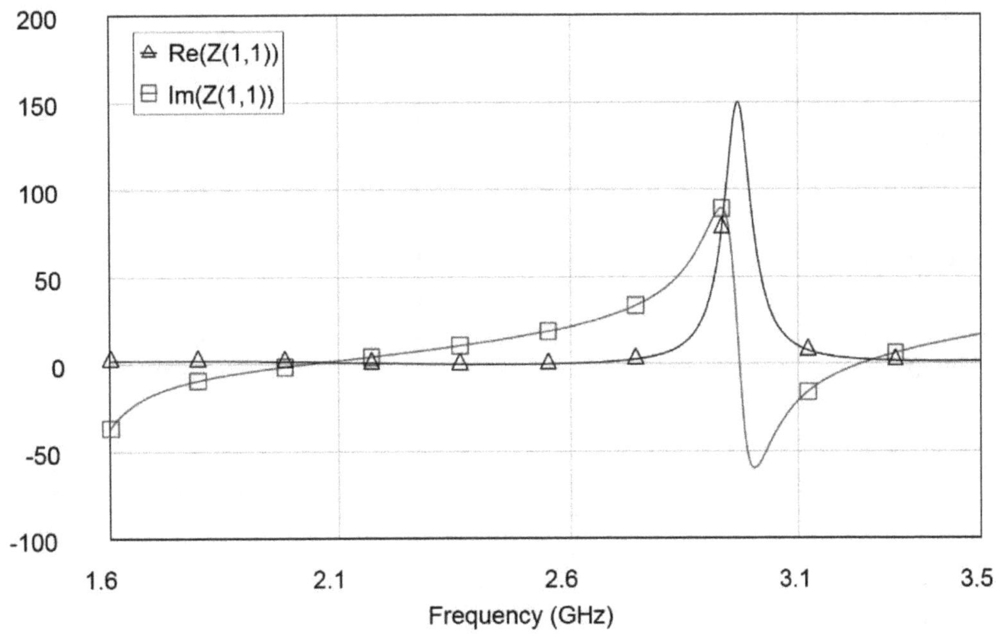

Figure 10. AI real and imaginary part.

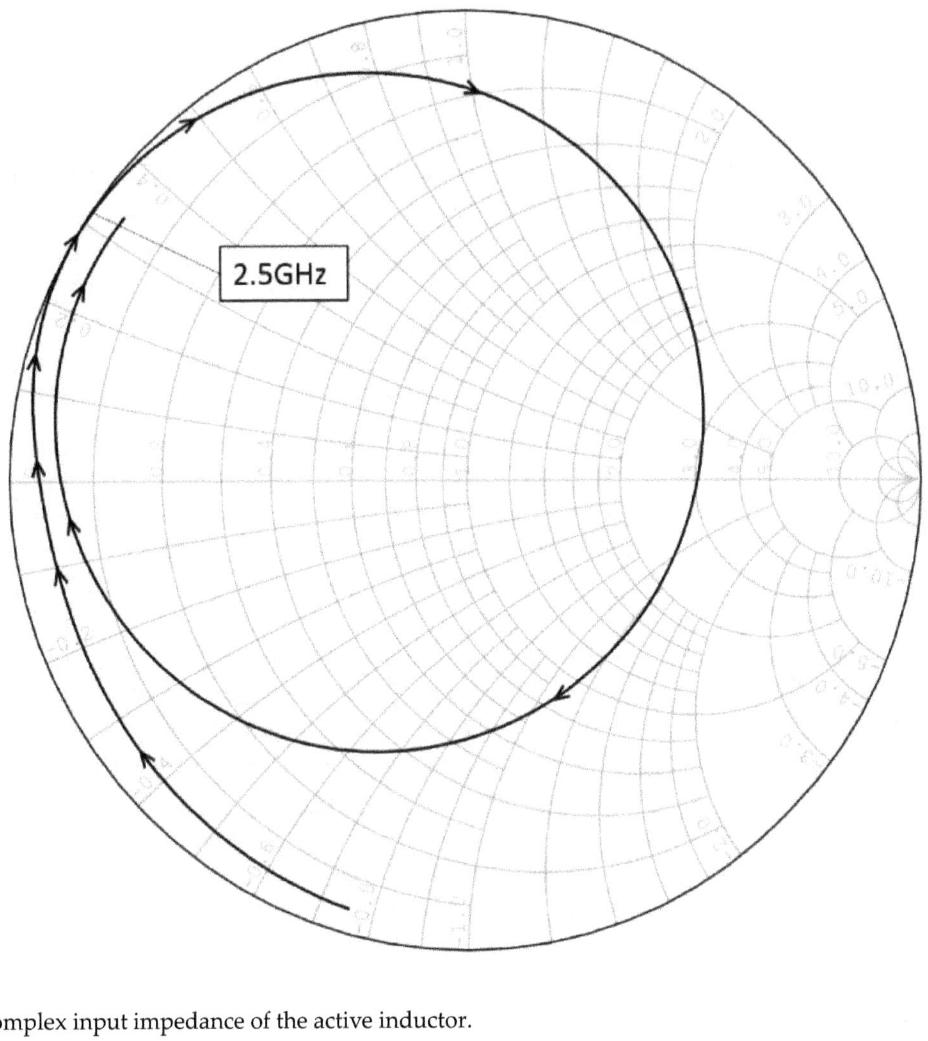

Figure 11. Complex input impedance of the active inductor.

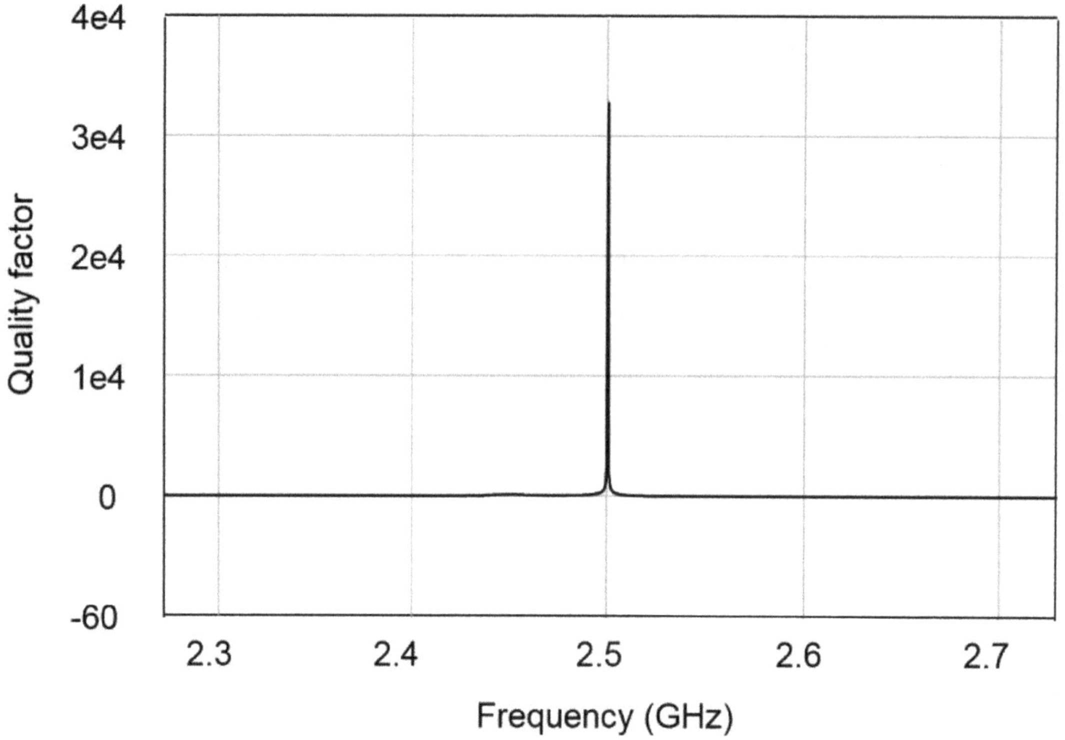

Figure 12. Quality factor of the active inductor.

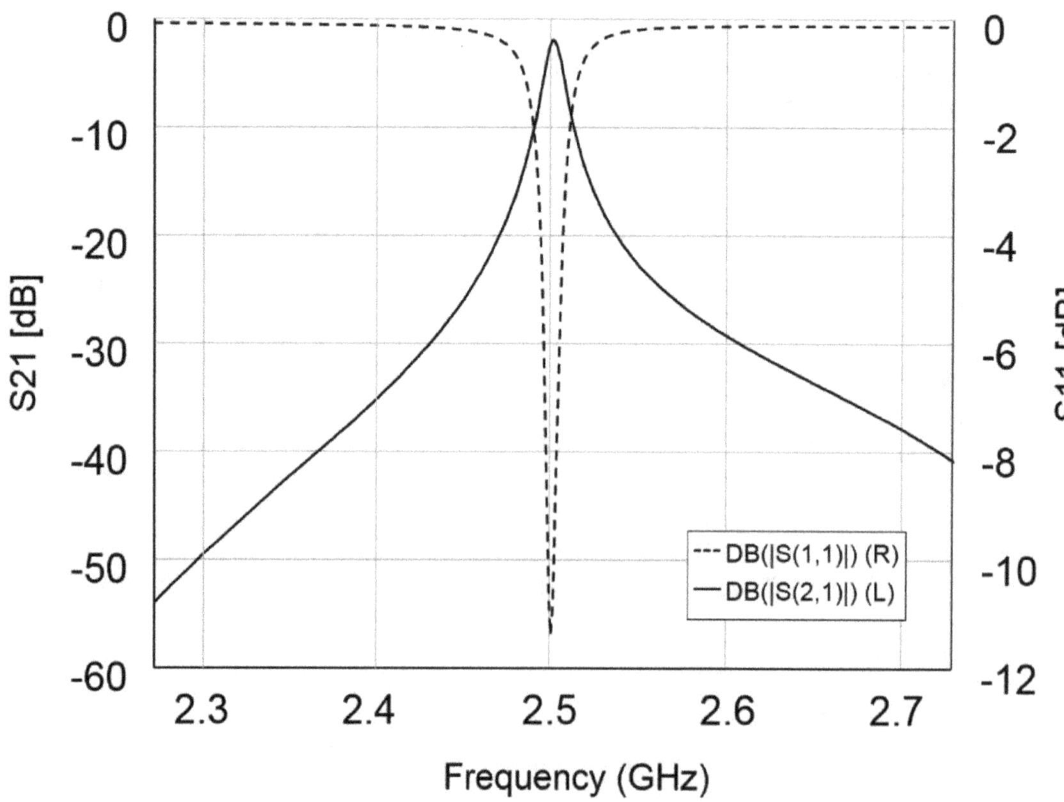

Figure 13. S-parameters of the designed active filter.

Author details

Vincenzo Stornelli*, Leonardo Pantoli and Giorgio Leuzzi

*Address all correspondence to: vincenzo.stornelli@univaq.it

Department of Industrial and Information Engineering and Economics, University of L'Aquila, L'Aquila, Italy

References

[1] Hara, S., Tokumitsu, T., Aikawa, M. "Lossless broad band monolithic microwave active inductor," IEEE Transactions on Microwave Theory and Techniques vol. 37, no. 12, pp: 1979–1984, December 1989.

[2] Wu, Y., Ding, X., Ismail, M., Olsson, H. "RF bandpass filter design based on CMOS active inductors" IEEE Transactions on Circuits and Systems-II: Analog and Digital Signal Processing vol. 50, no. 12, pp: 942–949, December 2003.

[3] Hara, S., Tokumitsu, T., Tanaka, T., Aikawa, M. "Broad band monolithic microwave active inductor and its application to miniaturized wide band amplifiers," IEEE Transactions on Microwave Theory and Techniques, vol. 36, no. 12, pp: 1920–1924, December 1988.

[4] Mukhopadhyay, R., Nuttinck, S., Woo, S. H., Kim, J. H, Lee, S. S., Lee, C. H., Laskar, J. Tunable active inductor, United States Patent Application Publication, no. US 2006/0170523 A1, appl. no. 60/649/221, Int. Cl. H03H 11/00, August 3, 2006.

[5] Kaunisto, R., Alinikula, P., Stadius, K., Porra, V. "A Low-Power HBT MMIC filter based on tunable active inductors," IEEE Microwave and Guided Wave Letters, vol. 7, no. 8, pp: 209–211, 1997.

[6] Wu, Y., Ding, X., Ismail, M., Olsson, H. "CMOS active inductor and its application in RF bandpass filter," IEEE Radio Frequency Integrated Circuits Symposium, pp: 655–658, 2004 .

[7] Campbell, C. F., Weber, R. J. "Design of a broadband microwave BJT active inductor circuit," Proceedings of the 34th Midwest Symposium on Circuits and Systems, vol. 1, pp: 407–409, 1991.

[8] He, X., Kuhn, W. B. "A 2.5-GHz low-power, high dynamic range self-tuned Q-enhanced LC ¯ filter in SOI," IEEE Journal of Solid-State Circuits, vol. 40, no. 8, 1618–1628, 2005.

[9] Kaunisto, R., Alinikula, P., Stadius, K. "Q-enhancing technique for high speed active inductors," IEEE International Symposium on Circuit and Systems ISCAS 1994, vol. 5, pp: 735–738, 1994.

[10] Hsiao, C., Kuo, C., Ho, C., Chan, Y. "Improved quality-factor of 0.18-μm CMOS active inductor by a feedback resistance design," IEEE Microwave and Wireless Components

Letters, vol. 12, no. 12. pp: 467–469, 2002.

[11] Lucyszyn, S., Robertson, I.D. "Monolithic narrow-band filter using ultrahigh-Q tuneable active inductors," IEEE Transactions on Microwave Theory and Techniques, vol. 42, no. 12, pp. 2617–2622, 1994.

[12] Thanachayanont, A. "Low-voltage Low-power high-Q CMOS RF bandpass filter" Electronics Letters, vol. 38, no. 13, pp. 615–616, 2002.

[13] Cho, Y., Hong, S., Kwon, Y. "A novel active inductor and its application to inductance-controlled oscillator," IEEE Transactions on Microwave Theory and Techniques, vol. 45, no. 8, pp. 1208–1213, 1997.

[14] Sharman, R., A'ain, A. K., Azmi, M., Huang, M. Z. "Design approach for tuneable CMOS active inductor," IEEE International Conference on Semiconductor Electronics ICSE 2004.

[15] Lu, L. H., Hsieh, H. H., Liao, Y. T., "A wide tuning-range CMOS VCO with a differential tunable active inductor," IEEE Transactions on Microwave Theory and Techniques, vol. 54, no. 9, pp. 3462 - 3468, 2006.

[16] Mukhopadhyay, R., Park, Y., Yoon, S. W., Lee, C. H., Nuttinck, N., Cressler, J. D., Laskar, J. "Active-inductor-based low-power broadband harmonic VCO in SiGe technology for wideband and multi-standard applications," IEEE MTT-S International Microwave Symposium Digest, Long Beach, CA. June 2005.

[17] van Vliet, F. E., van der Bogaart, F. L. M., Tauritz, J. L., Baets, R. G. F. "Systematic analysis, synthesis and realization of monolithic microwave active inductors," IEEE MTT-S International Microwave Symposium Digest, vol. 3, pp. 1659–1662, 1996.

[18] Thanachayanont, A., A. Payne, "VHF CMOS integrated active inductor," Electronics Letters, vol. 32, no. 11, pp. 999–1000, 1996.

[19] Georgescu, B., Finvers, I. G., Ghannouchi, F. "2 GHz Q-enhanced active filter with low passband distortion and high dynamic range," IEEE Journal of Solid-State Circuits, vol. 41, no. 9, pp. 2029 - 2039, 2006.

[20] Soorapanth, T., Wong, S. S., "A 0-dB IL 2140 § 30 MHz bandpass filter utilizing Q-enhanced spiral inductors in standard CMOS," IEEE Journal of Solid-State Circuits, vol. 37, no. 5, pp. 579 - 586, 2002.

[21] Gao, Z., Yu, M., Ma, J. "Wide tuning range of a CMOS RF bandpass filter for wireless application," IEEE Conference on Electron Devices and Solid-State Circuits, pp. 53–56, 2005.

[22] Allidina, K., Mirabbasi, S. "A widely tunable active RF filter topology," 2006 IEEE International Symposium on Circuits and Systems, pp. 879–882, 2006.

[23] Sae-Ngow, S., Thanachayanont, A. "A low-voltage, wide dynamic range CMOS floating active inductor," TENCON 2003. Conference on Convergent Technologies for Asia-Pacific Region, vol. 4, pp. 1460–1463, 2003.

[24] Mukhopadhyay, R., Yoon, S. W., Park, Y, Lee, C. H., Nuttinck, S., Laskar, J. "Investigation of inductors for digital Si-CMOS technologies," 2006 IEEE International Symposium on Circuits and Systems, pp. 3750–3753, 2006.

[25] Sussman-Fort, S.E. "Design concepts for microwave GaAs FET active filters," IEEE Transactions on Microwave Theory and Techniques, vol. 37, no. 9, pp. 1418–1424, 1989.

[26] Sujin, S., Namsik, R., Heungjae, C., Yongchae, J. "Novel high-Q inductor using active inductor structure and feedback parallel resonance circuit," IEEE Symposium on Radio Frequency Integrated Circuits (RFIC), vol. 1, pp. 467–470, 2007.

[27] Andriesei, C., Goras, L., Temcamani, F., Delacressoniere, B. "CMOS RF active inductor with improved tuning capability," International Semiconductor Conference, vol. 2, pp. 397–400, 2009.

[28] Chun-Lee, L., A'ain, A., Kordesch, A. V. "CMOS active inductor linearity improvement using feed-forward current source technique," IEEE Transactions on Microwave Theory and Techniques, vol. 57, no. 8, pp. 1915–1924, 2009.

[29] Yongho, C., Songcheol, H., Youngse, K. "Monolithic VCO using a novel active inductor," IEEE Microwave and Millimeter-Wave Monolithic Circuits Symposium, DOI: 10.1109/MCS.1996.506325, pp. 155–158, 1996.

[30] Rohde, U. L., Poddar, A.K. "Active inductor oscillator and noise dynamics," IEEE MTT-S International Microwave Symposium Digest (MTT), DOI: 10.1109/MWSYM.2011.5972933, pp. 1–4, 2011.

[31] Yushi, Z., Fei, Y. "Subthreshold CMOS active inductors with applications to low-power injection-locked oscillators for passive wireless microsystems," Circuits and Systems (MWSCAS), 2010 53rd IEEE International Midwest Symposium on, Aug. 2010.

[32] Gamm, G. U. "Low power wake-up receiver for wireless sensor nodes," IEEE Conference on Intelligent Sensors, Sensor Networks and Information Processing, vol. 1, pp. 121–126, 2010.

[33] Pantoli, L., Stornelli, V., Leuzzi, G. "A low-voltage low-power 0.25 μm integrated single transistor active inductor-based filter," Analog Integrated Circuits and Signal Processing, vol. 87, no. 3, pp. 463–469, 2016.

[34] Pantoli, L., Stornelli, V., Leuzzi, G. "Low-noise tunable filter design by means of active components," Electronics Letters, vol. 52, no. 1, pp. 86–88, 2016.

[35] Leuzzi, G., Stornelli, V., Del Re, S. "A tuneable active inductor with high dynamic range for band-pass filter applications," IEEE Transactions on Circuits and Systems II: Express Briefs, vol. 58, no. 10, pp. 647–651, 2011.

[36] Pantoli, L., Stornelli, V., Leuzzi, G. "Class AB tunable active inductor," Electronics Letters, vol. 51, pp. 65–67, 2015, ISSN: 0013-5194.

[37] Branchi, P., Pantoli L., Stornelli, V., Leuzzi, G. "RF and microwave high-Q floating active

inductor design and implementation," International Journal of Circuit Theory and Applications, vol. 43, no. 8, pp. 1095–1104, Aug. 2015. ISSN: 0098-9886.

[38] Leuzzi, G., Stornelli, V., Pantoli, L., Del Re, S. "Single transistor high linearity and wide dynamic range active inductor," International Journal of Circuit Theory and Applications, vol. 43, no. 3, pp. 277–285, March 2015. ISSN: 0098-9886.

[39] Pantoli, L., Stornelli, V., Leuzzi, G. "Tunable active filters for RF and microwave applications," Journal of Circuits, Systems, and Computers, vol. 23, pp. 1450088, 2014. ISSN: 0218-1266.

[40] Stornelli, V., Pantoli, L., Leuzzi, G. "High quality factor L-band AI-based band-pass filters," Journal of Circuits, Systems, and Computers, vol. 22, 3, 2013. ISSN: 0218-1266.

[41] Colucci, P., Leuzzi, G., Pantoli, L., Stornelli, V. "A third order integrable UHF bandpass filter using active inductors," Microwave and Optical Technology Letters, Microwave and Optical Technology Letters, vol. 54, no. 6, pp. 1426–1429, June 2012.

[42] Stornelli, V., Pantoli, L., Leuzzi, G. "An assessment on low voltage low power integrable single transistor active inductor design for RF filter applications" (Invited Paper), IEEE International Conference on Integrated Circuit Design and Technology (ICICDT), June 27–29, 2016.

[43] Pantoli, L., Stornelli, V., Leuzzi, G. "Class AB gyrator-based active inductor." In: IEEE Integrated Nonlinear Microwave and Millimetre-wave Circuits (INMMiC), 2015 International Workshop, Taormina, 29–30 September, 2015.

[44] Pantoli, L., Stornelli, V., Leuzzi, G. "RF active circuit simulating a floating inductance." In: IEEE, Integrated Nonlinear Microwave and Millimetre-wave Circuits (INMMiC), 2015 International Workshop, Taormina, 29–30 September, 2015.

[45] Pantoli, L., Stornelli, V., Leuzzi, G. "A wideband class-AB tunable active filter." In: IEEE, European Microwave Week 2015. EuMW 2015—Conference Proceedings; EuMIC 2015: European Microwave Integrated Circuits Conference, Parigi, 7–8 September, 2015.

[46] Pantoli, L., Stornelli, V., Leuzzi, G. "A wideband class-AB tunable active filter." In: IEEE, European Microwave Week 2015. EuMW 2015—Conference Proceedings; EuMC 2015: European Microwave Conference, Parigi, 7–8 September, 2015.

[47] Pantoli, L., Stornelli, V., Leuzzi, G., Di Carlofelice, A. "A single transistor post selector active tunable filter for radio receivers applications." In: IEEE Integrated Nonlinear Microwave and Millimetre-wave Circuits (INMMiC), 2014 International Workshop on. DOI: 10.1109/INMMIC.2014.6815106, 2014.

[48] Pantoli, L., Stornelli, V., Leuzzi, G. "Low voltage high-order agile active filter for microwave applications." In: IEEE. EuMC 2013: 43rd European Microwave Conference, pp. 1203–1206, 2013.

[49] Pantoli, L., Stornelli, V., Leuzzi, G. "A single-transistor tunable filter for bluetooth applications." In: EuMW 2012, Conference Proceedings—7th European Microwave Integrated

Circuits Conference, EuMIC 2012, Amsterdam, October 29–30, 2012.

[50] Pantoli, L., Stornelli, V., Leuzzi, G. "A single-transistor tunable filter for bluetooth applications." In: Conference Proceedings—42nd European Microwave Conference, EuMC 2012. pp. 1269–1272, ISBN: 978-287487027-9.

[51] Stornelli, V., Leuzzi, G., Pantoli, L., Del Re, S. "High dynamic range bandpass filters design based on active inductor." In: Microwave Integrated Circuits Conference (EuMIC), 2011 European, Manchester, October 10–11, pp. 168–171.

[52] Leuzzi, G., Stornelli, V., Del Re, S., Pantoli, L. "High quality factor integrable bandpass filter by using tunable active inductor." In: Integrated Nonlinear Microwave and Millimetre-Wave Circuits (INMMIC), 2011 Workshop on, Vienna, April 18–19, 2011.

[53] Leuzzi, G., Stornelli, V., Colucci, P., Pantoli, L. "Low-noise electronic circuit simulating the behavior of an inductance. 2014, US201414301526 20140611.

9

Modeling and Simulation Techniques for Microwave Components

Sayed Alireza Sadrossadat and Farnaz Mohammadi

Abstract

This chapter discusses about methods used in simulation and modeling of radio frequency (RF)/microwave circuits and components. The main topic that is discussed here is about one of the most powerful methods, that is, artificial neural networks. In this chapter, different types of neural network such as dynamic and recurrent neural networks will be discussed. Other techniques that are popular in the area of microwave components simulation and modeling are numerical techniques such as vector fitting, Krylov method, and Pade approximation. At the end of the chapter, vector fitting as an example of numerical methods will be discussed.

Keywords: artificial neural networks, circuit simulation and modeling, transient analysis, function approximation, RF/microwave circuits

1. Introduction

In recent years, ascending development of wireless communication products and huge trend for commercial market in this ground caused significant improvement in modeling and simulation approaches of radio frequency (RF) and microwave circuits. Such high-frequency circuits are leading to the development of a large variety of microwave models for passive and active devices and circuit components [1]. Modeling and computer-aided design (CAD) methods have an essential role in microwave designs and simulations [2]. The older approaches were mainly based on slow trial-and-error processes and an emphasis on performance at any price, but today seems to be a new era in high-frequency circuit design and modeling, since development in this ground has enabled microwave engineers to design larger, more efficient, and more complicated circuits than before [1, 3]. This complexity requires new materials and technologies that require not only new models but also new algorithms in computer-aided design [4] for RF/microwave circuits, antennas, and systems to keep up with the advancement of technology with emphasis on time-to-market and low-cost approaches [1, 3]. In addition to accurate parametric-modeling techniques to describe the behavior of the microwave device, a reliable description that explains the changes of its behavior against geometrical or physical parameters is also needed [5].

Also, since circuit models at high frequencies often lack fidelity, detailed electromagnetic (EM) simulation techniques are needed to improve design accuracy. Although EM simulation techniques are heavily used yet, they are computationally expensive, so there is a demand for design methodologies to be not only accurate but also fast. Another concerning problem today is optimization. To meet this purpose, computer-based algorithms that work with iterative circuit evaluation are needed; this process also needs a highly repetitive computational process. Another concerning issue according to Ref. [6] is the possibility of employing knowledge-based tools for initial design, that is, one of the steps toward designing and modeling process. It is hard to satisfy all these problems with the traditional CAD technologies [1, 3]. In conclusion, obviously there is a serious need for a powerful accurate and fast processing and modeling tool.

Neural networks (NNs), or artificial neural networks, are information-processing systems that can imitate the ability of human brain to learn from observation and generalize by abstraction to create complex models [7]. Neural network gives a great approximation of system regardless of linear or nonlinear correlation between the input data and can be used as knowledge-based tool (to be employed for initial design in RF/microwave applications) [1]. The ability of NN to be trained resulted in their use in many diverse fields such as pattern recognition, system identification, control, telecommunications, biomedical instrumentation, and many other grounds. Recently, many researchers in communication area are focusing on using neural network in their modeling and simulation, and NN has been recognized as a useful alternative to conventional approaches in microwave modeling [1, 3]. Neural network models are simple and fast, and they can enhance the accuracy of existing models. The basis of neural network is on the universal approximation theorem, which says that a neural network with at least one hidden layer can give an approximation of nonlinear multidimensional function to any intended accuracy [8]. This property makes neural network a favorite modeling tool for microwave engineers. Neural network approach is generic, that is, the same modeling technique that can be reused for passive/active devices/circuits. Another advantage of NN is the ease of updating neural models regarding changes in technology [2]. Neural network is now used in various microwave modeling and simulation applications, such as vertical interconnect accesses (Vias) and interconnects [9], parasitic modeling [10], coplanar waveguide (CWG) components [11], antenna applications, nonlinear microwave circuit optimization [12], power amplifier modeling, nonlinear device modeling, wave-guide filter, enhanced elemental method (EM) computation, and so on [2].

Artificial neural networks are classified into two main categories: static neural networks and dynamic neural networks. In this chapter, the first neural network structures will be presented, and then a general overview of static and dynamic neural networks and different types of them and their applications in microwave modeling will be discussed. The last part is devoted to another method called vector fitting (VF) that is a numerical technique used for system identification and macromodeling [13].

2. Training neural network

2.1. Basic structure of neural network

The idea behind neural network is similar to the function of human brain. A typical neural network structure has two types of basic components: the processing elements and interconnection between them. The processing elements are known as neurons, and the interconnections are called links. Each link is recognized with a corresponding weight parameter. Every neuron receives stimuli from neighbor neurons connected to it [3]. Input neurons receive stimuli from the outside of the network and the neurons that produce the output result are called output neurons, and neurons that not only send but also receive stimuli are called hidden neurons [1]. There are different ways to connect neurons to each other, so there are different neural network structures. A neural network structure defines how information is processed inside a neuron, and how the neurons are connected. In this chapter, we discuss the models that are more common in microwave simulations and modeling.

Generally, artificial neural networks have an input data vector, an output data vector, a vector including all the weight parameters, and a function that mathematically presents the neural network [14].

Assume N_i and N_o to represent the number of input and output neurons of the neural network, respectively, w to be the vector of weight parameters, and $y = y(x,w)$ to define the function that represents the neural network. A simple scheme of artificial neural network is shown in **Figure 1**.

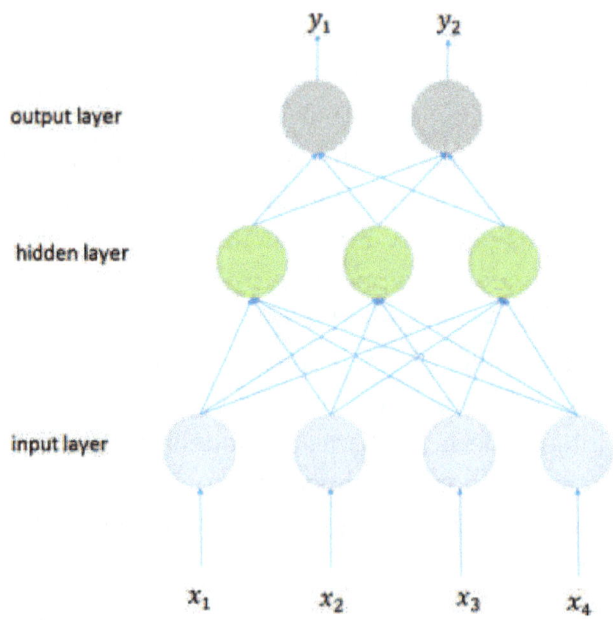

Figure 1. A general scheme of a neural network with four input neurons, two output neuron, and one hidden layer. Also, the links between the neurons are the weights.

Given a set of input and output data, a neural network can be constructed and trained. The network tries to estimate a function, so that it is able to give the closest result to the intended output. Commonly, a large percentage of input and corresponding output data are used as training data and the network will be trained by means of them. The act of training means identifying the weights, so that they reach the optimum values. The remaining percentage of data is used as validation and testing. Validation set is used to determine an approximation of generalized error and is a factor for determining when to stop the process to prohibit over-learning and under-learning [7]. Testing data obviously is used for checking the accuracy and correctness of the network after training is completed.

In each level of information processing, the output of each neuron is received by the next neuron, from input neurons to output neurons. An overview of information processing in layers is shown in **Figure 2**. The inputs of a neuron are first multiplied by the corresponding weight parameters individually, then the results are added to produce a weighted sum of γ, which then will pass through a neuron activation function σ(.) to produce the final output of the neuron. This output is the input of the neuron in the next layer, and this process is repeated for all the neurons, until it reaches the output layer.

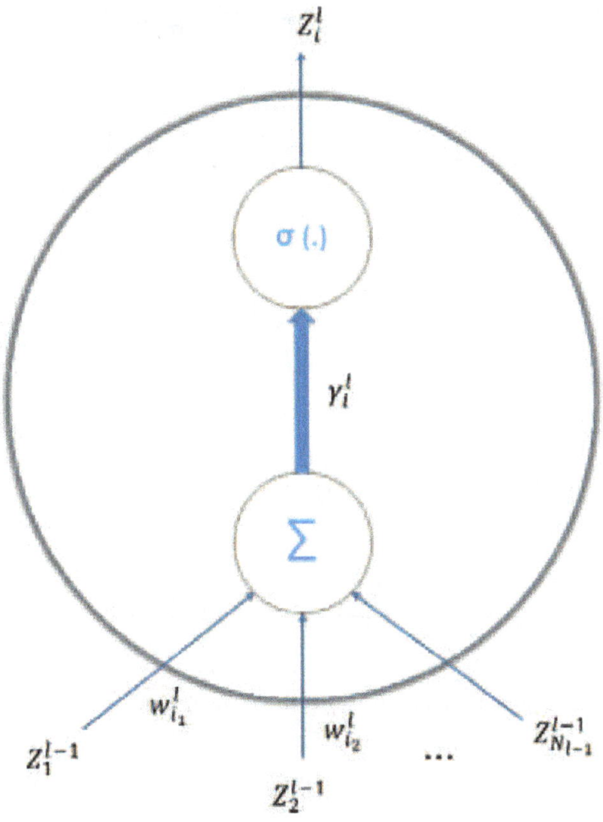

Figure 2. Information processing inside a neuron.

Let $w_{i0}{}^{l}$ to be the bias for i^{th} neuron of lth layer. So the vector of weights is $w = [w_{10}^2 w_{11}^2 w_{12}^2 \ldots w_{2N_1}^2 w_{10}^3 \ldots w_{N_L N_{L-1}}^L]^T$ and the output of the neuron is $z_i^l = \sigma(\gamma_i^l)$.

There are different types of training in neural network [1], here we explain each of them shortly:

1. Sample-by-sample (or online) training: each time a training sample is presented, the weights (w) are updated based on training error.
2. Batch-mode (or offline) training: after each epoch weights are updated based on training error from all the samples in training data set.
3. Supervised training: using x and y data for training process, where x is the input of the neural network and y is the output of the neural network.
4. Un-supervised training: using just x data for training process.

2.2. Activation functions

Activation function, also known as transfer function, is one of the most important units in a neural network structure, that is, a scalar-to-scalar function transforms a set of input signals into an output signal. Common types of activation functions are arctangent as shown in **Figure 3**, hyperbolic tangent shown in **Figure 4**, and sigmoid functions which are shown in **Figure 5** [15].

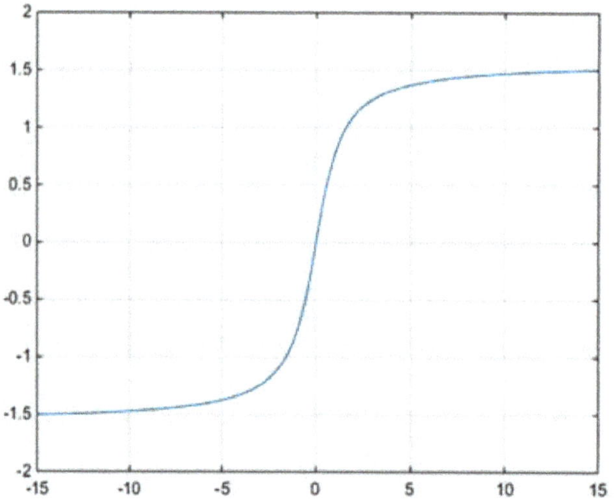

Figure 3. Mathematical graph of arctangent function $z = \sigma(\gamma) = \frac{2}{\pi} \arctan(\gamma)$.

Sigmoid function:

$$z = \sigma(\gamma) = \frac{1}{1 + e^{-\gamma}} \tag{1}$$

Arctangent function:

$$z = \sigma(\gamma) = \frac{2}{\pi} \arctan(\gamma) \tag{2}$$

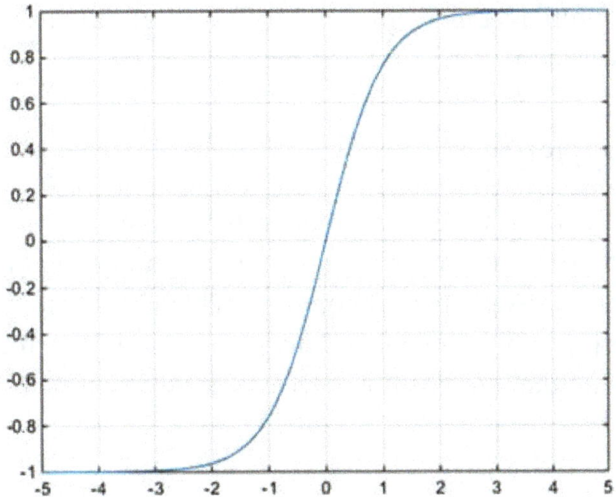

Figure 4. Mathematical graph of hyperbolic tangent function $z = \sigma(\gamma) = \frac{e^{\gamma}-e^{-\gamma}}{e^{\gamma}+e^{-\gamma}}$.

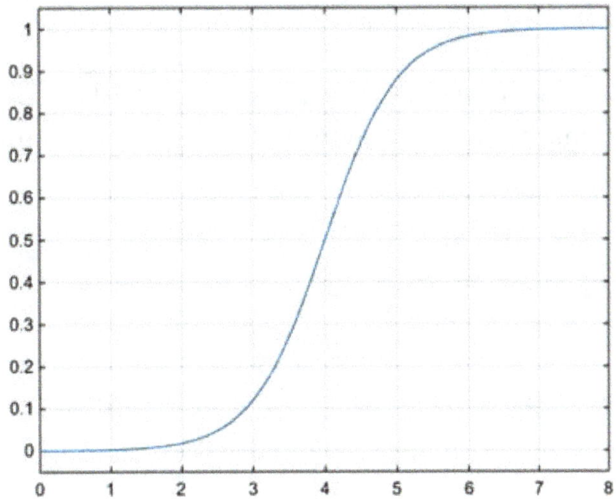

Figure 5. Mathematical diagram of sigmoid function $z = \sigma(\gamma) = \frac{1}{1+e^{-\gamma}}$.

Hyperbolic tangent function:

$$z = \sigma(\gamma) = \frac{e^{\gamma}-e^{-\gamma}}{e^{\gamma}+e^{-\gamma}} \tag{3}$$

3. Static neural networks

In the past few years, artificial neural networks have gained attention as a valuable computer-aided design tool for modeling high-frequency circuits. They can mainly be categorized as techniques for modeling frequency-domain response of components and time-domain response of them. For frequency-domain modeling, static neural networks are employed. Their main architectures are multilayer perceptron (MLP) and radial-basis function (RBF), which will be discussed in this section.

3.1. Multilayer perceptron (MLP)

Multilayer perceptron structure is the most frequently used structure in many areas including microwave modeling and optimization problems. This technique belongs to a subcategory of neural network called feed-forward neural network, which is able to approximate continuous and integrable functions [1], and their connectivity consists of layer groups that are only linked to adjacent layers, meaning that there is not a cycle or a recursive path [16].

3.1.1. MLP structure

In MLP structure, neurons are classified into different layers. A typical MLP neural network consists of one input layer, one or more hidden layers, and one output layer, as shown in **Figure 6**. Consider L as the total number of layers, layer 1 is the input layer, layer 2 to layer (L-1) are hidden layers, and layer L is the output layer. Also, suppose the number of neurons in l^{th} layer is N_l, $l = 2, 3, \ldots, L$.

Here, consider x_i as the ith input of the MLP, and z_i^l as the output of ith neuron of lth layer. Also, w_{ij}^l is the weight of the link between j^{th} neuron of $(l - 1)^{\text{th}}$ layer, and i^{th} neuron of lth layer $1 \le j \le N_{l-1}, 1 \le i \le N_l$.

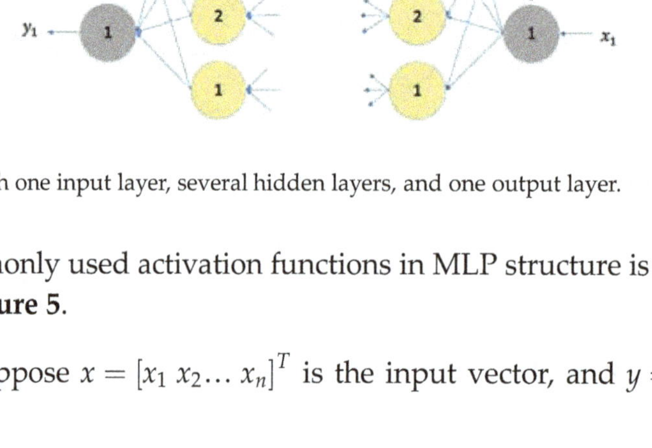

Figure 6. MLP structure with one input layer, several hidden layers, and one output layer.

One of the most commonly used activation functions in MLP structure is sigmoid function [1], which is shown in **Figure 5**.

In summary, if we suppose $x = [x_1\ x_2 \ldots\ x_n]^T$ is the input vector, and $y = [y_1\ y_2 \cdots\ y_m]^T$ is the output vector,

For

$$l = 1: \ z_i^l = x_i, \ i = 1, 2, \ldots, \ n = N_1 \tag{4}$$

For

$$l = 2,3,\ldots,L : \gamma_i^l = \sum_{j=0}^{N_{l-1}} w_{ij}^{l-1} z_j^{l-1}, z_i^l = \sigma(\gamma_i^l) \tag{5}$$

And

$$y_i = z_i^L, i = 1,2,\ldots,m, m = N_L \tag{6}$$

3.2. Radial-basis function (RBF) networks

Radial-basis function neural network, like MLP, is a subset of feed-forward neural network. It is used in a wide range of applications related to microwave transistors and high-speed integrated circuits, and modeling of intermodulation distortion behavior of MESFETs and HEMTs [1, 17].

3.2.1. RBF structure

A radial-basis function is a real-valued function whose value depends only on the distance from the origin, so that $\varphi(x) = \varphi(\|x\|)$, or alternatively on the distance from some other point *c*, called a *center*, so that $\varphi(x,c) = \varphi(\|x-c\|)$. Any function that satisfies the property $\varphi(x) = \varphi(\|x\|)$ is a radial function.

The main approach in this structure is based on approximation of a curve that best fits to the training data set in high-dimensional space by determining λ and c that are standard deviation and center of the activation functions, respectively, and are parameters of the function [7, 15]. The dimension of the hidden space is related directly to the accuracy of the approximated model [14]. In this structure, we have

$$y(x) = \sum_{i=1}^{N} w_i \varphi(\|x-x_i\|), \tag{7}$$

where *y(x)* is the approximating function that is the weighted sum of radial-basis functions [1].

In RBF structure, there is just one hidden layer, and the function of input and output layers stays the same like MLP structure. RBF uses the radial basis as activation function [7]. **Figure 7** shows a typical RBF neural network.

Radial-basis activation functions include Gaussian and multiquadratic functions.

Gaussian function:

$$\sigma(\gamma) = \exp\left(-\left(\frac{\gamma}{\lambda}\right)^2\right) \tag{8}$$

Multiquadratic function:

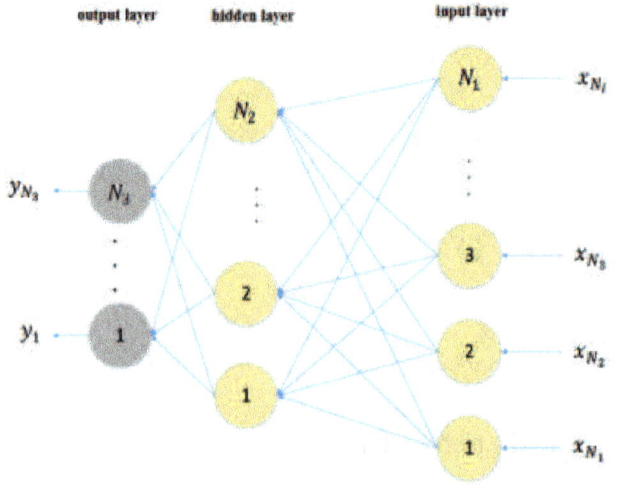

Figure 7. Structure of RBF neural network.

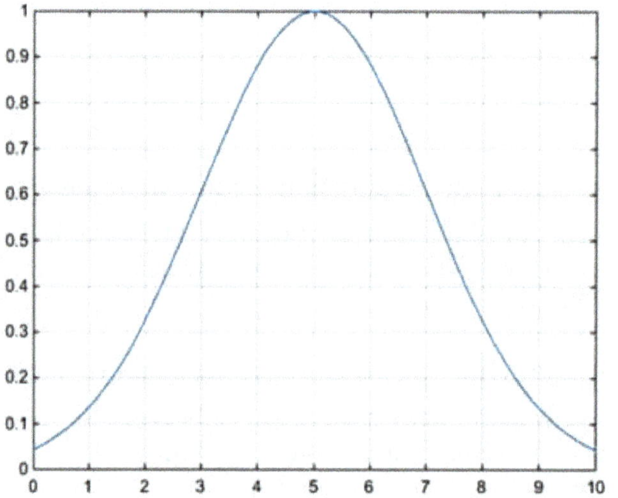

Figure 8. The Gaussian function.

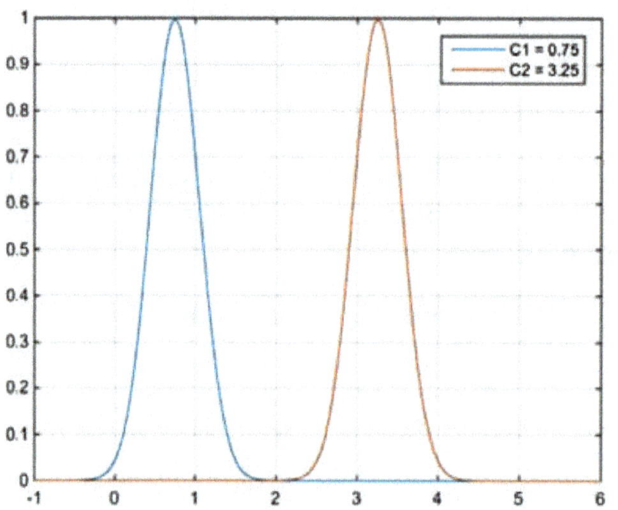

Figure 9. The multiquadratic function. *C* represents the value of centers.

$$\sigma(\gamma) = \frac{1}{(c^2 + \gamma^2)^{\alpha}}, \alpha > 0 \tag{9}$$

The Gaussian and multiquadratic functions are shown in **Figure 8** and **Figure 9**, respectively.

4. Time-domain neural networks

For time-domain modeling of components and systems, time-domain artificial neural network structures are usually employed in the literature. The main time-domain architectures are dynamic neural networks (DNNs) and recurrent neural networks (RNNs), which will be discussed in this section.

Most neural network structures used by engineers are feed-forward neural networks that are suitable for time-independent static input-output mapping [18]. In feed-forward neural networks, the flow of information is straight forward from the first neuron of the first layer to the last neuron of the output layer, and the procedure is not recursive, so the output of neurons does not have any effect on the input of the last neurons, although the stability of a neural network is the result of the absence of feedback in the network. In spite of static NNs, a dynamic neural network uses feedback between neurons in the same layer, or even neurons in different layers, also it provides more computational advantages [19]. Feedback-based neural networks are good approaches for modeling, identification, and control of systems, since most of systems in real world such as airplanes, rockets, and so forth are nonlinear dynamical systems [18, 20].

4.1. Recurrent neural networks (RNNs)

Recurrent neural network is a discrete time-domain neural network that allows time-domain behaviors of a dynamic system to be modeled [1]. Its structure is suitable for modeling tasks such as dynamic system control and finite-difference time-domain (FDTD) solutions in electromagnetic modeling [21]. The output of the neural network is a function of its present inputs and a history of its inputs and outputs [22]. The delayed outputs are fed back to the inputs and the feed-forward network along with the feedback delay constructs the recurrent neural network structure. In this architecture, we suppose the inputs and outputs to be a function of time, representing this functionality with parameter t, also τ which is the delay representing the effect of history of the neural network inputs and outputs.

Suppose the external single input of the neural network to be $x(t)$ so the history of it would be $x(t-\tau)$, $x(t-2\tau)$, $x(t-3\tau)$, ..., $x(t-\alpha\tau)$ where α is the maximum number of delay steps for x, and suppose the single output of the RNN to be $y(t)$ with history of it demonstrating as $y(t-\tau)$, $y(t-2\tau)$, $y(t-3\tau)$, ..., $y(t-\beta\tau)$ that β is the maximum number of delay steps for y. The architecture of the RNN is shown in **Figure 10**. The corresponding formulation is

$$y(t) = f\left(y(t-\tau), y(t-2\tau), \ldots, y(t-\beta\tau), x(t), x(t-\tau), x(t-2\tau), \ldots, x(t-\alpha\tau)\right) \tag{10}$$

Suppose a three-layer discrete-time MLP neural network as above, with activation function $z = \sigma(\gamma)$. Applying the delays to the process, the output of the ith neuron at t is

$$y_i(t) = \sigma\left(\gamma_i(t-\tau)\right) \tag{11}$$

in which,

$$\gamma_i(t) = x_i(t) + \sum_{j=1}^{T} w_{ij} y_j(t) \tag{12}$$

where T is the total number of neurons, x_i is the external input, $i = 1, 2, 3, \ldots, T$, y_i is the output of neuron itself, also y_j is the output of other neurons, $j = 1, 2, 3, \ldots, T$, $i \neq j$.

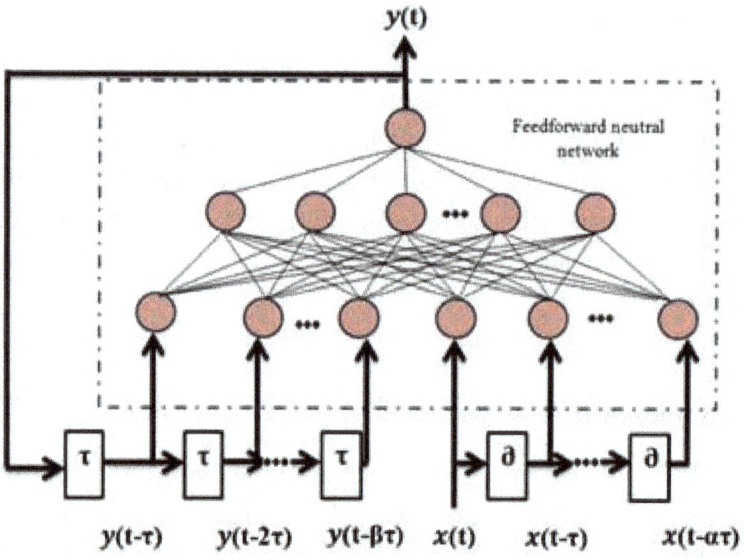

Figure 10. Recurrent neural network.

5. Dynamic neural networks (DNNs)

Dynamic neural network is a continuous time-domain neural network that is one of the best formulations for modeling nonlinear microwave circuits [9]. DNN is highly efficient in theory and practice. It is suitable for a wide range of needs in nonlinear microwave simulations, for example, it is suitable for both time- and frequency-domain applications, multitone simulations, and so on [12]. In comparison with other neural network methods, DNN provides a faster and more accurate network modeling that is significantly required in today's efficient CAD algorithms in high-level and large-scale nonlinear microwave designs. DNN also can be developed directly from input-output data without a need to depend on internal details of the circuit [12]. In DNN, the outputs are a function of inputs and their derivatives, and also a function of derivatives of outputs. **Figure 11** shows the architecture of dynamic neural network and the process occurring at each level.

In **Figure 11**, $y^{(n)}(t)$ is the nth derivative of *y(t)* and is integrated and fed back as an input to the system along with the inputs *x(t)* and their derivatives, $x^{(i)}(t)$. $y^{(n)}(t) = f\left(y^{(n-1)}(t), y^{(n-2)}(t), \ldots, y(t), x^{(n)}(t), x^{(n-1)}(t), \ldots, x(t)\right)$, in which f (.) represents the MLP nonlinear function and $y^{(i)}(t)$ represents i^{th} derivative of *y(t)*.

The DNN model can represent a nonlinear circuit when trained and tested with an appropriate data set, measured or obtained from the original circuit.

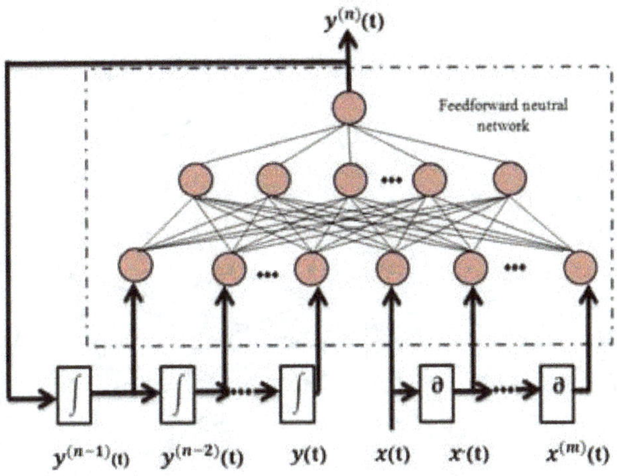

Figure 11. Dynamic neural network.

5.1. State-space dynamic neural network (SSDNN)

State-space dynamic neural network (SSDNN) is a technique for modeling nonlinear transient behaviors especially in high-speed IC and nonlinear circuits. The SSDNN-modeling technique is based on DNN structure and is a combination of DNN and state-space concept, which expands continuous DNN into a more general and flexible approach for nonlinear transient modeling and design with good accuracy [23].

Let $v \in \mathbb{R}^N$ be the transient input signal of a nonlinear circuit, and let $y \in \mathbb{R}^M$ be the transient output signal of a nonlinear circuit, where N and M are the number of inputs and outputs of the circuit, respectively, w also is the weight parameter matrix that is divided into three matrixes: w_v, w_s, w_o, which are weights connecting to the inputs (v), weights connecting to the state variables, and weights connecting the hidden neurons of the hidden layer to the outputs, respectively. Also, η is a constant scaling parameter. SSDNN model formulation can be represented by the equations as follows:

$$\left\{ \begin{array}{c} \dot{x}(t) = -x(t) + \eta g_{\text{ANN}}\Big(v(t), x(t), w\Big) \\ y(t) = Bx(t) \end{array} \right\} \tag{13}$$

where $x = [x_1, \ldots, x_L]^T \in \mathbb{R}^L$ and is a vector of state variables, with initial condition $x(0) = x_0$ and L is the dimension of the state space, that is, the order of the model, $g_{\text{ANN}} = [g_{\text{ANN-1}}, \ldots, g_{\text{ANN-}L}]^T$ is a representation of feed-forward MLP neural network, which has $N + L$ input neurons and L output neurons, also $B = [b_{ij}] \in \mathbb{R}^{LM}$ is the matrix which maps state space into output space [23]. A simple structure of SSDNN is shown in **Figure 12**.

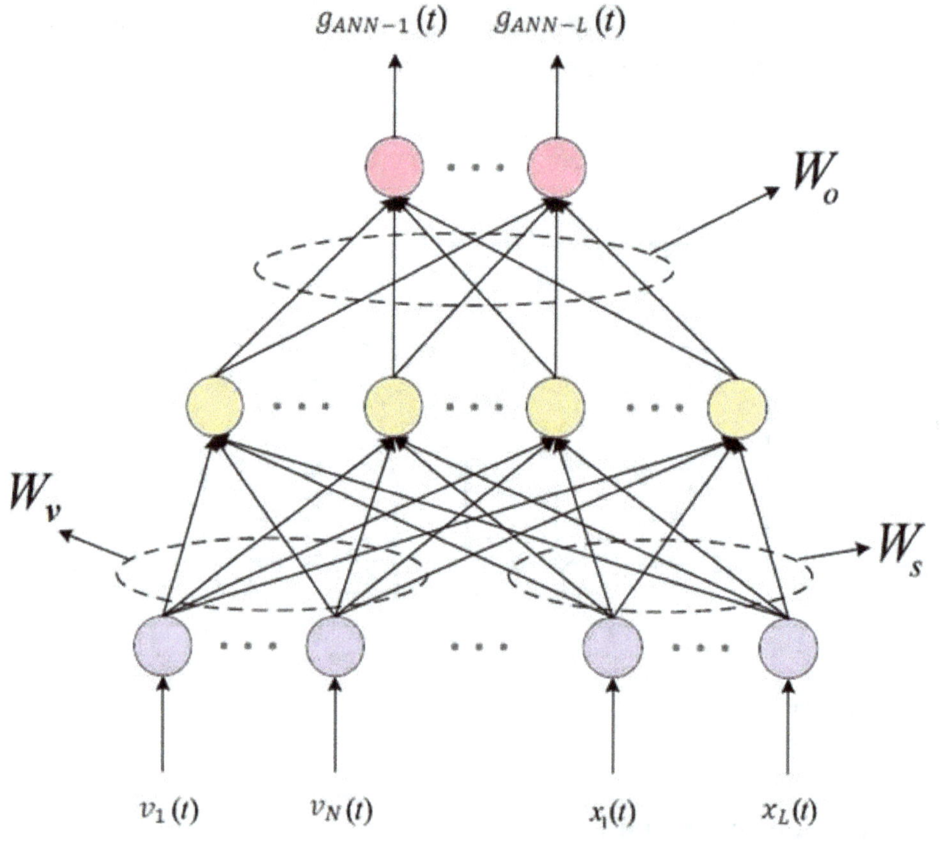

Figure 12. Structure of SSDNN model.

5.1.1. Adjoint state-space dynamic neural network (ASSDNN)

Adjoint state-space dynamic neural network (ASDNN) method, like SSDNN method, is used for modeling the transient behavior of nonlinear electronic and photonic components. It is an extension of SSDNN technique that is capable of adding the derivative information of the output to the training patterns of nonlinear components simultaneously, so that the training process can be done more efficient requiring less data without sacrificing model accuracy and efficiency [23, 24]. It has been shown in Ref. [24] that testing error from the model trained by ASSDNN method is much less than that obtained from SSDNN. Here is the formulation of ASSDNN using notation similar to SSDNN mentioned already, and an overview of structure of ASSDNN is shown in **Figure 13**.

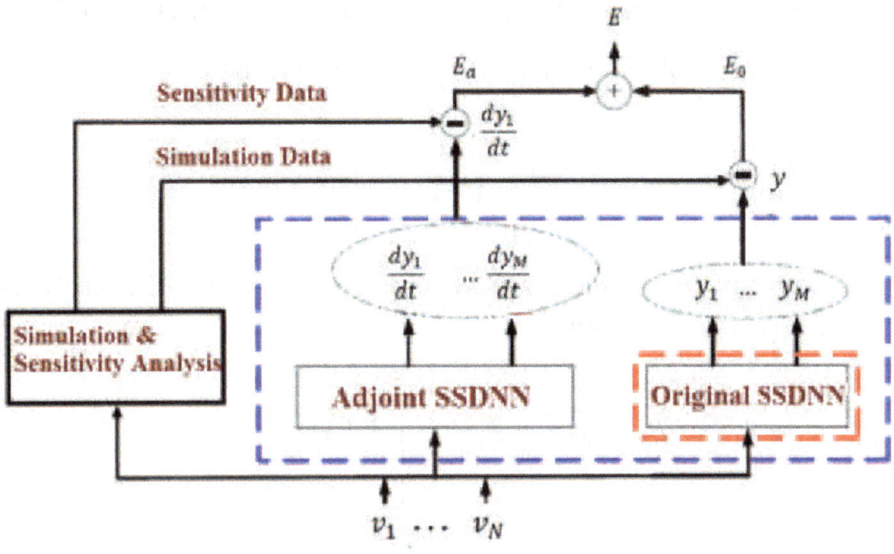

Figure 13. Structure of ASSDNN-based model.

$$\left\{ \begin{array}{c} \dot{x}(t) = -x(t) + \eta g_{\text{ANN}}\Big(v(t), x(t), w\Big) \\ y(t) = Bx(t) \\ \dot{y}(t) = B\dot{x}(t) \end{array} \right\} \tag{14}$$

6. Other methods in microwave modeling

There are other methods which are not based on neural networks for modeling microwave components such as Krylov method [25], finite element [26], mode-matching method [27], vector-fitting method, and so forth [28]. In the preceding section, we present vector-fitting method. This technique has been used in many microwave simulations and modeling researches [29–31].

6.1. Vector-fitting method

Vector fitting (VF) is a robust numerical technique for rational approximation of transfer functions and *s*-parameter in the frequency domain, especially in microwave devices using poles and residues [32]. It allows calculating multiport models directly from measured or computed frequency responses. The resulting approximation has guaranteed stable poles that are real or come in complex conjugate pairs, and the model can be converted directly into a state-space model [13].

Basically, vector fitting is a pole relocation method where the poles are improved in an iterative manner. This is achieved by repeatedly solving a linear problem until convergence is achieved [13]. The VF formulation avoids the ill-conditioning problems encountered with some alternative approaches, as the formulation is given in the form of simple fractions instead of polynomials. Unstable poles are flipped into the left-half plane to enforce stable poles. This makes VF applicable to high-order systems and wide frequency bands [33].

Mathematical representation of vector-fitting method is presented briefly in this section.

Let $\{p_n\}$ be a set of unknown poles, and $\{r_n\}$ be residues, H(s) is the given rational function:

$$H(s) = \left(\sum_{n=1}^{M} \frac{r_n}{s - p_n} \right) \tag{15}$$

in which M is the order of the macromodel. The poles are identified by solving the linear problem shown in Eq. (16) for i^{th} iteration,

$$\sum_{n=1}^{M} \frac{r_n^i}{s - p_n^i} \approx \left(\left(\sum_{n=1}^{M} \frac{\gamma_n^i}{s - p_n^i} \right) + 1 \right) H(s) \tag{16}$$

in which $r_n^{\ i}$ is r_n for i^{th} iteration, the same is $p_n^{\ i}$, and $\gamma_n^{\ i}$ is found in matrix x. In Eq. (6), we call $\left(\left(\sum_{n=1}^{M} \frac{r_n^i}{s - p_n^i} \right) + 1 \right)$ as σ^i(s), and $\sum_{n=1}^{M} \frac{r_n^i}{s - p_n^i}$ as σH^i(s) which are unknown rational functions with given poles. By writing Eq. (16) for several frequency points, we have an overdetermined linear problem with a frequency-sampled data point f_t:

$$A_t x = H_t(f_t), \tag{17}$$

in which t = 1, 2, …, N_f, $A_t = \left[\frac{1}{s+p_1^i} \cdots \frac{1}{s+p_n^i} 1 \frac{-H(f_t)}{s+p_1^i} \cdots \frac{-H(f_t)}{s+p_n^i} \right]$, and $x = [r_1^i \ldots r_N^i 0 \gamma_1^i \ldots \gamma_N^i]$ that can become a linear equation problem that is,

$$[A_1^T A_2^T \ldots A_{N_s}^T]^T x = [H_1 H_2 \ldots H_{N_f}]^T \tag{18}$$

It can be proven that the poles of $H(s)$ are equal to the zeroes of σ^i(s), also the zeroes of σ^i(s) can be calculated by solving an eigenvalue problem as shown in Eq. (19) [33]

$$\varphi = \begin{bmatrix} p_1^i & & & & \\ & p_2^i & & & \\ & & p_3^i & & \\ & & & \ddots & \\ & & & & p_N^i \end{bmatrix} - \begin{bmatrix} 1 \\ 1 \\ 1 \\ \vdots \\ 1 \end{bmatrix} \cdot \begin{bmatrix} \gamma_1^i \\ \gamma_2^i \\ \gamma_3^i \\ \vdots \\ \gamma_N^i \end{bmatrix}^T \tag{19}$$

Also for initialization there are different approaches. Basically, initial poles should be complex with weak attenuation and can be obtained by a simple calculation such as Prony method [13] or simply can be spaced within the desired range of frequency, for example, between 50 Hz and 1 MHz [17], and the advantage here is even if the starting poles were selected poorly, the result does not change significantly [32]. By solving Eq. (19), new set of poles are identified. After identifying all the poles, residues are calculated by solving Eq. (16) which is again a linear problem. As a conclusion, VF method samples the given function with an appropriate sample rate, and in this way a summation of partial fractions can be found, that is, the discrete-function approximation of the original transfer function.

7. Related work

This part of the chapter briefly discusses about the application of computer-aided design (CAD) techniques in modeling and simulation of RF and microwave-passive components. Neural network-based modeling approaches have been widely used for modeling variety of RF and microwave-passive components such as coupled-line filters, coplanar waveguides, Vias and multilayer interconnects, and some other passive components.

7.1. General procedure of modeling

Here, we provide a brief review of procedure used in neural network-based modeling of RF and microwave-passive components.

For modeling microwave components in frequency domain [1], first input and output parameters of the components should be selected in a wide range of frequencies. In most ANN models, it is desired to represent the parameters in terms of scattering parameters (*S*-parameters). The next step is data generation. For passive component models, electromagnetic simulation approach is widely used for generating data. EM simulators, which are used in the process of developing ANN models, produce *S*-parameter for the components. After data training, there should be a criterion for deliberation of the accuracy of the model, so error of the model in different formulations is measured. In most EM-ANN models, the absolute average and standard deviation of error is measured for each output.

After training and verification of the EM-ANN model, based on the usage they can be used either in stand-alone mode or in integrated mode along with microwave circuit simulators. In integrated mode, there is a linear model subroutine that connects models to the simulator. This subroutine returns *S*-parameter of a component back for further simulation. When the simulation is running, the simulator passes parameters such as frequency of a component. For computation of *S*-parameter, there is a feed-forward ANN subroutine that receives input variables and also it holds the algorithm for finding the output of the ANN model. Besides, models can be connected to the circuit simulator as a group, where these collections of models are called libraries [34].

7.2. Parametric modeling of a coupled-line filter

In this example, we demonstrate the use of ANN techniques to develop a model for a family of coupled-line filters [35]. Here, S_1 and S_2 are the spacing between lines and D_1, D_2, D_3 are offset distances from the ends of each coupled lines to the corresponding fringes. This model has six inputs that are $x = \{S_1, S_2, D_1, D_2, D_3, \omega\}$ and four outputs as *S*-parameters RS_{11}, IS_{11}, RS_{12}, IS_{12} which are real and imaginary parts of *S*-parameters S_{11} and S_{12}. The testing and training data were obtained from CST microwave studio [36]. **Table 1** shows the final testing and training results after developing the ANN model for coupled-line filter. In the table, 6-40-4 as neural network structure means a neural network with four inputs, 40 hidden neurons, and four outputs that was used to develop the model for this component. As it can be seen from the table, the ANN model matches the desired data obtained from the simulation tool with very good accuracy, which validates the usage of this method for model creation.

Model type	Neural network structure	Average training error (%)	Average testing error (%)
ANN model using 120 sets of training data	6-40-4	0.897	0.989
ANN model using 40 sets of training data	6-35-4	1.073	4.357

Table 1. Final testing and training results of ANN model (from Ref. [35]).

7.3. EM-ANN models for CPW components

The use of coplanar waveguides (CPWs) in RF and microwave-integrated circuits has brought many advantages. Accurate modeling of these components is necessary for accurate simulation of circuits. One of the fields that experts recently have been working on is toward the development of accurate and efficient methods for EM simulation of CPW discontinuities, but the challenge of using these tools for iterative CAD and circuit optimization [37] is the time-consuming nature of EM simulation. To overcome this problem, EM-ANN models have been suggested [38]. The models include CPW transmission line, short- and open-circuit stubs, step-in width discontinuities, and T-junctions. These EM-ANN models are linked to microwave circuit simulators and allow for the accurate and very fast EM circuit optimization in the framework of circuit simulator [1]. A general schematic of a coplanar waveguide is shown in **Figure 14**. In this figure, *W* is the center conductor width, *G* is the spacing between conductor and ground plane, and *L* is the center conductor length.

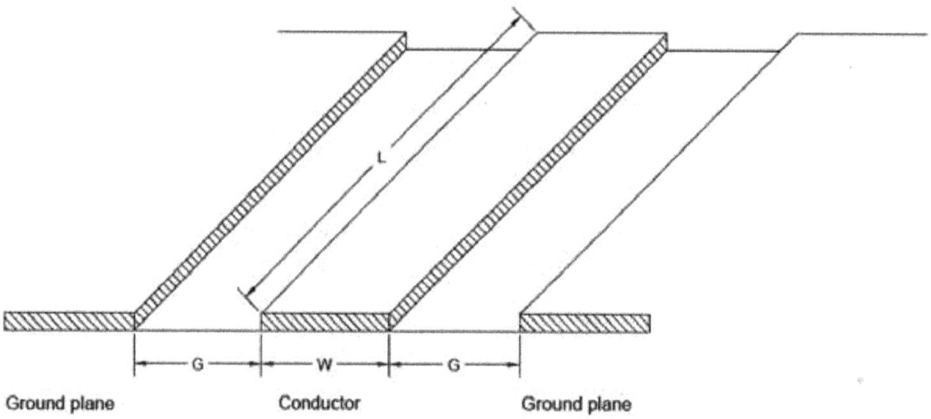

Figure 14. Simple coplanar waveguide.

7.4. Vias elements in microstrip circuits and multilayer Vias connectors

Progress in technology caused merging large number of microwave circuits and creating multilayer complexities that leads to investing much effort on optimizing and lowering the cost and weight of these circuits. Besides, accuracy and efficiency are important factors that should be satisfied in designs to have desirable simulation results. EM-ANN-based methodology despite other solutions that have been suggested was enormously successful in modeling Vias elements in microstrip circuits and multilayer Vias connectors [39]. Some other suggestions had limitations such as heavily computational expenses or limited range of frequency. As an example, microstrip transmission line model is one of the implemented ANN-based models in this case. In this model, input parameters are frequency which is in the range, log $\left(\frac{W_I}{H_{sub}}\right)$ in which W_I is the microstrip width and H_{sub} is the substrate height which varies between -1 and 1, and ε_r relative dielectric constant of the substrate in the range of 2–13. A simple schematic of microstrip transmission line is shown in **Figure 15**.

Output parameter is Z_0 which is the characteristic impedance, and ε_{eff} is the effective dielectric constant. A total of 155 data were used as training, 100 for validation, and 10 hidden neurons to create the model. To test the model, standard deviation and average error criteria were used. Error results for microstrip transmission line are shown in **Table 2**.

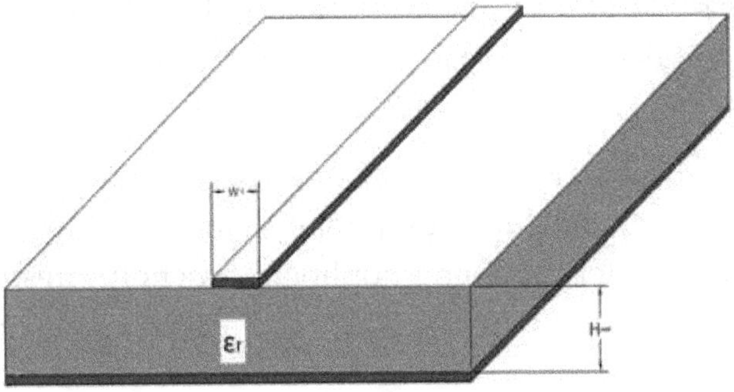

Figure 15. Simple schematic of microstrip transmission line.

	Z_0(%)	τ_{eff}(%)
Training data average error	1.161	0.377
Training data SD	1.157	0.376
Validation data average error	0.774	0.293
Validation data SD	0.875	0.223

Table 2. Average and standard deviation (SD), for absolute error training (from Ref. [1]).

8. Conclusion

In this chapter, a review of some tools commonly used in RF/microwave simulation and modeling has been presented. In the last few decades, high-frequency effects have become an important factor in RF/microwave area. These effects can be found in all levels of design from tiny chips to packaging structures. In order to capture these effects, it is common to use physics-based models or electrical models which lead to large equations and large computational efforts for solving and simulating them, which is extremely time-consuming and expensive. Artificial neural networks recently have become popular among computer-aided design tools. The main topic in this chapter was a discussion on neural network which was mentioned as a powerful tool in modeling and simulation areas, also two main types of neural network structures including static and dynamic neural networks and their different types has been presented. In static neural network section, we talked about multilayer perceptron (MLP) and radial-basis function (RBF) structure, and in time-domain part we discussed recurrent neural network (RNN), dynamic neural network (DNN), state-space dynamic neural network (SSDNN), and adjoint state-space dynamic neural network (ASSDNN) methods. Other than neural network, as mentioned already, there are several numerical methods that are being used in the procedure of simulation and modeling microwave components such as Krylov method, finite-difference time-domain (FDTD), finite-element time-domain (FEDT), and vector fitting (VF). Here, we presented vector-fitting method that is widely used for modeling microwave and electromagnetic components with good performance. VF despite other system identification methods avoids ill-conditioning calculation, and because of this, it works more efficiently. Also, this method is very robust; it performs well even for high-order fitting and does not disturb by poorly selected starting poles. VF technique is very easy to implement in a computer program, since it is constructed upon matrices from simple fractions, and the problems in this case are easy to solve.

As a conclusion, the ANN-based methodologies and other mentioned methods are capable of applying to RF/microwave modeling and components simulation and are shown to have both speed and accuracy advantage for modeling nonlinear functions, despite many other conventional techniques.

Author details

Sayed Alireza Sadrossadat* and Farnaz Mohammadi

*Address all correspondence to: Alireza.sadr@ece.ut.ac.ir

Department of Electrical and Computer Engineering, University of Tehran, Tehran, Iran

References

[1] Zhang Q. J., Gupta K. C. Neural Networks for RF and Microwave Design. Boston: Artech House; 2000. 369 p.

[2] Zhang Q. J., Gupta K. C., Devabhaktuni V. K. Artificial Neural Networks for RF and Microwave Design – From Theory to Practice. IEEE Trans. Microwave Theory Tech. 2003; 51(4): 1339–1350.

[3] Kabir H., Zhang L., Yu M., Aaen P. H., Wood J., Zhang Q. J. Smart Modeling of Microwave Devices. IEEE Microwave Mag. 2010; 11(3): 105–118.

[4] Burrascano P., Fiori S., Mongiardo M. A Review of Artificial Neural Networks Applications in Microwave Computer-Aided Design. Int. J. RF Microwave CAE. 1999; 9: 158–174.

[5] Steer M. B., Bandler J. W., Snowden C. M. Computer-Aided Design of RF and Microwave Circuits and Systems. IEEE Trans. Microwave Theory Tech. 2002; 50(3): 996–1005.

[6] Gupta K. C. ANN and Knowledge-Based Approaches for Microwave Design. In Directions for the Next Generation of MIMIC Devices and Systems, N. K. Das and H. L. Bertoni, Eds. NY: Plenum. 1996; 389–396.

[7] Engelbrecht A. P. Computational Intelligence: An Introduction. John Wiley & Sons, New York; 2007[23]. 581 p.

[8] Hornik K., Stinchcombe M., White H. Multilayer Feed-Forward Networks Are Universal Approximators. Neural Netw. 1989; 2: 359–366.

[9] Watson P. M., Gupta K. C. EM-ANN Models for Microstrip Vias and Interconnects in Dataset Circuits. IEEE Trans. Microwave Theory Tech. 1996; 44(12): 2495–2503.

[10] Sen P., Woods W. H., Sarkar S., Pratap R. J., Dufrene B. M., Mukhopadhyay R., Lee C., Mina E. F., Laskar J. Neural-Network-Based Parasitic Modeling and Extraction Verification for RF/Millimeter-Wave Integrated Circuit Design. IEEE Trans. Microwave Theory Tech. 2006; 54(6): 2604–2614.

[11] Nihad D., Jehad A., Amjad O. CAD Modeling of Coplanar Waveguide Interdigital Capacitor. Int. J. RF Microwave Computer-Aided Eng. 2005; 15(6): 551–558.

[12] Xu J. J., Yagoub M. C. E., Ding R., Zhang Q. J. Neural-Based Dynamic Modeling of Nonlinear Microwave Circuits. IEEE Trans. Microw. Theory Tech. 2002; 50(12): 2769–2780.

[13] Lei C., Wang Y., Chen Q., Wong N. On Vector Fitting Methods in Signal/Power Integrity Applications. Proc. Int. Multi-Conf. Eng. Comp. Scient. 2010; 2: 1407–1412.

[14] Haykin S., Lippmann R. Neural Networks, a Comprehensive Foundation. International Journal of Neural Systems; 1994. 823 p.

[15] Karlik B., Vehbi O. A. Performance Analysis of Various Activation Functions in Generalized MLP Architectures of Neural Networks. Int. J. Artificial Intellig. Expert Syst. 2011; 1 (4): 111–122.

[16] Dorffner G. Unified Frameworks for MLP and RBFNs: Introducing Conic Section Function Networks. Cybern. Syst. 1994; 25: 511–554.

[17] Gustavsen B. Improving the Pole Relocating Properties of Vector Fitting. IEEE Trans. Power Deliv. 2006; 21(3): 1587–1592.

[18] Sinha N. K., Gupta M. M., Rao D. H. Dynamic Neural Networks: An Overview. Proc. IEEE Int. Conf. Industr. Technol. 2000; 2: 491–496.

[19] Bertsekas D. P., John N. T. Neuro-Dynamic Programming: An Overview. Proc. 34th IEEE Conf. 1995; 1: 560–564.

[20] Narendra K. S., Parthasarthy K. Identification and Control of Dynamical Systems Using Neural Networks. IEEE Trans. Neural Netw. 1990; 1(1): 4–27.

[21] Wu C., Nguyen M., Lima J. On Incorporating Finite Impulse Response Neural Network with Finite Difference Time Domain Method for Simulating Electromagnetic Problems. IEEE AP-S Antennas Propag. Int. Symp. 1996; 3: 1678–1681.

[22] Aweya J., Zhang Q. J., Montuno D. A Direct Adaptive Neural Controller for Flow Control in Computer Networks. IEEE Int. Conf. Neural Netw. (Anchorage, Alaska). 1998; 1: 140–145.

[23] Cao Y., Ding R., Zhang Q. J. State-Space Dynamic Neural Network Technique for High-Speed IC Applications: Modeling and Stability Analysis. IEEE Trans. Microwave Theory Techn. 2006; 54(6): 2398–2409.

[24] Sadrossadat S. A., Gunupudi P., Zhang Q. J. Nonlinear Electronic/Photonic Component Modeling Using Adjoint State-Space Dynamic Neural Network Technique. IEEE Trans. Comp. Packaging Manufact. Technol. 2015; 5(11): 1679–1693.

[25] Chaniotis D., Pai M. A. Model Reduction in Power System Using Krylov Subspace Methods. IEEE Trans. Power Syst. 2005; 20: 888–894.

[26] Singh T. Design and Finite Element Modeling of Series-Shunt Configuration Based RF MEMS Switch for High Isolation Operation in K-Ka Band. J. Comput. Electr. 2015; 14(1): 167–179.

[27] Kordiboroujeni Z., Bornemann J., Sieverding T. Mode-Matching Design of Substrate-Integrated Waveguide Couplers. Asia-Pacific Symp. Electromag. Compatib. Singapore. 2012; 1: 701–704.

[28] Bai Z. Krylov Subspace Techniques for Reduced-Order Modeling of Large-Scale Dynamical Systems. Appl. Numer. Math. 2002; 43(1): 9–44.

[29] Li E. P., Liu E. X., Li L. W., Leong M. S. A Coupled Efficient and Systematic Full-Wave Time-Domain Macromodeling and Circuit Simulation Method for Signal Integrity Analysis of High-Speed Interconnects. IEEE Trans. Adv. Packag. 2004; 27(1): 213–223.

[30] Antonini G. SPICE Equivalent Circuits of Frequency-Domain Responses. IEEE Trans. Electromagn. Compat. 2003; 45(3): 502–512.

[31] Cai Y., Mias C. Faster 3D Finite Element Time Domain-Floquet Absorbing Boundary Condition Modeling Using Recursive Convolution and Vector Fitting. IET Microwaves Antennas Propag. 2009; 3(2): 310–324.

[32] Gustavsen B., Semlyen A. Rational Approximation of Frequency Domain Responses by Vector Fitting. IEEE Trans. Power Deliv. 1999; 14(3): 1052–1061.

[33] Semlyen A., Gustavsen B. Vector Fitting by Pole Relocation for the State Equation Approximation of Nonlinear Transfer Matrices. Circuits Syst. Signal Process. 2000; 19(6): 549–566.

[34] Cao Y., Wang G., Gunupudi P., Zhang Q. J. Parametric Modeling of Microwave Passive Components Using Combined Neural Networks and Transfer Functions in the Time and Frequency. Int. J. RF Microwave Comp. Aid. Eng. 2013; 23: 20–33.

[35] Sadrossadat S. A., Cao Y., Zhang Q. J. Parametric Modeling of Microwave Passive Components Using Sensitivity-Analysis-Based Adjoint Neural-Network Technique. IEEE Trans. Microwave Theory Techn. 2013; 61(5): 1733–1747.

[36] CST Microwave Studio. CST AG, Darmstadt, Germany, 2010 [Online]. Available at: http://www.cst.com.

[37] Doerner R., et al. Modelling of Passive Elements for Coplanar SiGe MMIC's. Microwave Symp. Digest IEEE MTT-S Int. Orlando, FL, USA. 1995; 3: 1187–1190.

[38] Watson P. M., Kuldip C. G. Design and Optimization of CPW Circuits Using EM-ANN Models for CPW Components. IEEE Trans. Microwave Theory Techn. 1997; 45(12): 2515–2523.

[39] Cao Y., Zhang Q. J. Neural Network Techniques for Fast Parametric Modeling of Vias on Multilayered Circuit Packages. IEEE Electr. Design Adv. Pack. Syst. Symp. 2010; 1: 1–4.

10

Manifold Multiplexer

Imran Mohsin

Abstract

In wireless communications, bandwidth is a valuable resource that can be smartly shared by multiple users simultaneously utilizing multiplexers. This chapter offers a short review and brief impression of the working principle and the design methodology of the multiplexers in RF and microwave systems. Predominantly used different multi-plexer design patterns are discussed here, however the compact manifold multiplexer is discussed in details with an example. It is designed by using advanced design system (ADS) software and implemented utilizing Microstrip technology for its low cost and simplicity.

Keywords: manifold multiplexer, diplexer, microwave filters, advanced design system, microstrip technology

1. Introduction

The idea of multiplexing has been used in different areas. In telegraph systems, it was used for the first time in 1870s, after few decades it was employed in telephony [1]. Multiplexing is a useful process applied in different wireless communication systems where signals from different users (channels) can be multiplexed or demultiplexed by utilizing the multiplexers in order to use the primary recourse (frequency spectrum) intelligently and to increase the speed of the data transmission. **Figure 1** shows the basic block diagram of the transmitter-receiver pair in the wireless communication system. It consists of antenna, RF front-end and baseband system. Usually, the RF front-end is the set of circuit component after antenna which down converts the RF frequency into the intermediate frequency for further processing in the baseband [2]. The well-known components in the front-end are low noise amplifier (LNA), band pass filter (BPF), power amplifier (PA), local oscillator (LO) and mixers. Microwave multiplexers can also be found in front-end [3–5].

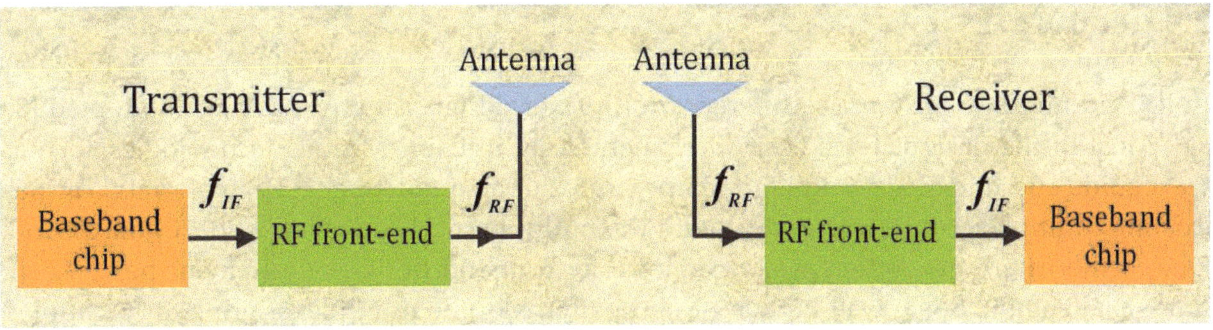

Figure 1. Basic block diagram of wireless communication system.

Frequencies more than 1 GHz are typically known as microwave frequencies in which most of the systems operate [3]. High-frequency circuit designs are different compared to the low-frequency circuit design because at high frequency, wave nature of current is needed to be considered (it will be further discussed in microstrip transmission line technology). After the announcement of a wide range of ultra-wide band (UWB) for public use in 2002 by Federal Communication Commission (FCC), many designs dealing with high frequency and wide bandwidth became centre of attraction [4]. With the evolution of the technology, design and implementation of multiplexing networks become more sophisticated. **Figure 2** shows the scheme of sharing the bandwidth among N channels using multiplexing.

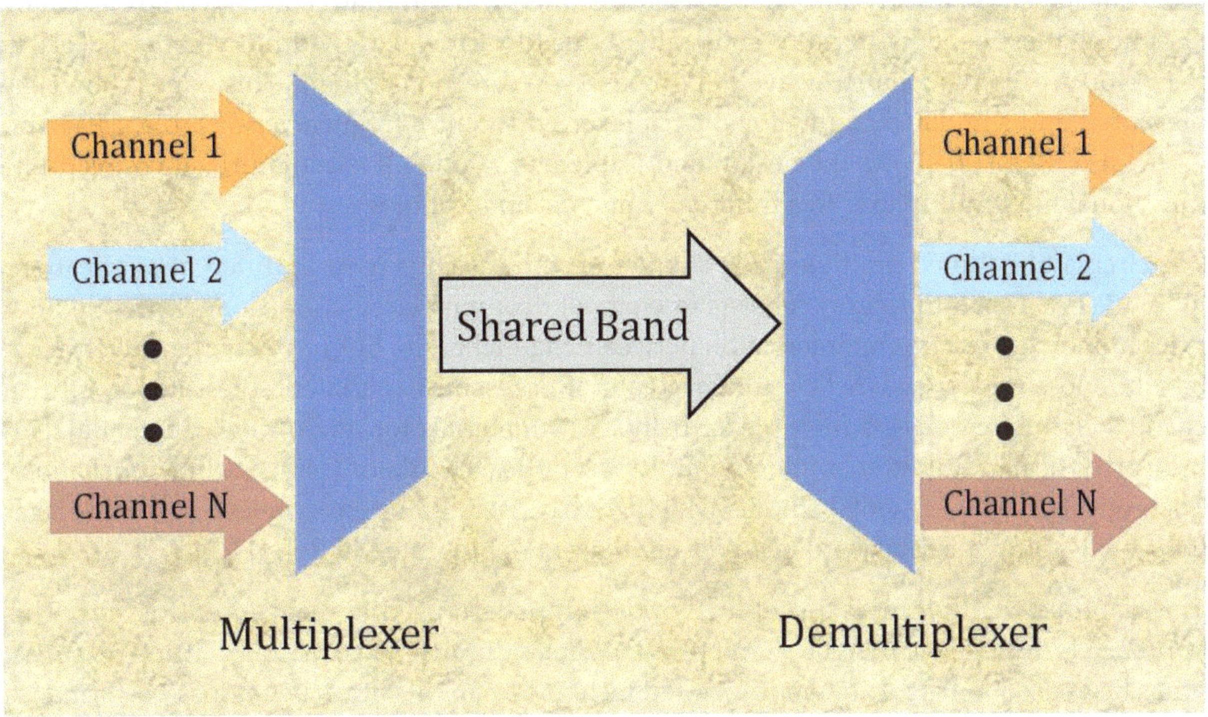

Figure 2. N channels multiplexing to share the bandwidth.

2. Literature review

In RF and microwave systems, different multiplexer design patterns have been adopted [3–24]. Most of the designs have been implemented by utilizing various technologies such as waveguide and microstrip, and high-temperature superconducting (HTS) materials. Various existing multiplexers are able to deal with the multiple sub-bands with a short guard band. Due to the future generation system, amenable front-end design requirement has become more challenging. Some well-known design patterns of the multiplexer are hybrid-coupled multiplexer, directional filter multiplexer, circulator coupled multiplexer and manifold multiplexer [5–7].

The hybrid-coupled multiplexer design pattern has been adopted by many researchers [8–10]. This design approach is easy to tune and there is no interaction between channel filters, so it can accommodate any change in channel frequencies or addition of new channel. Hybrid multiplexers are comparatively large due to the presence of two hybrid and two filters for each channel. Jonathan et al. [8], designed hybrid coupled multiplexer using microstrip technology. Mansour et al. [9] studied the possibility of building HTS using hybrid-coupled multiplexers. Rubin [10] represented the frequency multiplexer with the hybrid-coupled configuration by two-port analysis.

In another design pattern of multiplexer, four port directional filters are used [6, 11, 12]. By keeping one port terminated the other three ports are used for incident input, filtered output and reflected signals. Unlike hybrid-coupled multiplexers, it uses one filter per channel which can be designed separately and can be easily modified without changing the whole design [6]. Despite all these benefits, its use is constrained due to narrow bandwidth. Volker et al. [11], in their study, presented a microwave design based on directional filters. This paper shows the measured and simulated results for the triplexer with centre frequencies of 1.472, 2.944 and 4.416 GHz. Ref. [12] shows the design and implementation of the miniaturized coupled-line directional filter multiplexer in a multilayer microstrip technology.

The circulator-coupled multiplexer is another design which is used in different applications [6, 7, 13, 14]. This multiplexer consists of channel dropping circulators along with one filter per channel. It provides no interaction between channel filters but have comparatively high insertion loss in later channel circulator. Raafat [7] presents experimental results for a three-channel circulator-coupled multiplexer using HTS materials for microwave. Chen et al. [13] used an optical multiplexer employing multiport optical circulator (MOC) using optical fibre. Theoretical and experimental studies which were carried out upon a waveguide multiplexer that consists of two branches joined by a circulator are presented in Ref. [14].

The manifold multiplexer is a well-known design pattern used to realize multiplexers. This approach is the focus of this chapter. It provides better insertion loss and amplitude response. It has been used in HTS thin film integrated technology and it shows better loss performance but is sensitive to the material defects [15]. On the other hand, tuning this multiplexer is little complex because of its need to deal with all filters at the same time. So, it is not flexible to frequency changes or additional channel frequency and in the case of any modification in frequency arrangement, the whole multiplexer need to be redesigned.

This topology has been presented in many research papers using waveguides and microstrip technology. Morinil et al. [15] presented an improvement in the dual manifold multiplexer to design the reconfigurable multiplexer. Ho and Battensby [16] discuss the application of microwave active filter manifold in multiplexing. Ten-channel manifold multiplexer by using waveguide is presented with contiguous and non-contiguous channels [17]. In Ref. [18], the theory of manifold multiplexer using helical filter is discussed. Although the reconfigurable filter technology still not fully developed, a theoretical study of tunable manifold multiplexer has been carried out with five channels [19].

Figure 3 illustrates the most adopted structural patterns of the manifold multiplexer with one filter for each channel and these are the most compact design approach patterns [6, 7]. Several manifold multiplexer designs with all filters on the same side are presented in **Figure 3(a)** [20]. An alternate filter design as shown in **Figure 3(b)** is presented in Refs. [21, 22]. Both this approaches have been adopted according to requirements. Cameron and Yu [6] classified the manifold multiplexer design patterns into three categories namely comb, herringbone and end-fed (can be applicable in both **Figure 3(a)** and **(b)**, where one filter feed the manifold design).

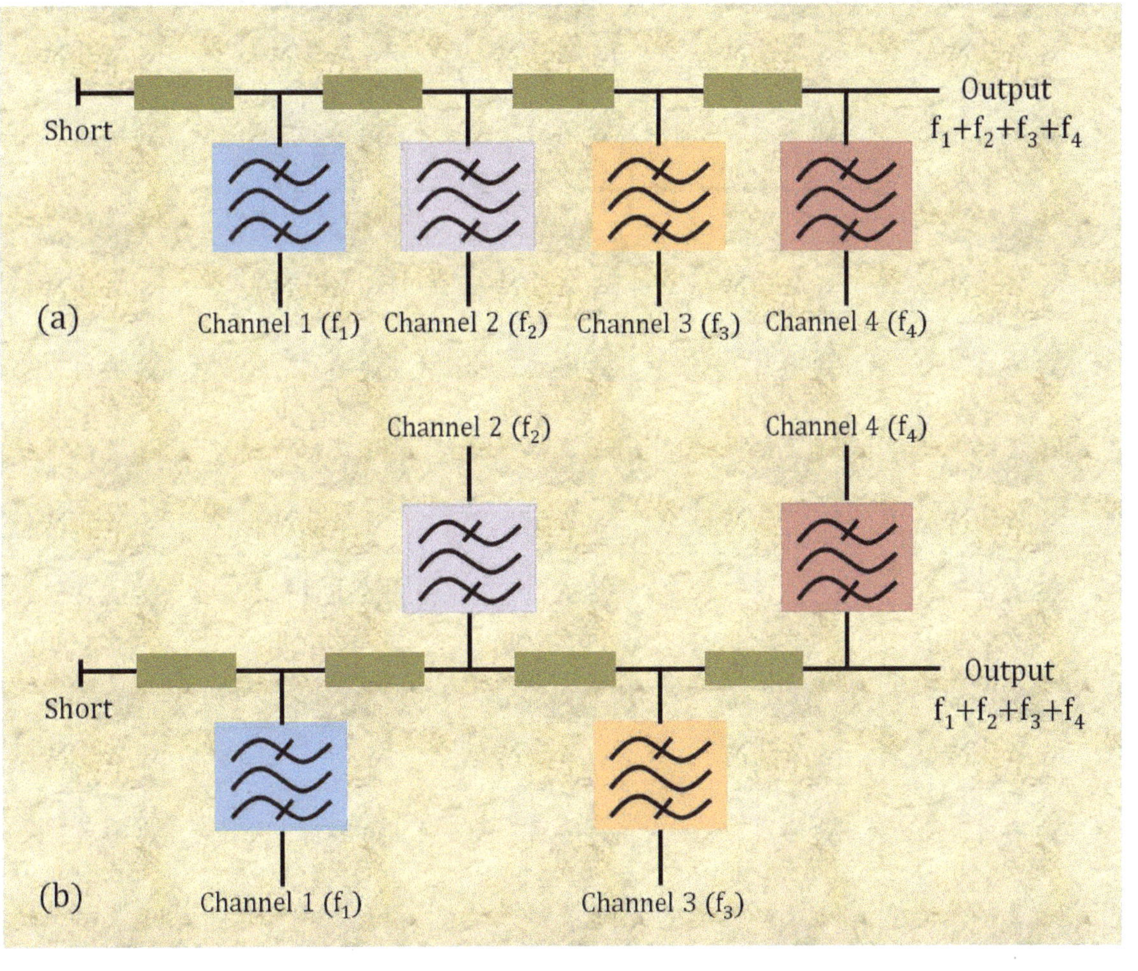

Figure 3. Manifold multiplexer design pattern, (a) all channels on one side, and (b) filters on alternate sides.

Compared to other multiplexer patterns, manifold has no isolation between channels, which make it act as a whole circuit but at the same time, which make it more complicated when the number of channels are increased. However, it gives small losses due to the absence of the lossy isolation. That is why manifold design is the preferred choice of the circuit designer aiming for small losses and miniaturized circuit. With the help of circuit design software, complexities in the design can be overcome.

3. Frequency multiplexing network

Figure 4 shows the working principle of the frequency multiplexing network (FMN) using manifold approach for *N* number of channels with one filter per channel. Filters are interconnected by using quarter wavelength transformer and horizontal transmission line. The

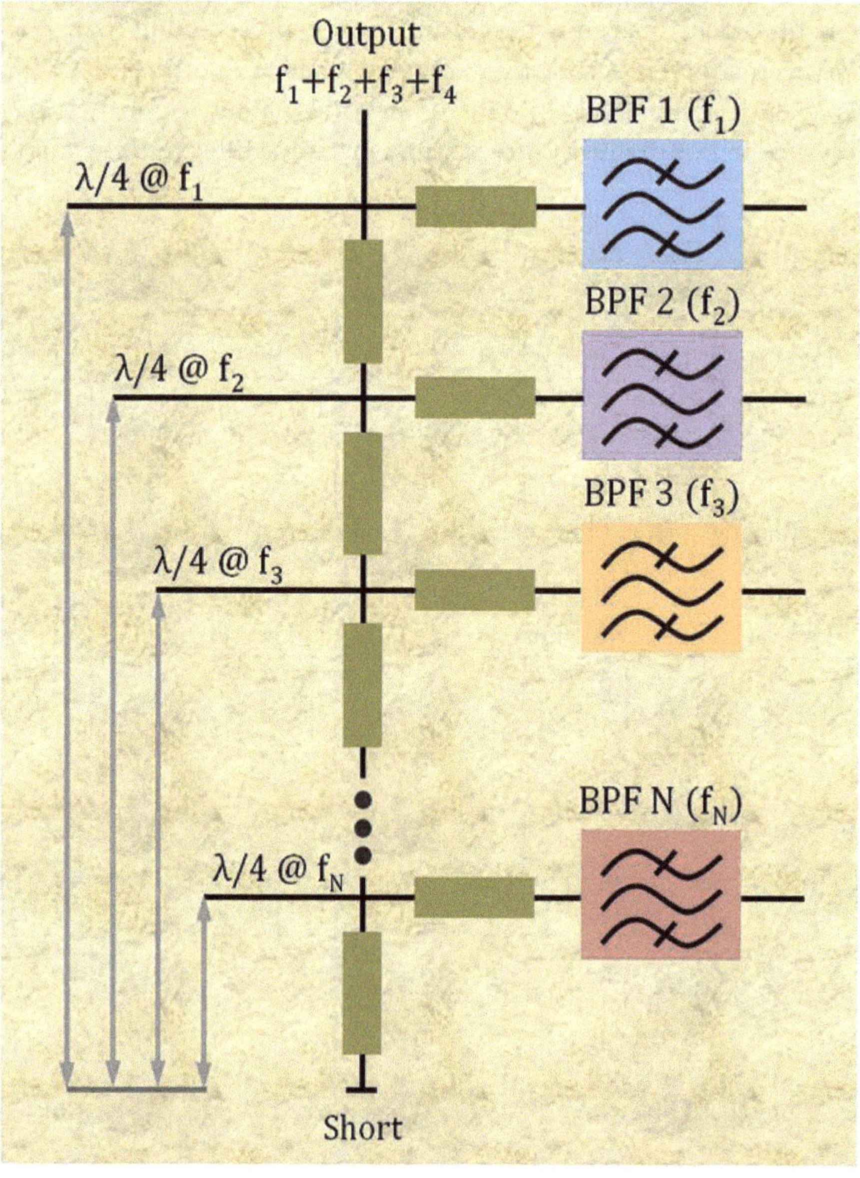

Figure 4. Frequency multiplexing network [23].

available frequency bandwidth is divided into N channels with frequencies $f_1, f_{2, \ldots,} f_N$, where f_1 is lowest frequency band and f_N is the highest frequency band. Quarter wave transmission lines provide the high impedance to the corresponding frequency band, where as horizontal transmission lines help in tuning that band. FMN is a passive network that can be used as a multiplexer in the transmitter and demultiplexer in the receiver. The presence of a suitable guard band can prevent overlapping of the sub-bands. The minimum guard band is always preferred as this band will not be used in transmission. Normally the guard band of less than 10% relative bandwidth is used most of the time [4].

4. Microstrip transmission line technology

Transmission line is a special purpose medium intended to transmit high-frequency signals which have short wavelengths. According to the rule of thumb, signal carrying medium will be considered as the transmission line, if the wavelength of the signal is less than the ten times the circuit component [24]. **Figure 5** shows the microstrip transmission line, which has two conductors separated by the dielectric material of permittivity ε_r and substrate height h. Here L is the length, W is the width and t is the thickness of the microstrip. It can be manufactured utilizing a double-sided printed circuit board (PCB). Input impedance (Z_{in}) of a terminated transmission line can be represented as [24]:

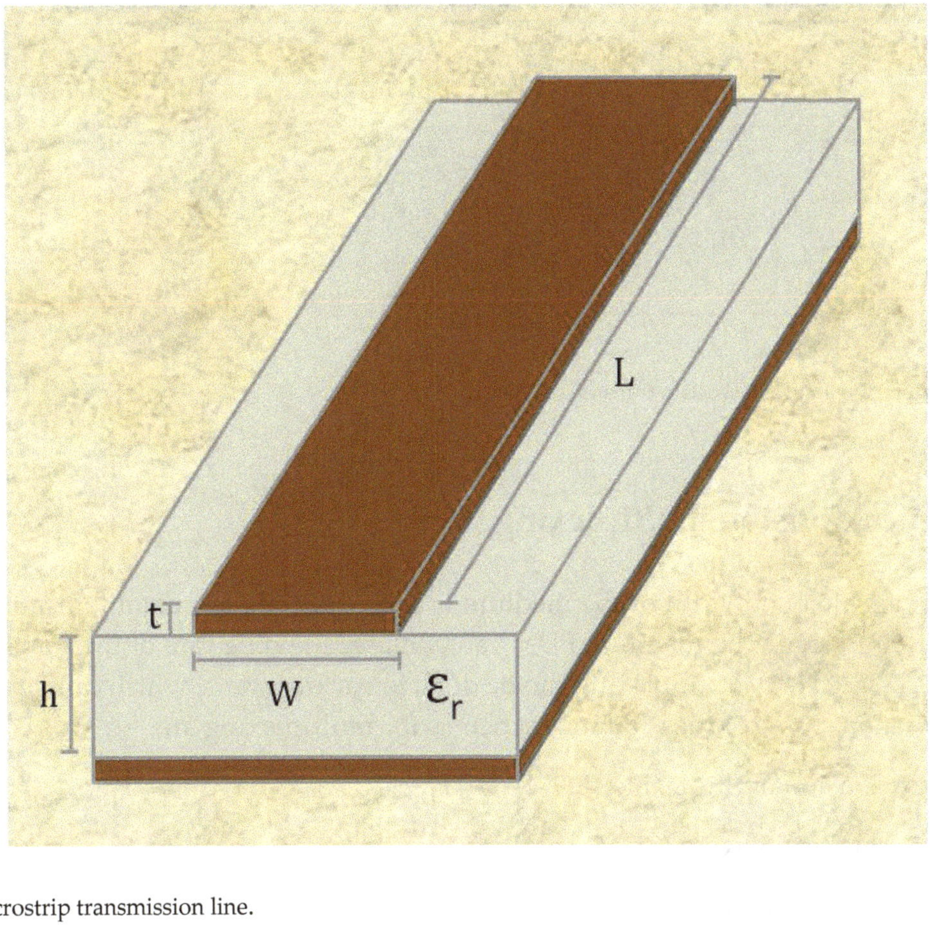

Figure 5. Microstrip transmission line.

$$Z_{in}(d) = Z_o \frac{Z_L + j Z_o \tan(\beta d)}{Z_L - j Z_o \tan(\beta d)} \tag{1}$$

where Z_L is load impedance, Z_o characteristic impedance, d is the length of transmission line and β is the wave number.

In order to achieve the maximum power transfer we need to match the source and load power, which can be done by using a passive network known as matching network. Many types of matching networks are using discrete components as a cheapest solution in circuit designing such as LC network, T network and Pi network in low-frequency applications. A combination of lumped and discrete component matching network can be used for comparatively high frequencies and purely discrete components have been used in high frequencies of the order GHz. Quarter wavelength transmission line, single and double stub are examples of discrete matching networks.

Quarter wavelength transmission line or transformer is a simple matching network. The impedance of the quarter wavelength transmission line can be calculated for matching load and source. **Figure 6** shows the quarter wavelength microstrip transmission line.

From Eq. (1) for (Z_S=Z_L)

$$Z_{in}\left(d = \frac{\lambda}{4}\right) = \frac{Z_o^2}{Z_L} \text{ or } Z = \sqrt{Z_{in} Z_L} \tag{2}$$

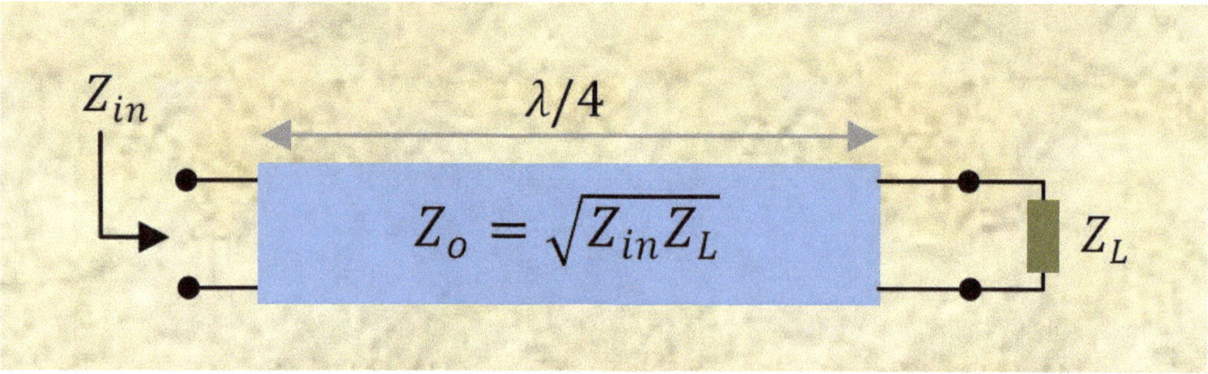

Figure 6. Quarter wavelength microstrip transmission line.

5. Frequency division multiplexing

When the available bandwidth of the medium is more than that of communicating devices, then it is better to share the medium. FDM can enhance the data rate of the transmission by parallel processing [4, 5]. Multiplexing can be done in various ways, which can depend on the target application. Some of these famous schemes for multiplexing are:

- Frequency division multiplexing (FDM)
- Time division multiplexing (TDM)
- Wavelength division multiplexing (WDM)

In wireless communications, different schemes are applied to efficiently use the valuable bandwidth. FDM is mostly used in the radio frequency band. It is the analog multiplexing schemes which divide the available bandwidth into two or more frequency bands (channels). **Figure 7** shows the schematic of frequency division for the multiple channels utilizing FDM. The guard band (narrow portion of the band spectrum which is not used in communication but used to protect sub-bands from interference) is present between each neighbouring channel. One antenna pair is sufficient for the communication of all the users of the sub-bands.

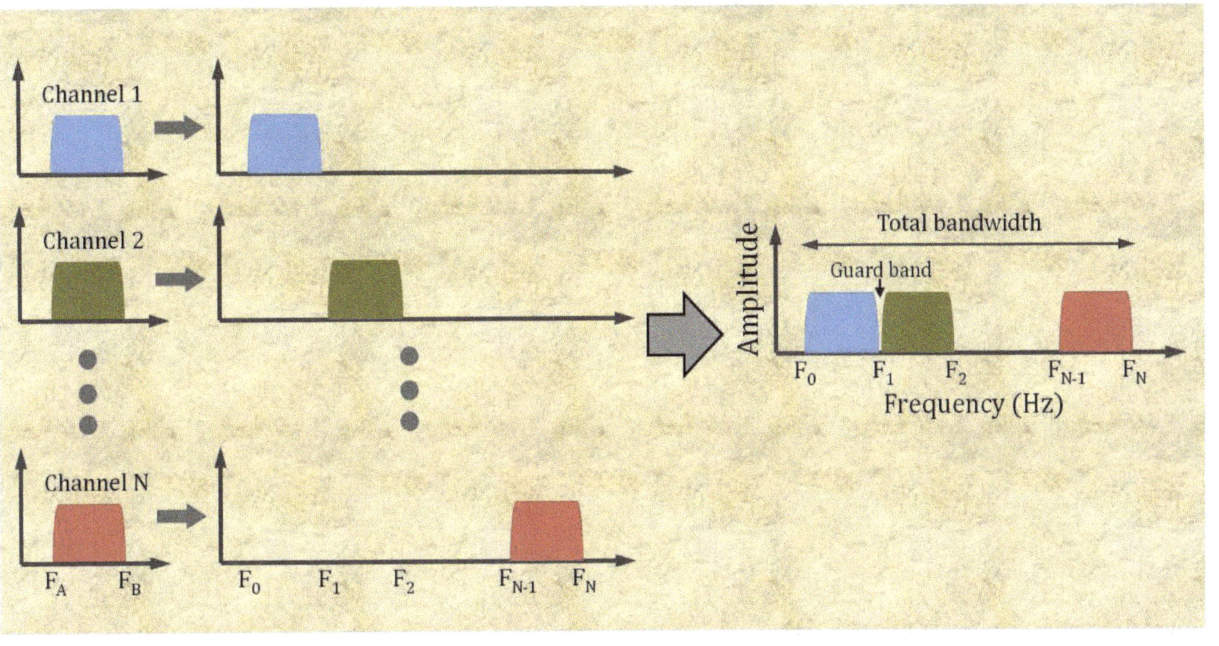

Figure 7. Multiple sub-bands (channels) in available bandwidth using FDM scheme.

6. Design and implementation of manifold multiplexer

This section will give the detailed designing process of the diplexer (an example of manifold multiplexer) using advanced design circuit (ADS) from Agilent's simulation software and implementation on a double-sided PCB. Finally, the prototype will be measured by a two-port vector network analyser.

6.1. Frequency diplexer network (manifold multiplexer)

By using the working principle of the FMN (discussed earlier), a diplexer is designed. **Figure 8** shows the block diagram of the frequency diplexer network (FDN). It has two channels in the UWB frequency range from 6 to 9 GHz. The guard band of 200 MHz is present between these two sub-bands (6–7.4 and 7.6–9 GHz). As discussed earlier all the filters in the manifold design are needed to be tuned at the same time. However, individual channel filters are optimized and then combined with a matching network which is a systematic approach to make the designing process simple.

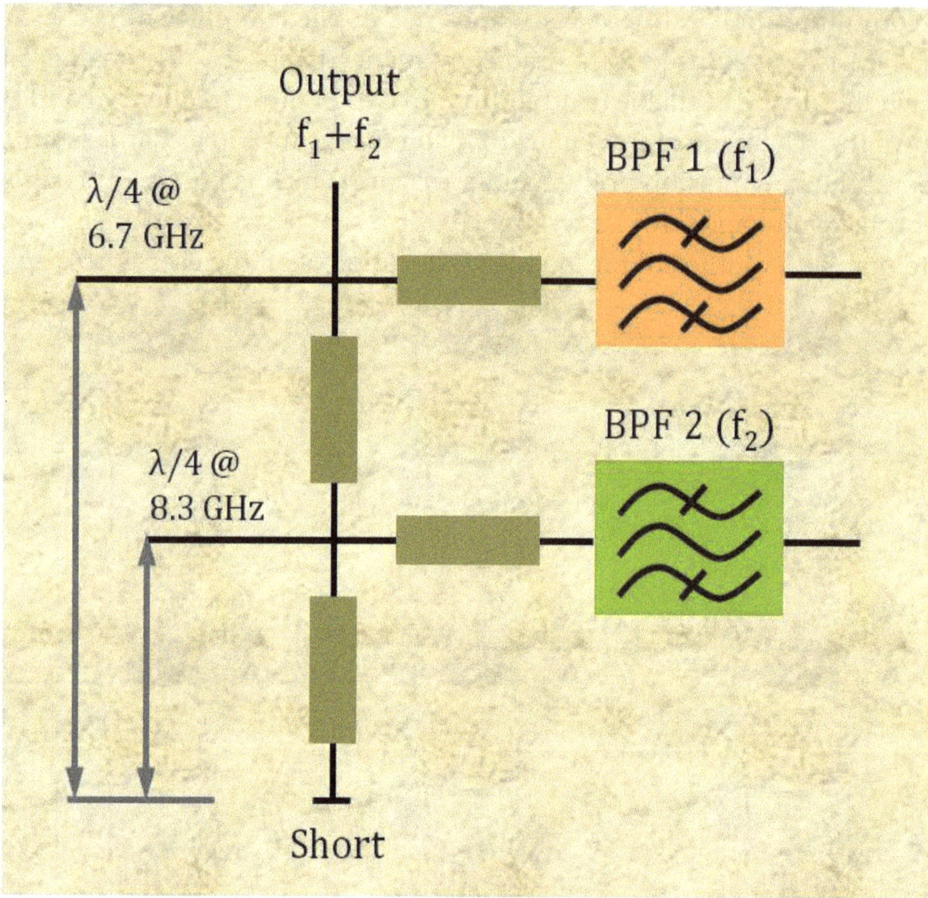

Figure 8. Frequency diplexer network [21].

6.2. Systematic processes

In order to understand the process of design and implementation it is categorised into the four following steps [4].

1. Schematic designing by using selected substrate properties of PCB to show the ideal behaviour.
2. Layout of the schematic design to see the real behaviour by running momentum tool.
3. Analysis of schematic and momentum results together by using layout in schematic for final simulation results.
4. Implementation of the approved design on to the PCB and comparison of the measured results with simulation results.

6.2.1. First step

Rogers 4350B is used in many high frequency electronic circuits. It is a double-sided PCB and selected here for the design implementation. **Table 1** shows the substrate properties of PCB.

Dielectric thickness	254 μm
Relative dielectric constant	3.48
Metal thickness	25 μm
Metal conductivity	58 MS/m

Table 1. Substrate properties of Rogers 4350B.

Diplexer can be divided in three blocks as first channel filter, second channel filter and matching network. **Figure 9** shows the schematic view of the first filter in ADS. MSub and S-parameters show the substrate properties and frequency plan, respectively.

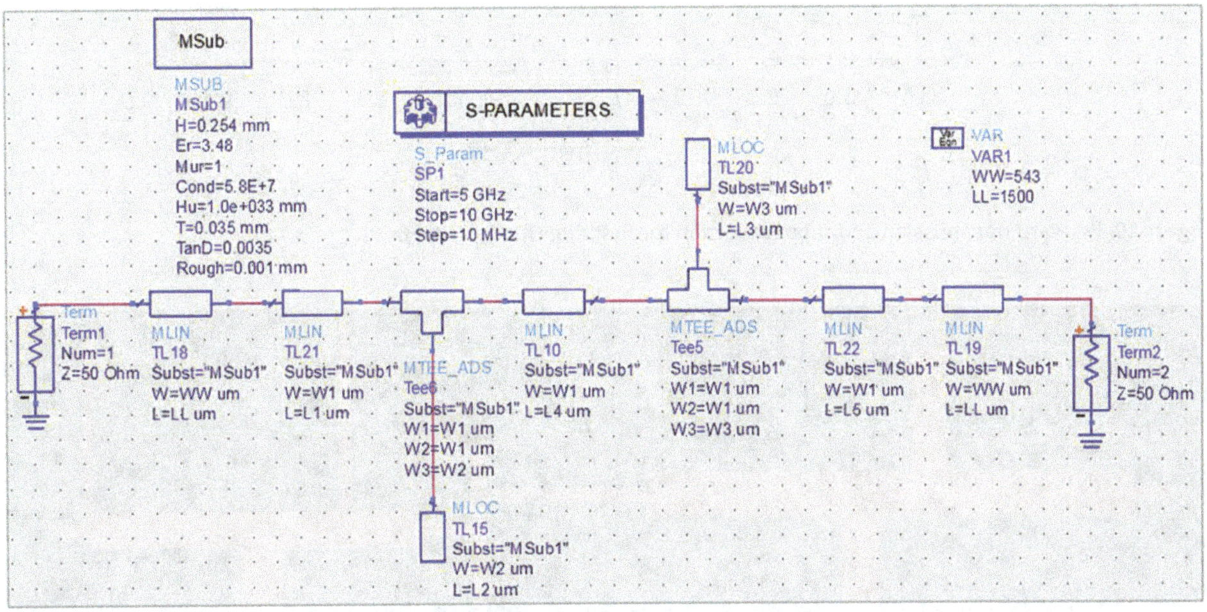

Figure 9. Schematic diagram of first filter for 6–7.4 GHz.

Figure 10 presents the simulated forward transmission and input reflection result for the first filter in S-parameter. In a similar way, schematic of the second filter can be designed. After the optimization of the filters, both can be combined by the transmission line network as shown in the FDN figure to form the diplexers. Now tune the whole circuit together to get the optimization. Different tuning tools are available in ADS which can be used to tune the circuit.

6.2.2. Second step

After getting the optimized diplexer's schematic, second step is to convert the whole schematic into a layout as shown in **Figure 11**. It has three inputs/outputs and one connected to ground, so four ports are assigned to this FDN. Now run momentum, to see the real behaviour of the design which is a simulation engine use for the layout. It has two modes, RF and microwave (microwave mode gives full electromagnetic simulation with radiation effect and RF mode is quick simulation for the designs which do not radiate).

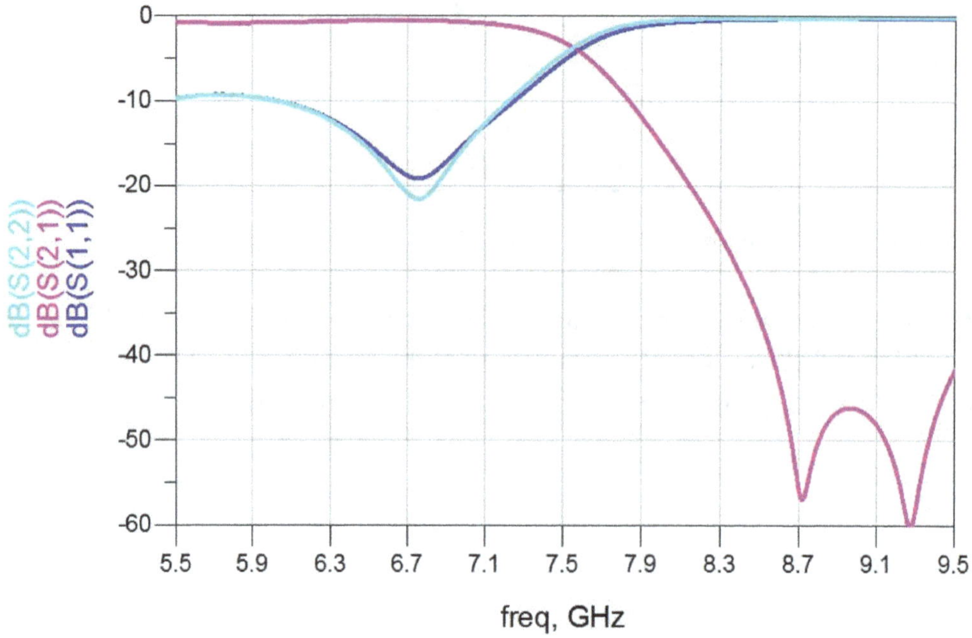

Figure 10. Forward transmission and input reflection for first filter for 6–7.4 GHz.

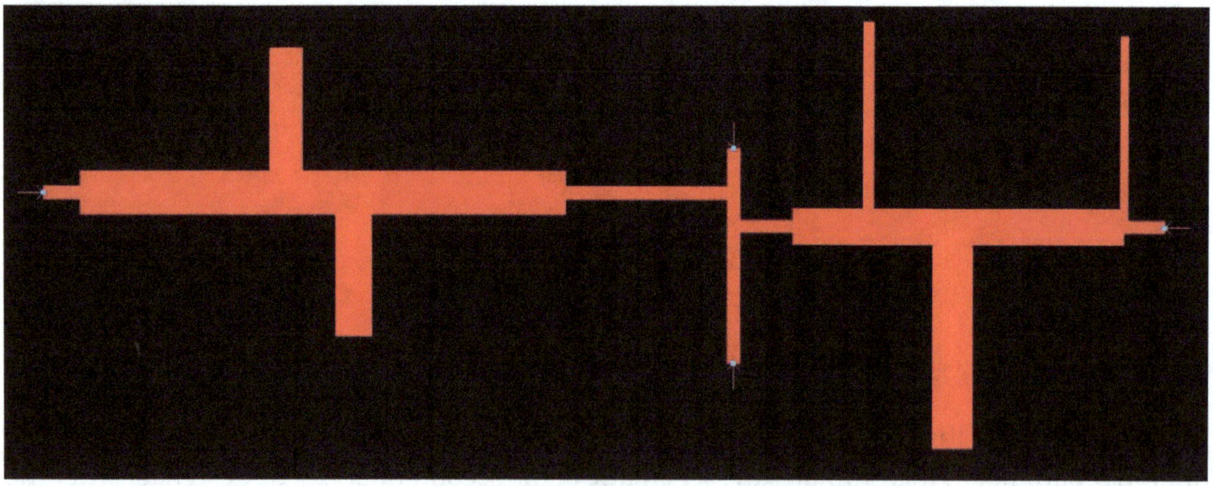

Figure 11. Layout of the FDN.

6.2.3. Third step

Third step is to convert the layout of the FDN into the component and call this component into the schematic as shown in **Figure 12**. Here schematic and layout simulate together and provide final simulation results in terms of ideal and real physical design.

Figure 13 shows the final simulated results of the FDN in terms of the S-parameters. Forward transmission (S21, S31) and input reflection (S11, S22, S33) show flat response in their respective bands jamming others. Appropriate values of group delay can be observed from the results that show small variations.

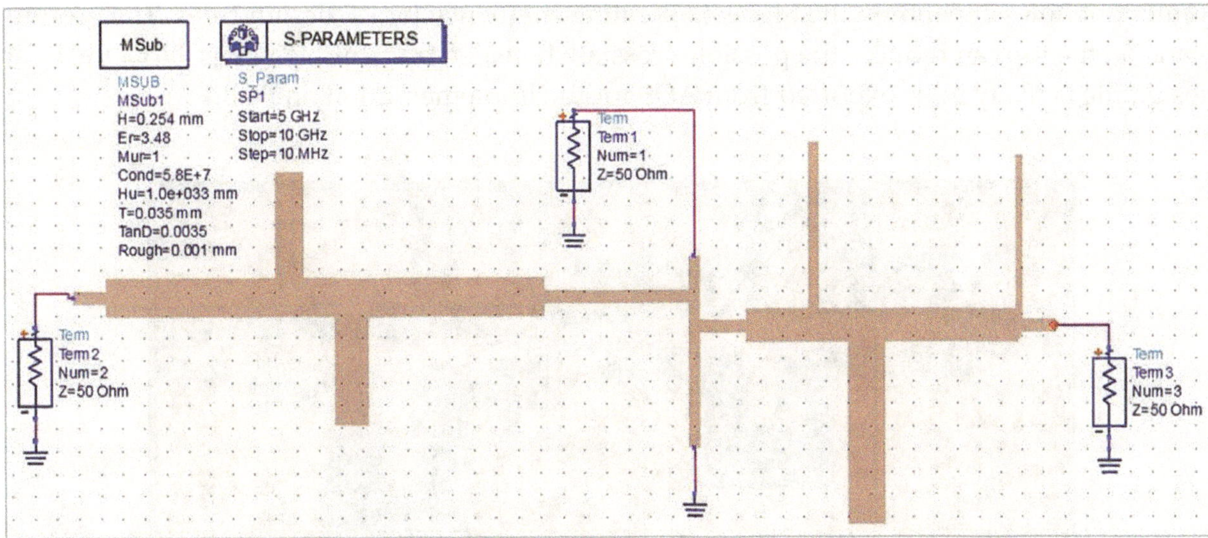

Figure 12. FDN as layout component in schematic representation.

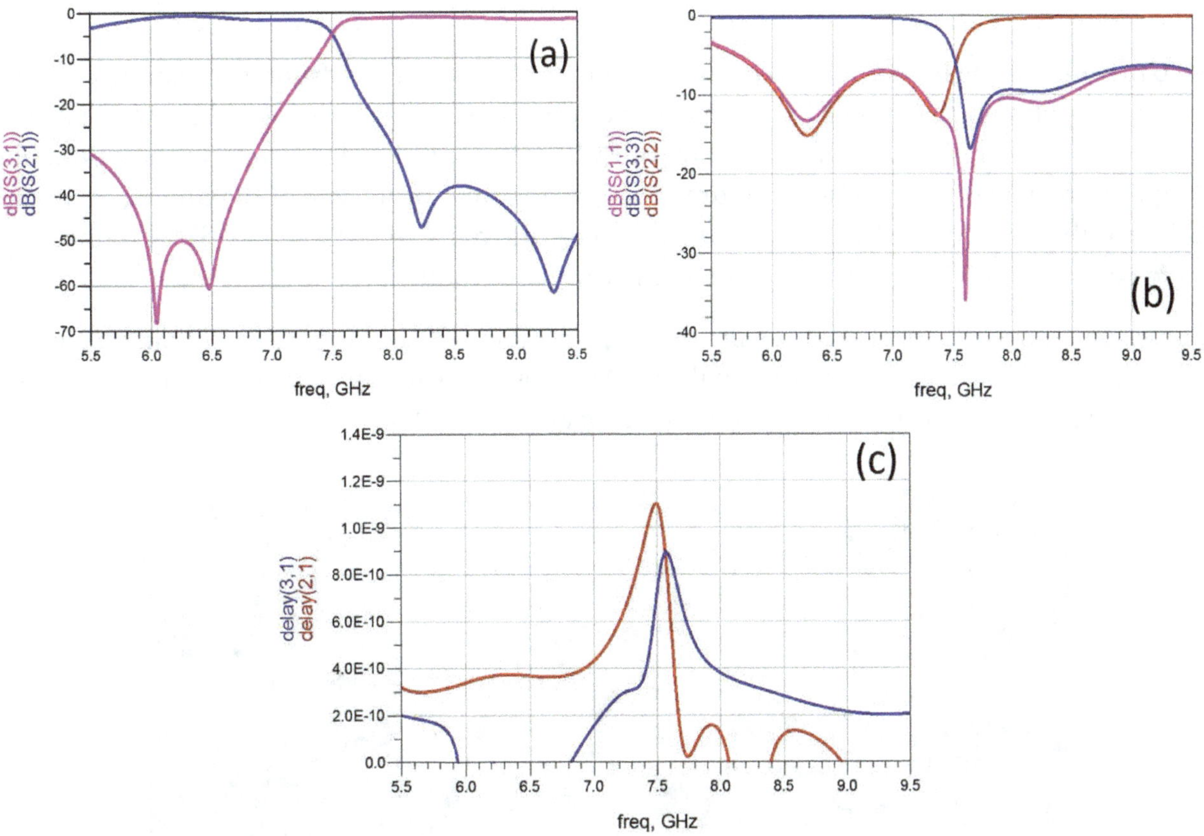

Figure 13. Simulated results for FDN, (a) forward transmission, (b) input reflection, and (c) group delay.

6.2.4. Fourth step

Finally, the layout of the PCB is prepared for the implementation on the double-sided PCB. **Figure 14** shows the top ground plane connected with the bottom ground plane through ground via hole. Small diameter circles are placed in already defined layer for holes. The number of holes depend on the highest operating frequency by RF design rules. The ground plane on the top can reduces the possible crosstalk between transmission lines. Now the final design file is ready to be exported from ADS for the implementation on PCB.

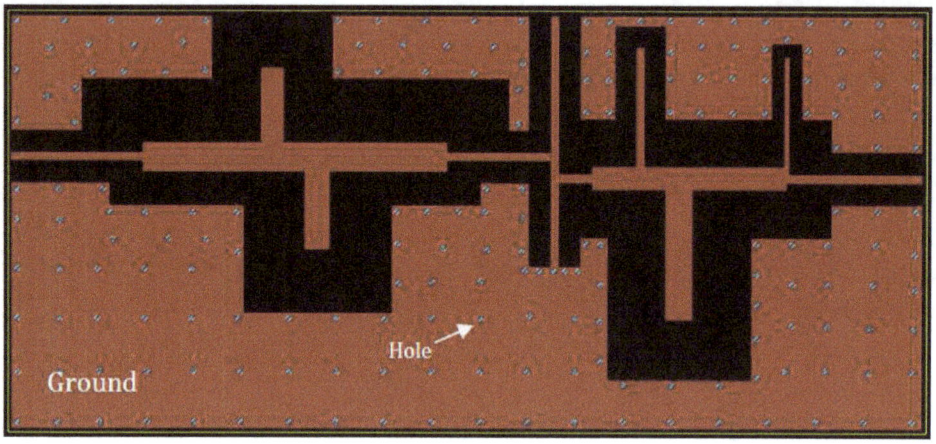

Figure 14. Layout of FDN for PCB.

PCB manufacturing has various steps, such as film processing, CNC drilling, brushing, plating, laminating photoresist on PCB, UV-exposure of photoresist, development of photoresist and etching. After etching, a visual inspection of PCB is needed in order to remove any possible copper remained during etching. **Figure 15** shows the final prototype of the diplexer with 50 Ω line at the input/output extended for connecting the big size SMA. Three SMA connectors have been soldered in the PCB in order to measure the results.

Figure 16 shows the measured results of the prototype which are measured using a Rhode and Schwartz ZVM vector network analyser. Any possible difference between measured and the simulated results can be due to the soldering and SMA [25].

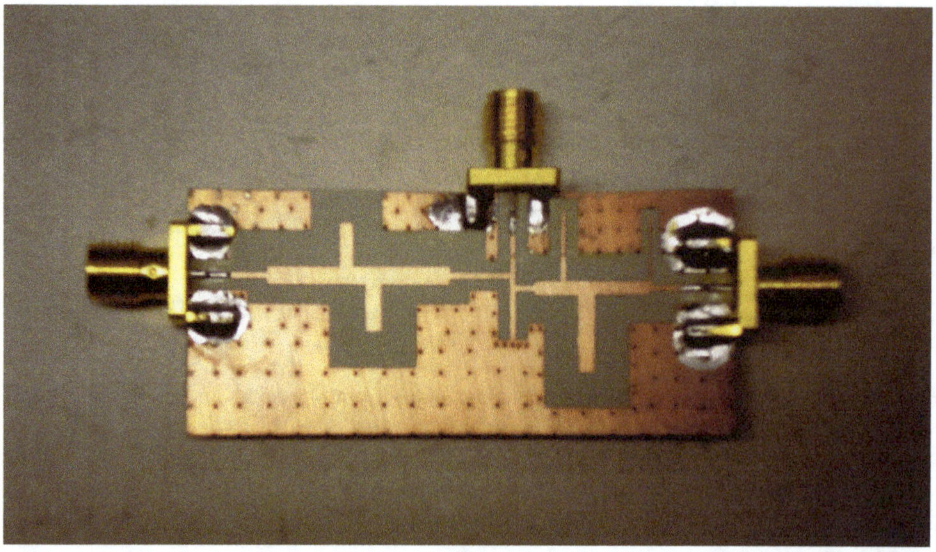

Figure 15. Prototype of FDN.

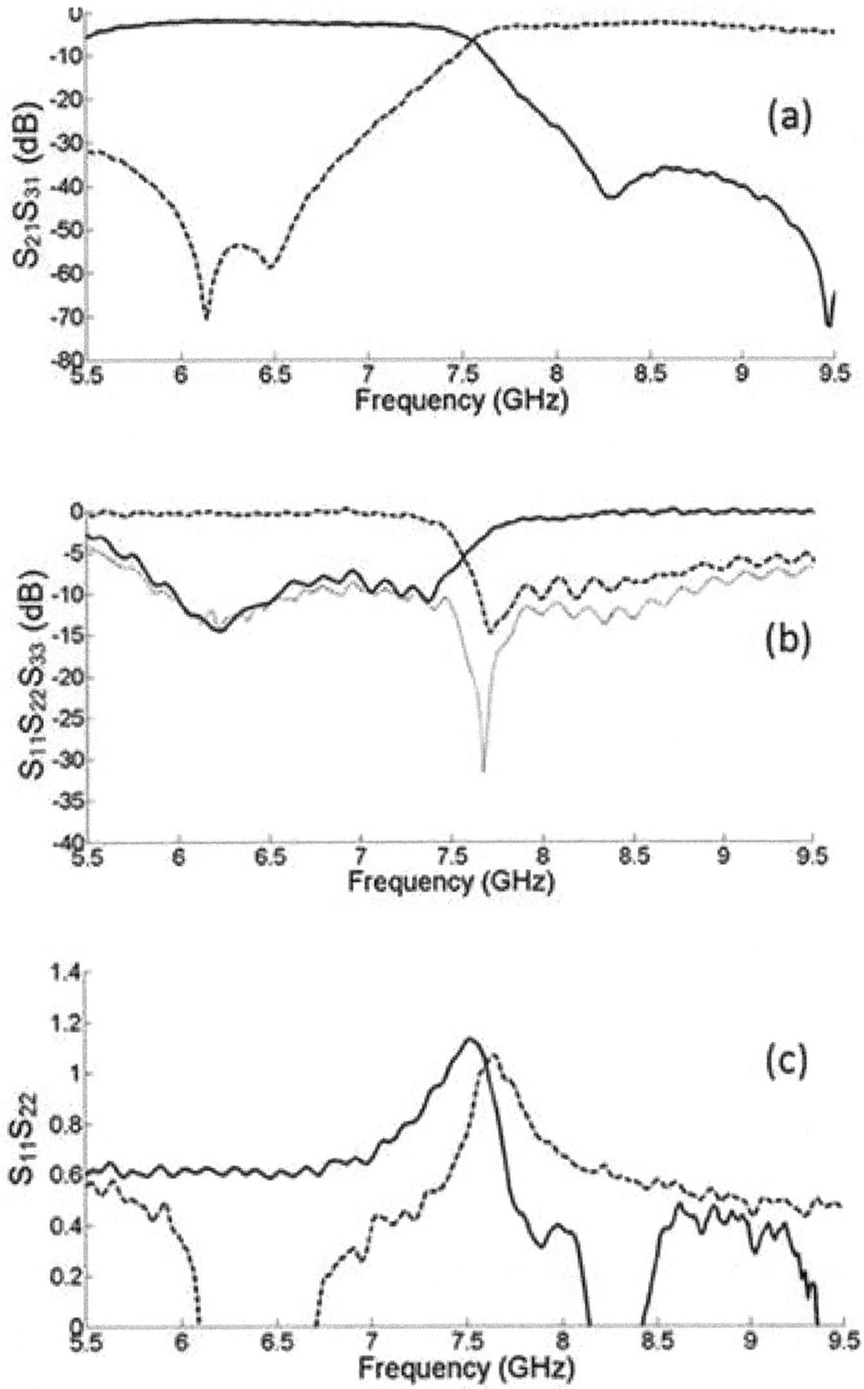

Figure 16. Measured results for FDN, (a) forward transmission, (b) input reflection, and (c) group delay.

Note: Here a diplexer is designed and implemented on the basis of FMN (**Figure 4**) and to see the design and implementation of the triplexer with a different BPF, see Ref. [22].

Acknowledgements

The author thanks Magnus Karlsson and Gustav Knutsson at the Department of Science and Technology Linköping University Sweden, for the assistance in PCB manufacturing.

Appendices and nomenclatures

ADS	Advanced Design System
BPF	Band Pass Filter
FCC	Federal Communications Commission
FDM	Frequency Division Multiplexing
FDN	Frequency Diplexer Network
FMN	Frequency Multiplexing Network
HTS	High-Temperature Superconductor
LO	Local Oscillator
LNA	Low Noise amplifier
MOC	Multiport Optical Circulator
PA	Power Amplifier
PCB	Printed Circuit Board
PPM	Pulse Position Modulation
PSM	Pulse Shape Modulation
SMA	Sub-Miniature version A
TDM	Time Division Multiplexing
UWB	Ultra-Wide Band
WDM	Wavelength Division Multiplexing

Author details

Imran Mohsin

Address all correspondence to: imranmoh27@gmail.com

1 Shenzhen Institutes of Advanced Technology (SIAT), Chinese Academy of Sciences, Shenzhen, China

2 Shenzhen College of Advanced Technology, University of Chinese Academy of Sciences, Shenzhen, China

References

[1] Angwin A S. The Alexander Graham Bell centenary lecture. Graham Bell-pioneer: An era of outstanding developments in world communication. Journal of the Institution of Electrical Engineers—Part I: General. 2010; 94(78): 269–274. DOI: 10.1049/ji-1.1947.0084

[2] Joseph Carr. The Technician's Radio Receiver Handbook: Wireless and Telecommunication Technology. 1st ed. Oxford: Newnes; 2000. 23 p. DOI: 9780080518596

[3] Karlsson M. Ultra-wideband antenna and radio front-end system. [thesis]. Sweden: Department of Science and Technology, Linkoping University; 2007.

[4] Mohsin I. Design of a smart six-port circuit and wireless parallel data transmission. [thesis]. Sweden: Department of Science and Technology, Linkoping University; 2010.

[5] Håkansson P. High speed wireless parallel transmission and six-port transmitters and receivers. [thesis]. Sweden: Department of Science and Technology, Linkoping University; 2008.

[6] Cameron R I, Yu M. Design of manifold-coupled multiplexers. IEEE Microwave Magazine. 2007; 8(5): 46–59. DOI: 10.1109/MMM.2007.904715

[7] Raafat R M. Design of superconductive multiplexers using single-mode and dual-mode filters. IEEE Transactions on Microwave Theory and Techniques, 1994; 42(7): 1411–1418. DOI: 10.1109/22.299738

[8] Jonathan C, Jerry F, Stephen B, Duane H. Microstrip hybrid coupled input multiplexer design for satellite communications. In: IEEE MTT-S International Microwave Symposium (IMS2014); 1—6 June 2014; FL.

[9] Mansour R, Rammo F, Dokas V. Design of Hybrid-Coupled Multiplexers and Diplexers using Asymmetrical Superconducting Filters. In: Microwave Symposium Digest, IEEE MTT-S International; 14–18 June 1993; IEEE; 2002. pp. 1281–1284.

[10] Rubin D. Millimeter-wave hybrid coupled reflection amplifiers and multiplexer. IEEE Transactions on Microwave Theory and Techniques. 1982; 30: 2156–2162. DOI: 10.1109/TMTT.1982.1131400.

[11] Volker N, Markus M, Gottfried M. A Novel Low Loss Microwave Multiplexer Design Based on Directional Filters. In: Radio and Wireless Conference, RAWCON; 11–14 August 2002; IEEE; 2002. pp. 257–260.

[12] Jakub S, Ilona P, Slawomir G, Krzysztof W. Miniaturized Directional Filter Multiplexer for Band Separation in UWB Antenna Systems. In: International Symposium on Antennas and Propagation (ISAP); 9–12 November 2015; IEEE; 2016. pp. 1–4.

[13] Chen K Y, Hu C J, Lee C C, Feng K M, Lu M K, Chang C H, Tu Y K, Tzeng S L. Low-crosstalk and compact optical add–drop multiplexer using a multiport circulator and fiber Bragg gratings. IEEE Photonics Technology Letters. 2000; 12(10): 1394–1396. DOI: 10.1109/68.883841.

[14] Andrzejewski N. K, Pietraszewy-ski E J. Waveguide Multiplexer with Eight Contiguous Channels X Band. In: Microwave Conference, 9th European; 17–20 Sept. 1979; IEEE; 1979. pp. 392–396.

[15] Morinil A, Baldellil M, Venanzonil G, Cifolal L, Farinal M, Iglesias P M, Emst C, Macchiarella G. Improvement of Dual-Manifold Architecture for the Design of

Reconfigurable Diplexers. In: IEEE MTT-S International Microwave Symposium; 17–22 May 2015; IEEE; 2015. pp. 1–4.

[16] Ho R Y C, Battensby B. Subminiature Microwave Active Filter Manifolds. In: Microwave Symposium, G-MTT 1970 International; 11–14 May 1970; IEEE; 2010. pp. 62–65.

[17] María B, Santiago C, Mónica M, Pablo S, Vicente E B, Marco G. Design of Waveguide Manifold Multiplexers with Dual-Mode Filters Using Distributed Models. In: IEEE MTT-S International Microwave Symposium (IMS2014); 1–6 June 2014; IEEE; 2014. pp. 1–4.

[18] Liang P L, Tao S, Zhi P R., Bian W. Design of Helical Filters Manifold Multiplexer Using A Novel Equivalent Circuit Model. In: International Symposium on Signals, Systems and Electronics; 17–20 Sept. 2010; IEEE; 2010. pp. 1–4.

[19] Uwe R, Dietmar R, Wolfgang R, Dieter W. Tunable Manifold Multiplexers - A New Possibility For Satellite Redundancy Philosophy. In: Microwave Conference, 18th European; 12–15 Sept. 1988; IEEE; 2007. pp. 870–875.

[20] Santiago C, Pablo S, Vicente E. B, Marco G, María B, Benito G, and David R. Efficient design of waveguide manifold multiplexers based on low-order EM distributed models. IEEE Transactions on Microwave Theory and Techniques. 2015; 63(8): 2540–2549. DOI: 10.1109/TMTT.2015.2442990

[21] Mohsin I, Karlsson M, Gong S. Frequency diplexer network for wireless parallel data transmission and ultra-wideband systems utilizing manifold technique. Microwave and Optical Letters. 2014; 56(8): 1869–1871. DOI: 10.1002/mop.28464

[22] Karlsson M, Håkansson P, Gong S. A frequency triplexer for the ultra-wideband systems utilizing combined broadside- and edge-coupled filters. IEEE Transactions on Advanced Packaging. 2008; 31(4): 794–801. DOI: 10.1109/TADVP.2008.2004415

[23] Håkansson P, Huynh A, Gong S. A Study of Wireless Parallel Data Transmission of Extremely High Data Rate up to 6.17 Gbps per Channel. In: Asia-Pacific Microwave Conference; 12–15 Dec. 2006; IEEE; 2008. pp. 975–978.

[24] Ludwig R, Bretchko P. RF Circuit Design, Theory and Application. 2nd ed. Upper Saddle River, NJ: Prentice Hall; 2000. ISBN:0-13-095323-7

[25] Mohsin I, Karlsson M, Owais O, Gong S. Design and implementation of a UWB six-port correlator for 6–9 GHz frequency band. Microwave and Optical Letters. 2012; 55(1): 190–193. DOI: 10.1002/mop.27235

11

Multiple Person Localization Based on their Vital Sign Detection using UWB Sensor

Dušan Kocur, Daniel Novák and Mária Švecová

Abstract

In the past period, great efforts have been made to develop methods for through an obstacle detection of human vital signs such as breathing or heart beating. For that purpose, ultra-wideband (UWB) radars operating in the frequency band DC-5 GHz can be used as a proper tool. The basic principle of respiratory motion detection consists in the identification of radar signal components possessing a significant power in the frequency band 0.2–0.7 Hz (frequency band of human respiratory rate) corresponding to a constant bistatic range between the target and radar. To tackle the task of detecting respiratory motion, a variety of methods have been developed. However, the problem of person localization based on his or her respiratory motion detection has not been studied deeply. In order to fill this gap, an approach for multiple person localization based on the detection of their respiratory motion will be introduced in this chapter.

Keywords: emergency events, person localization, respiratory motion, signal processing, ultra-wideband (UWB) sensor/radar, vital signs detection of persons

1. Introduction

At the beginning of the twenty-first century, the human society faced quite a number of specific social trends. The increasing density of population of towns and town agglomerations, criminality growing and political tensions producing terrorism, and growth of the percentage of elder people can be ranked among them.

A high density of the population at emergency events such as earthquakes or building collapses results in a high number of injured and trapped persons. At these disasters, many persons can survive the most critical moments. However, they have often limited capability of the motion or they are unconscious, what makes more difficult to save them. There is, however, necessary to emphasize that the capability and rapidity of the victim localization is the most critical for their lifesaving.

People localization is very important not only for post-disaster rescue but also for military and security operations [1]. Here, reservoirs, power plants, and other critical infrastructures are extremely vulnerable to terrorist attack. Therefore, the request for monitoring of the mentioned critical areas and for the detection of unauthorized intrusion is still needed. The information about the number of persons and their positions can be very useful for security troops to take the right decisions. In the case of the mentioned disasters and law enforcement operations, people to be detected and localized are often situated behind nonmetallic obstacles (e.g. walls). That is the reason why conventional sensors (e.g. cameras) cannot be applied for person monitoring for such scenarios.

It is expected that the number of seniors will increase in the future. It will result in dramatic challenges to the society since elder people need additional care and assistance. It is well known that many people aged above 65 years fall down in their dwelling every year and above 3% of them are not able to rise again without external help. They have to stay in their uncomfortable situation over hours or even days before they get help. As a result, a number of people are dying in consequence of such accidents [2]. In order to save life of the seniors, they can be monitored, for example, by conventional cameras. However, this simple solution represents a serious invasion to their private life, and hence it is not well accepted. On the other hand, a solution enabling to localize seniors at their home taking into account their privacy and at the same time detection of emergency events is really very needful. It has been shown in Ref. [2] that ultra-wide frequency band radars (UWB sensors) can provide solutions of the outlined problem.

The analyses of the requirements for person localization according to above outlined scenarios have shown that short-range high-resolution radars emitting electromagnetic waves with ultra-wide frequency band (UWB radars) using relatively low frequencies can be used with advantage for such applications. The exploitation of the ultra-wide frequency band provides the high resolution of the radar. On the other hand, the electromagnetic waves emitted in the frequency band DC-5GHz can penetrate through standard building materials (e.g. wood, concrete, bricks, etc.) with acceptable attenuation. Therefore, UWB sensors employing this frequency band are capable to detect a target situated not only for line-of-sight but also for a target located behind a nonmetallic obstacle. Moreover, UWB radar systems can be constructed to be lightweight and small.

UWB radars could be available as handheld sensor systems consisting of antenna system, radio frequency (RF) block, analog-to-digital (AD) convertors, and digital circuitry including a computer and software. On the other hand, UWB sensors provide also privacy requested at people monitoring at their home, and hence they can be implemented as a part of complex sensor networks applied for senior monitoring in ambient-assisted living (AAL) programs as well.

The key part of the UWB sensor system software consists in the implementation of UWB radar signal processing procedure intent on human being detection, localization, and tracking. The deeper analyses of person detection using UWB sensors have shown that it is necessary to distinguish between the detection of moving and static persons. In this context, a person moving within the monitored area in such a way that their coordinates are changing is referred to as a moving person. On the other hand, a living person is considered as static person, if he or she is situated but not moving within the monitored area in such a way that their coordinates are changing. Then, while a typical form of moving person movement is, for example, his or her walking or running, typical forms of static person movement are his or her respiration motions or heart beating, which is commonly referred to as person's vital signs. As examples of static persons, unconscious person located beneath the rubble of the building or avalanche, bound people (prisoners of terrorists) and sleeping persons, and so on can be given.

In this chapter, we focus on UWB radar signal processing procedure for localization of static persons situated behind a nonmetallic obstacle. First, we state the basic principle of the static person detection based on the detection of his or her respiratory motions. Then, the the state-of-the of UWB radar signal processing concerning static person detection and localization will be presented in Section 3. The core of the chapter will be presented in Section 4. Here, a Welch periodogram method for multiple static person localization based on their vital sign detection will be given. As the vital signs, person respiratory motions (breathing) will be considered. The particular phases of this method will be illustrated by corresponding experimental results. Finally, some conclusions will be summarized in Section 5.

2. Respiratory motion detection: problem statement and its fundamental solution

Figure 1 symbolizes the generic measurement arrangement illustrating the basic principle of respiratory motion detection [2]. A transmitting antenna *Tx* converts the signal $p(t,\tau)$ into an electromagnetic wave, which propagates toward the body of the person. There, it will be partially reflected and travels back to the receiving antenna *Rx*, which converts the wave back into signal $r(t,\tau)$ that has to be captured by appropriate radar electronics. Then, the radar device converts the received signals by a proper pulse compression method into impulse responses $h(t,\tau)$ of the environment through which the signal emitted by the radar is

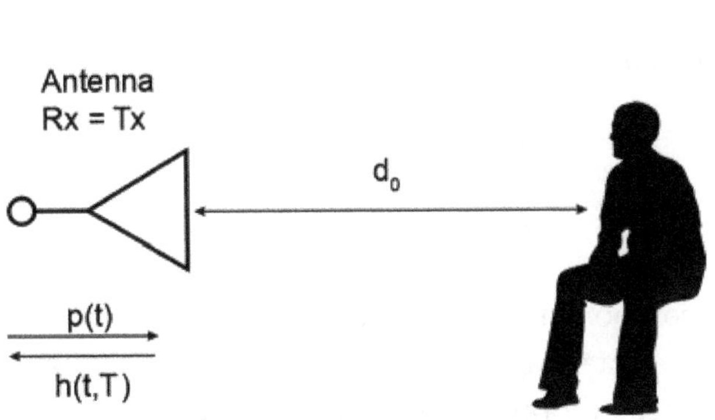

Figure 1. Basic measurement arrangement.

propagated. The variables t and τ, and the set of $h(t,\tau)$ represent fast-time (propagation time), slow-time (observation time), and raw radar data usually referred to as radargram, respectively.

Let us assume for this moment that the same antenna is used for the signal transmission and reception, and a static person is located in front of this antenna in the distance d_0. Under such conditions, signal $h(t,\tau)$ can be expressed as follows [2, 3]:

$$h(t,\tau) = \sum_{i=1}^{N} A_i p(t-t_i) + A_0 p\left(t-t_d(\tau)\right) = h_b(t,\tau) + h_0(t,\tau). \tag{1}$$

In this expression, the constant term A_i stands for the path gain (or loss) of the ith signal path. The signal components $A_i p(t-t_i)$ for $i\neq 0$ represent the reflections of the transmitted signal from static objects (i.e. not reflections due to static person). Then,

$$h_b(t,\tau) = \sum_{i=1}^{N} A_i p(t-t_i) \tag{2}$$

is the component of $h(t,\tau)$ representing static clutter and noise (e.g. noise due to antenna and other electronic components of radar device, etc.). On the other hand, signal

$$h_0(t,\tau) = A_0 p\left(t-t_d(\tau)\right) \tag{3}$$

expresses the component of $h(t,\tau)$ due to static persons. The time delay $t_d(\tau)$ (associated with vital signs) represents the time of arrival (TOA) of the static person. It covers the time signal required to travel across *Tx-target-Rx* trajectory. The time delay $t_d(\tau)$ varies depending on the distance of the target from the antenna system. Slight but periodic changes of TOA are caused by a movement of the target chest due to periodical chest movement caused by breathing process (with a frequency f_b) and heartbeats (with frequency f_h) [2, 3]. Then, $t_d(\tau)$ can be approximately expressed as

$$t_d(\tau) = 2d(t_0,\tau)/c \tag{4}$$

where $c = 3\text{x}10^8 \text{ms}^{-1}$ is the speed of light and

$$d(t_0,\tau) = d_0 + m_b(\tau) + m_h(\tau) = d_0 + m_b \sin(2\pi f_b \tau) + m_h \sin(2\pi f_h \tau) \tag{5}$$

$$m_b(\tau) = m_b \sin(2\pi f_b \tau) \tag{6}$$

$$m_h(\tau) = m_h \sin(2\pi f_h \tau) \tag{7}$$

The term $d_0(t_0,\tau)$ is a component expressing the reflection from static object situated in distance d_0 from the radar antenna, and m_b and m_h are the breathing and heart amplitude, respectively. The peak-to-peak thorax motion due to breathing of an adult person ranges from 4 to 12 mm, while the peak-to-peak motion due to the heartbeat is about 0.5 mm. Then,$m_b >> m_h$ and the signals $d(t_0,\tau)$ and $t_d(\tau)$ can be expressed as

$$d(t_0,\tau)\approx d_o + m_b \sin(2\pi f_b \tau) \tag{8}$$

$$t_d(\tau) = 2d(t_0)/c + (2m_b/c)\cdot \sin(2\pi f_b \tau). \tag{9}$$

Within a long-term observation, breathing can be qualified as a periodic process with its own frequency f_b. According to [4], this frequency can take values from 0.2 Hz caused by slow, calm breathing up to 0.7 Hz caused by fast, stressful breathing. Then, taking this knowledge and Eqs. (1), (3), and (8) as well into account, it can be concluded that the impulse response $h(t,\tau)$ consists of the clutter and noise Eq. (2) and a periodical signal $h_0(t,\tau)$. The signal $h_0(t,\tau)$ due to the respiratory motion of a person has a DC component (depending on d_0) and period $T_b = 1/f_b$. The frequency of its fundamental harmonic component is $f_b \in B$, where $B = \langle 0.2Hz,\ 0.7\ Hz\rangle$.

The abovementioned comments summarizing the detailed analyses presented in Refs. [2, 3] indicate that a static person can be detected based on the detection of a periodical signal components of the radargram $h(t,\tau)$ corresponding to a periodical motion with frequency $f_b \in B$ with regard to the slow-time variable (τ) for a constant fast-time instant (t_0). If such a component is detected for $t = t_0$, then the bistatic range of the target (i.e. the length of the trajectory distance *Tx-target-Rx*) is ct_0. And finally, if a proper number of bistatic ranges of the target for the properly defined positions of *Tx* and *Rx* are estimated, then the target coordinates can be estimated, for example, by a trilateration or multilateration method.

As it follows from the previous paragraph, the key element of the static person detection consists in the detection of the periodical components of the radargram with the frequency form the interval $B = \langle 0.2Hz,\ 0.7\ Hz\rangle$. Here, it should be stressed that the target echo to-noise-and clutter ratio (SNCR) and is usually low. The target detection and then its localization can be onward complicated by a complex environment where the target is situated (strong attenuation due to obstacle, multipath electromagnetic wave propagation, shadowing effect, narrowband interference due to wireless communication systems, etc.). Therefore, there are no simple solutions to be applied for static person detection and localization. The published solutions differ in their approach to clutter and noise suppression, to power spectrum estimation, and to target detection methods. Moreover, there are only a few papers dealing not only with the person detection but also with person localization. The state-of-the art in this field is shortly outlined in the next section.

3. Static person detection and localization: state-of-the art

During the last decade, a variety of the papers intent on human being detection by UWB sensors have been published. These contributions were initially focused mainly on UWB sensor system construction and on basic principle of human being detection (e.g. [5, 6]). Then, the research in the field of UWB sensor systems was devoted also to the development of UWB radar signal processing methods to be applied for moving and static person detection and localization. As we have shown in the previous section, the basic principle of the static person detection consists in the spectrum analyses of the radargram. This fact allows modifying the methods originally designed for static person detection for estimation of the breathing and heart-beating rate (e.g. [3, 7]). In the next section, we deal with the problem of radar signal processing for the detection and localization of static person only.

This section briefly highlights the approaches and the outcomes of the research in the detection and localization of the static persons but also points out their respective deficiencies, which altogether give ground for motivation of the research presented herein.

Deep analyses of the problem of the human beings detection based on the detection of their respiratory motions have been provided in Refs. [2, 8]. In these works, a very useful model of raw radar data has been introduced. It has been shown that the raw radar data can be modeled as an additive mixture of signal of interest (target echo), stationary and nonstationary clutter, jamming and random noise. What is important, the components of the target echo along the slow-time axis are represented by periodical signals with the fundamental harmonic located in the frequency interval $B = \langle 0.2Hz,\ 0.7\ Hz \rangle$. The further analyses of the raw radar data have shown that target echo to-noise and clutter ratio SNCR is usually very small. Therefore, the process of the static person detection can be decomposed at least into two phases, which include clutter and noise suppression and spectrum analyses of the radargrams with the suppressed clutter and noise.

The approach chosen for respiratory motion detection proposed in Refs. [8, 9] is divided into three stages: stationary-clutter removal (range-profile subtraction, mean subtraction, linear-trend subtraction), SNCR improvement (filtering in range dimension, slow-time frequency-domain windowing, and nonstationary-clutter suppression by singular value decomposition (SVD)), and threshold-based decision- and range-estimation stage (results of SVD decomposition processing). For the power spectrum estimation, discrete Fourier transform (DFT) is applied. The proposed approach has been successfully demonstrated for the detection of a single person located under rubble.

On the other hand, a bit simpler approach for a static person detection has been introduced in Refs. [2, 5]. In these papers, an exponential averaging and DFT or Welch periodogram has been proposed for background subtraction and power-spectrum estimation, respectively. The obtained results have shown that also this alternative approach can be successfully used for the static person detection under some scenarios.

The analyses of some several other contributions (e.g. [2, 4, 10, 11]) have shown that moving target indicator method (MTI), SVD, combination of correlation analyses and a Curvelet transformation, and CLEAN algorithm can be used for the SNCR improvement. Besides conventional methods of spectrum analyses (e.g. DFT or Welch periodogram), the application of some new methods for the detection of the periodical components of radargram has been proposed. The methods such as Hilbert-Huang transformation (e.g. [11, 12]), S-transform (e.g. [13]), MUSIC algorithm (e.g. [7]) and wavelet transformation and wavelet entropy (e.g. [14]) can be rated among them. The fundamentals of Hilbert-Huang transformation, S-transform, and MUSIC methods can be found in Refs. [15–17], respectively.

The results of the detection phase are represented usually by two-dimensional (2D) figures expressing a propagation rate (fast-time) versus a frequency axis (respiration rate), where high-level components (so-called "hot spots") of these figures indicate a presence of the targets. It means that the output of these methods cannot provide any information about target coordinates. In order to localize the static person, their bistatic range or TOA has to be estimated and associated and only then the target can be localized using a proper localization method.

The mentioned approach has been applied in Refs. [18, 19]. In this papers, a sensor network of bistatic UWB radars for multiple person localization was described. The signal processing procedure described in these studies consists of SNCR improvement (using a simple variant of MTI), detection (a power-spectrum estimator application), initial screening (it is conducted via frequency analysis of the detected signals), data association and parameter estimation (maximum likelihood observation-target association technique) and target localization (including determination of number of the target). The sensor network presented in Ref. [18] provides not only very good localization of a set of static persons for a line-of-sight scenario, but it provides the estimation of the frequency of their breathing, too.

The present short but comprehensive review of some papers devoted to the static person detection and localization indicate that there are only a few papers (e.g. [18, 19].) dealing with the problem of multiple static person localization. In order to fill this gap, an approach for multiple person localization based on the detection of their respiratory motion and employing a single UWB sensor only will be introduced in the next section.

4. WP-STAPELOC method

Taking into account the model of raw radar signal $h(t, \tau)$, the basic principle of static person detection introduced in Section 2, and the findings summarized in the previous section as well, the static persons can be localized using the following procedure of UWB sensor signal processing. The procedure consists of a set of phases of signal processing such as background subtraction, target echo enhancement, target detection, TOA estimation and TOA association, and target localization, where each phase is implemented by proper methods of signal processing. As it will be shown later, the detection of persons' respiratory motions is the key phase of the procedure. The input signal of the detectors employed in this phase will be formed using power spectrum of the radargram components. Because Welch periodogram will be used for the estimation of the mentioned power spectrum, the signal processing procedure for static person localization introduced in this section will be referred as WP-STAPELOC method (a method for STAtic PErson LOCalization based on power-spectrum estimation using Welch Periodogram).

In the next paragraphs of this section, the importance and implementation of the particular phases of WP-STAPELOC method will be presented. With the intention to provide a bit deeper insight into the procedure, the importance of its particular phases, the performance of the signal processing methods applied within these phases and the performance of the WP-STAPELOC method as a whole will be illustrated using the experimental data obtained by measurements intent on through-the-wall localization of three static persons using UWB sensor system (the so-called illustrative scenario). Following the outlined concept of this section, the illustrative scenario will be first described, and then the particular phases of WP-STAPELOC method will be introduced.

4.1. Illustrative scenario

The illustrative scenario was intent on through-the-wall localization of three static persons using an UWB sensor. The measurement scheme for the illustrative scenario is outlined in **Figure 2**. The monitored area was represented by the basement room. The UWB sensor (radar) was situated behind the structural concrete wall of thickness 0.5 m (**Figure 2**). Three persons to be localized stood in the positions P1, P2, and P3. The only observable movements of the persons to be localized were their respiratory motions.

The raw radar data analyzed in this contribution were acquired by means of M-sequence UWB radar systems equipped with one transmitting and two receiving horn antennas (**Figure 3**). The radar antenna positions are outlined in **Figure 2**. The distance between Tx and Rxi for $i = 1, 2$ was set to 0.45 m. The system clock frequency and operational bandwidth of the applied UWB sensor were 4.5 GHz and DC-2.25 GHz, respectively. The order of the M-sequence emitted by the radar was nine, that is, the impulse response covers 511 samples regularly spread over 114 ns. Then, the unambiguous range of the UWB radar was about 17 m.

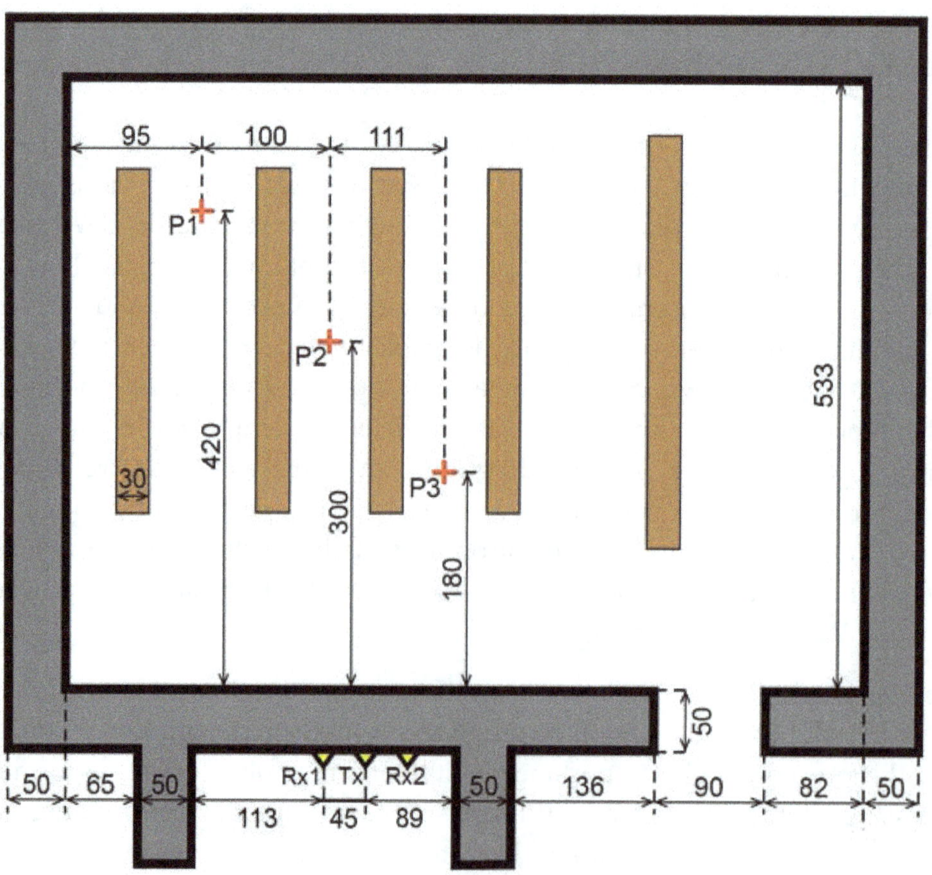

Figure 2. Measurement setup.

Figure 3. UWB radar system.

The radar measurement rate was approximately 13.5 impulse responses per second. The total length of the measurement was 74 s, which corresponds to 1000 impulse responses. The total power transmitted by the particular radar was about 1 mW.

The raw radar data obtained for the illustrative scenario are given in **Figures 4** and **5**. The strongest components identified in these figures correspond to the direct waves between Tx and Rxi (the so-called antenna coupling) and the reflections due to the wall. Unfortunately, the target-echo levels are too small, and hence no reflections due to persons are visible in **Figures 4** and **5**. These figures demonstrate very clearly that a further processing of the radargram is necessary to detect and localize the persons. In the next paragraph of this section, we describe WP-STAPELOC method, which can be applied for that purpose.

4.2. Phases of WP-STAPELOC method

As has been stated at the beginning of this section, the WP-STAPELOC method is UWB radar signal processing procedure consisting of the set of the five signal processing phases such as background subtraction, target-echo enhancement, target detection, TOA estimation and TOA association, and target localization, where each phase is implemented by proper methods of signal processing. Now, the particular phases will be presented in the next paragraphs of this section.

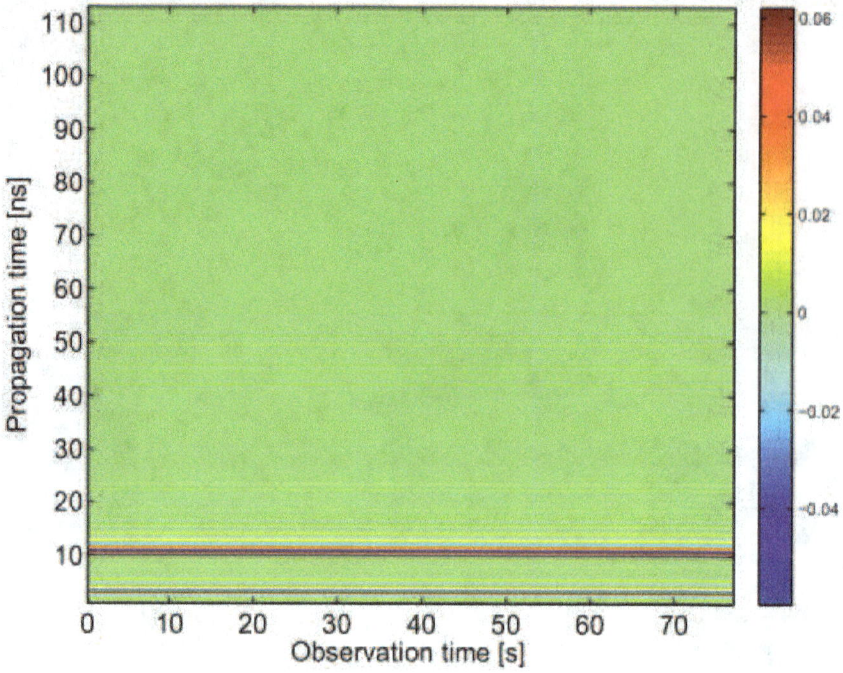

Figure 4. Raw radar data. Receiving channel Rx1.

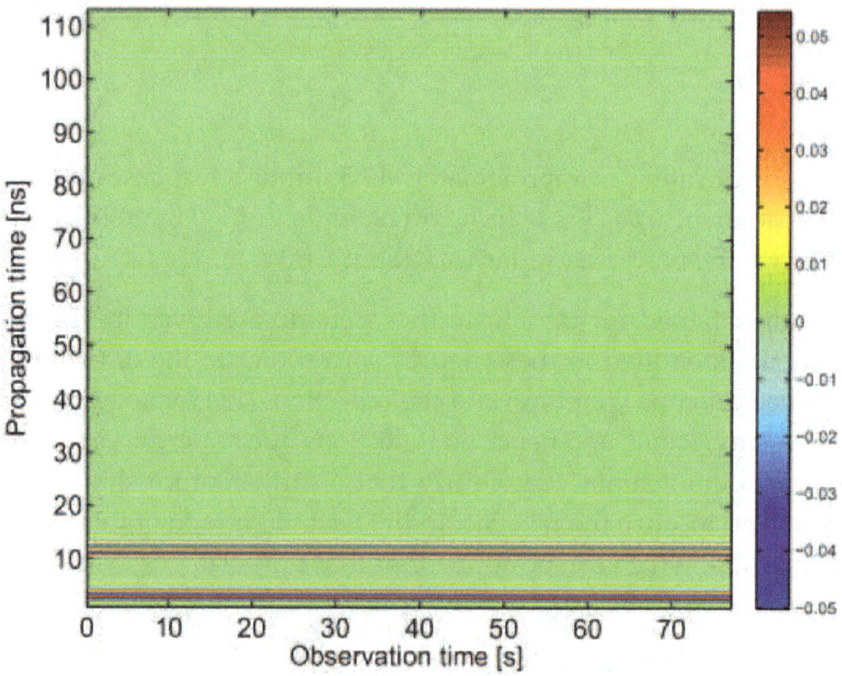

Figure 5. Raw radar data. Receiving channel Rx2.

4.2.1. Phase 1: background subtraction

The analysis of raw radar data has shown that it is impossible to directly identify any static persons in the obtained radargrams. This effect is due to the very small ratio of the signals reflected by a target (nonstationary components of the radargram) to noise and clutter (stationary components of radargram mainly due to reflections of electromagnetic waves from static objects). In order to detect a target, this ratio has to be increased. For that purpose, background subtraction methods can be used. In Ref. [20], a variety of methods of the background estimation and subtraction have been described. The mentioned methods differ in relation to assumptions concerning the clutter properties, as well as by their computational complexity and convenience for online signal processing. Because of a good performance, high robustness, and low computational complexity, the method of exponential averaging belongs to the most popular and often used methods of background subtraction [20]. According to this method, the background estimate $\hat{b}(t,\tau)$ is obtained by

$$\hat{b}(t,\tau) = \alpha\hat{b}(t,\tau{-}1) + (\alpha{-}1)h(t,\tau), \quad 0 < \alpha < 1 \tag{10}$$

where α is a constant weighting factor setting the length of background estimator memory. Then, the radargram with a subtracted background can be obtained by the expression

$$h_b(t,\tau) = h(t,\tau){-}\hat{b}(t,\tau) \tag{11}$$

The radargrams with the subtracted background for the illustrative scenario considered in this chapter are given in **Figures 6** and **7**, respectively. Three high-level artifacts similar to lines parallel to the slow-time axis are located in the interval ⟨20 ns, 40 ns⟩ of fast-time axis. These high-level signals represent the components of the radargram due to static persons.

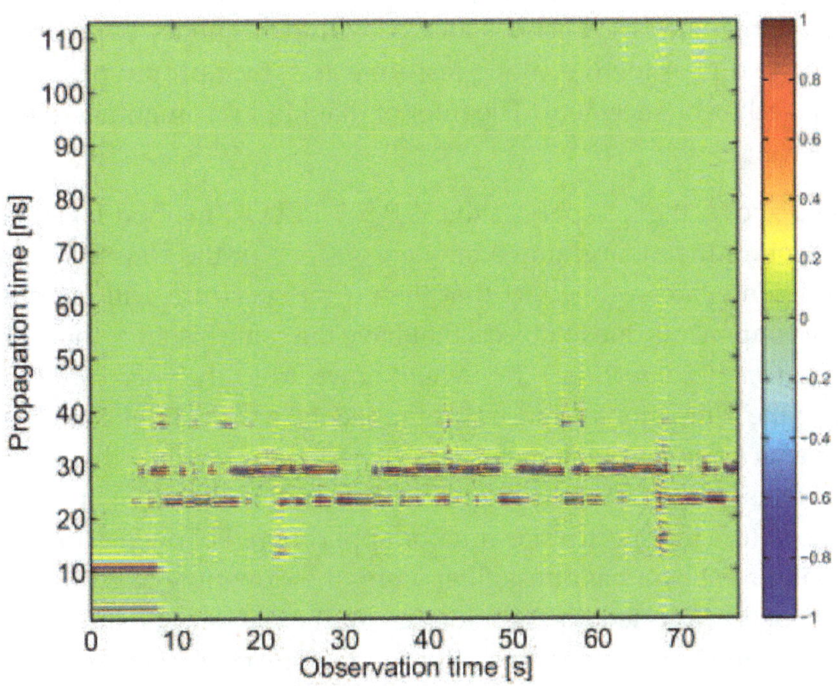

Figure 6. Radargram with subtracted background. Receiving channel Rx1.

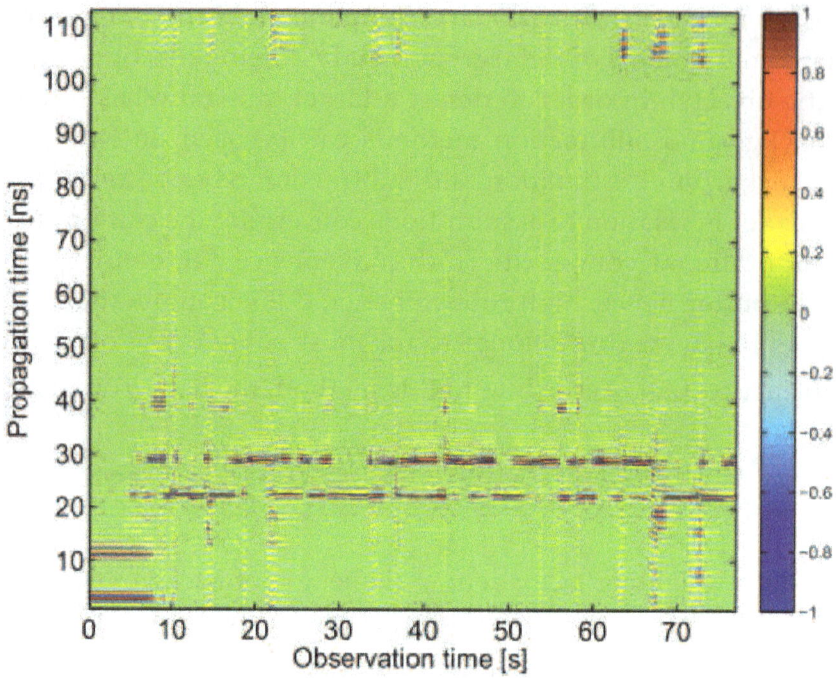

Figure 7. Radargram with subtracted background. Receiving channel Rx2.

4.2.2. Phase 2: target echo enhancement

SNCR can be improved using not only a background subtraction method but also a proper method of the target echo enhancement. An interesting approach for the improvement of SNCR with the special stress for the static person detection has been presented in, for example [8]. Here, the analyses of the impulse responses obtained at monitoring of a static person have shown that the target echo is located within frequency band 0.4–1.4 GHz. Then, filtering in the range dimension (i.e. along fast-time axis) by a proper pass-band filter can be used for low-level echo enhancement. The filter of that kind is sometimes also referred to as a range filter.

As we mentioned above, the key part of the WP-STAPELOC method consists in the power-spectrum estimation of the radargram components $h(t, \tau)$ for the constant values of t (i.e. for $t = t_k$). This condition expresses the fact that the power spectrum will be estimated only for those radargram components due to objects that have the same bistatic range with regard to Tx and Rx (i.e. *the distance Tx-target-Rxi*). It is well known from the spectrum estimation theory that in the case of the estimation of spectral components, which we like to detect (in our case, spectral components from the frequency band $B = \langle 0.2Hz,\ 0.7\ Hz \rangle$), the spectrum estimator performance depends usually on useful signal-to-noise ratio. Therefore, it would be helpful to increase this ratio before the spectrum estimator application. Following this idea, we propose to filter the signal $h(t_k, \tau)$ by a bandpass filter along the slow-time axis. The passband of this filter should be properly selected taking into account the frequency band $B = \langle 0.2Hz,\ 0.7\ Hz \rangle$ (signal components due to static persons) and the rate of measurement (i.e. frequency sampling of $h(t, \tau)$ along slow-time axis). In the next, a filter applied for this purpose will be referred to as a slow-time filter. The response of this filter will be denoted as $h_2(t, \tau)$.

In the case of the illustrative scenario, the low cutoff frequency of the Tx and Rxi antennas was 750 MHz and the operational bandwidth of UWB sensor was about DC-2.25 GHz. Because the practical operational bandwidth of UWB sensor system (radar device and antennas) used in the illustrative scenario was comparable with the frequency band 0.4–1.4 GHz reported in Ref. [8], no range filtering was applied for low-level echo enhancement. As the slow-time filter, the Cauer filter of the second order with the cutoff frequencies 0.1 and 0.8 Hz was applied. The obtained results are presented in **Figures 8** and **9**. If we compare the radargrams with the subtracted background and the corresponding slow-time filter outputs $h_2(t, \tau)$, we can see that the useful signal components (three high-level artifacts similar to lines parallel to slow-time axis located in the interval $\langle 20\,\text{ns}, 40\,\text{ns}\rangle$of fast-time axis) were enhanced, while some unwanted signal components were suppressed [21].

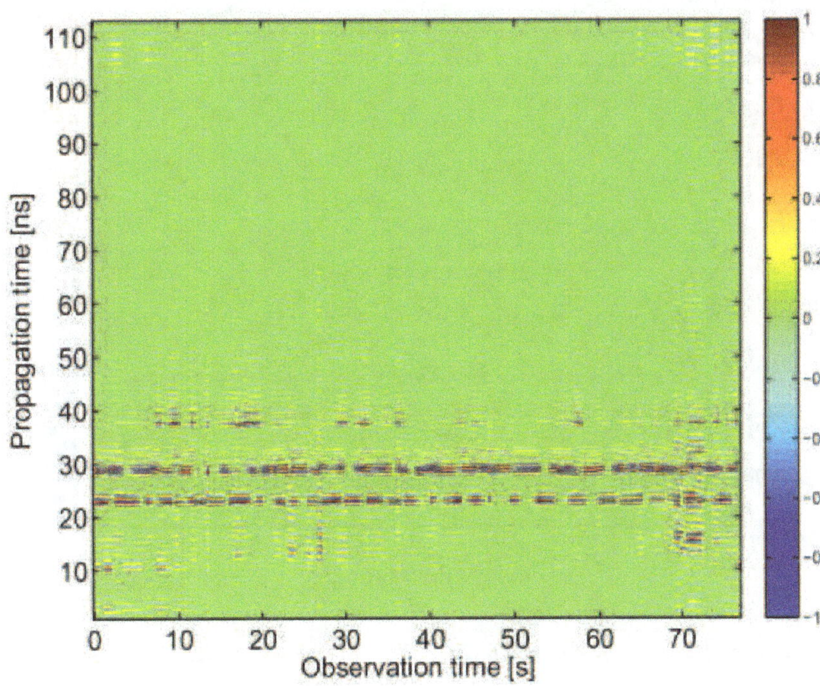

Figure 8. Slow-time filter output. Receiving channel Rx1.

4.2.3. Phase 3: target detection

Detection is the next phase of the radar signal processing procedure, which comes after the background subtraction and the target echo enhancement. Detection methods analyze the radargram with the subtracted background processed by range- and slow-time filters to reach the decision whether a signal scattered by a static person is present or absent in the analyzed radargram. Let the output of Phase 2 be denoted as $h_2(t, \tau)$. According to the conclusion of Section 2, the static person detection can be based on the detection of periodical components of signal $h_2(t_k, \tau)$ (i.e. a constant fast-time instant is considered). For those components, it is necessary to be located in the frequency band $B = \langle 0.2Hz,\ 0.7\,Hz\rangle$ and to have a sufficient level of their power as well. The former condition follows from the fact that the frequency band $B = \langle 0.2Hz,\ 0.7\,Hz\rangle$ covers the frequencies of human being respiration, while the latter condition is essential to ensure that we can distinguish between the level of background noise and signal components caused by a static person. Moreover, we have to consider the fact that a person is the so-called distributed target, that is, not the only one but usually several reflections can be received by the sensor from the same person. Then taking into account this knowledge, static persons can be detected by the detector consisting of power-spectrum estimator employing Welch periodogram [17], order statistic constant false alarm rate detector (OS-CFAR) [22] and a simple threshold detector (TD). In the next, this kind of the detector will be referred to as a two-stage detector [23]. Its operation can be described as follows.

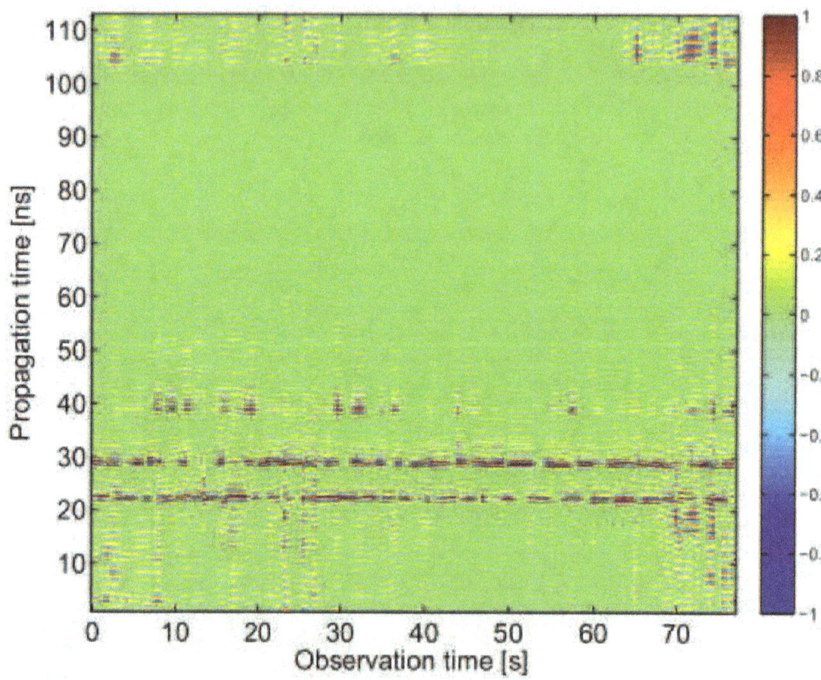

Figure 9. Slow-time filter output. Receiving channel Rx2.

First, the power spectra of the sequences $h_2(t_k, \tau)$ for $\tau \in \langle \tau_1, \tau_2 \rangle$ and for each time instant t_k are estimated using Welch periodogram [17]. Here, we would like to stress that the interval $\langle \tau_1, \tau_2 \rangle$ should be large enough to get a good power-spectrum estimation of the frequency components covering the frequency interval $B = \langle 0.2Hz,\ 0.7\,Hz \rangle$. In the next step, the total power of $h_2(t_k, \tau)$ contained in the frequency band $B = \langle 0.2Hz,\ 0.7\,Hz \rangle$ denoted as $X_1(t)$ is computed by the integration (summation) of the periodogram components located in this frequency band. Then, the variable $X_1(t_k)$ expresses a measure of the total power of $h_2(t_k, \tau)$ allocated in the frequency interval B including the power of signals reflected by all objects having the bistatic range $d(t_k) = ct_k$ (i.e. the distance *Tx-object-Rx*) with regard to the UWB sensor. As we stated before, a significant level of $X_1(t_k)$ indicates that in the bistatic range $d(t_k) = ct_k$ a static person can be located.

With the intention to decide if the level of $X_1(t_k)$ is significant to indicate the presence of a static person, OS-CFAR detector is employed with advantage as the first stage of the two-stage detector. The description of the OS-CFAR is beyond this paper. A reader interested in this topic can find the detailed description of OS-CFAR detector, for example, in Ref. [22]. The input-output relation of the OS-CFAR can be described by the expression

$$h_1(t) = \begin{cases} 0 & \text{if } X_1(t) \leq \gamma_1(t) \\ 1 & \text{if } X_1(t) > \gamma_1(t) \end{cases} \tag{12}$$

where the level of the threshold $\gamma_1(t)$ is controlled by the OS-CFAR algorithm. The detector output $h_1(t)$ is a binary signal, where "1" ("0") indicates the presence (absence) of the target in the bistatic range $d(t) = ct$.

In the case of person detection by a high-resolution radar, the radar-range resolution is much finer than the size of the target. Then, a human target has to be considered as a distributed target. It means that several reflections due to the same target can be detected. In order to avoid a multiple detection of the same target, the second detection stage is applied within the detection phase. The input signal of this detector denoted as $X_2(t)$ is formed by the sequential summation of $X_1(t_k)$ over the interval with the length V, where V represents approximately the maximum number samples of $h(t, \tau)$ containing the reflections from the same target. Then, the signal $X_2(t)$ can be expressed as follows:

$$X_2(t) = \sum_{l=k}^{k+V-1} h_1(t_l) \tag{13}$$

Now, the operation of the second stage of the detector (TD) can be described by the expression

$$h_2(t) = \begin{cases} 0 & \text{if } X_2(t) \leq \gamma_2 \\ 1 & \text{if } X_2(t) > \gamma_2 \end{cases} \tag{14}$$

where γ_2 is a constant threshold. The TD output $h_2(t)$ is a binary signal, where "1" ("0") indicates the presence (absence) of a static person in the bistatic range $d(t) = ct$. The signal $h_2(t)$ represents also the output of the detection phase.

Now, the performance of the above-described two-stage detector can be illustrated using the data obtained at our illustrative scenario. First, the power spectra of slow-time filter outputs are given in **Figures 10** and **11**. Here, the frequency components due to the static persons located in the frequency band $B = \langle 0.2Hz,\ 0.7\ Hz\rangle$ and within the fast-time interval ⟨20 ns, 40 ns⟩ can be identified. The performance of OS-CFAR detector is outlined in **Figures 12–15**. These figures indicate that OS-CFAR detector is able to detect the static persons (three target detected in the fast-time interval ⟨20 ns, 40 ns⟩). We can observe also in these figures that several isolated reflections from the same person were detected by OS-CFAR detector. After the application of TD, the number of the detected targets is decreased. Targets are mostly detected in the fast-time interval ⟨20 ns, 40 ns⟩ (**Figures 16–19**).

As it might be expected, some false alarms can be observed in the OS-CFAR and TD outputs. The total number of the false alarms produced by TD is smaller than that created by OS-CFAR detector. Hence, we can see that TD application decreases the total number of false alarms observed at the output of the detection phase. We will show in the following paragraphs that these false alarms will be eliminated by the TD output processing within TOA estimation and TOA association phase.

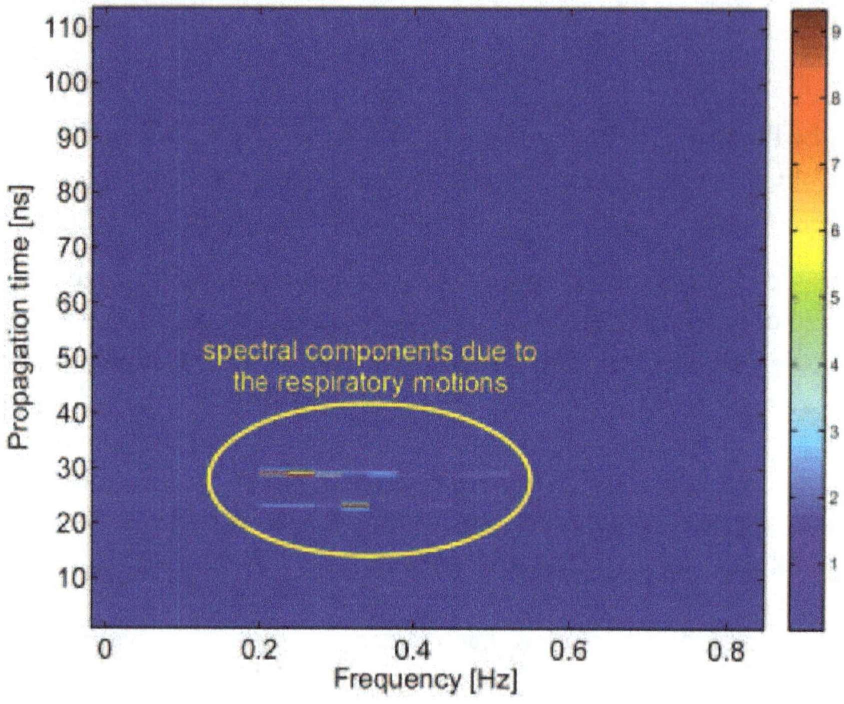

Figure 10. Power spectrum of slow-time filter output. Receiving channel Rx1.

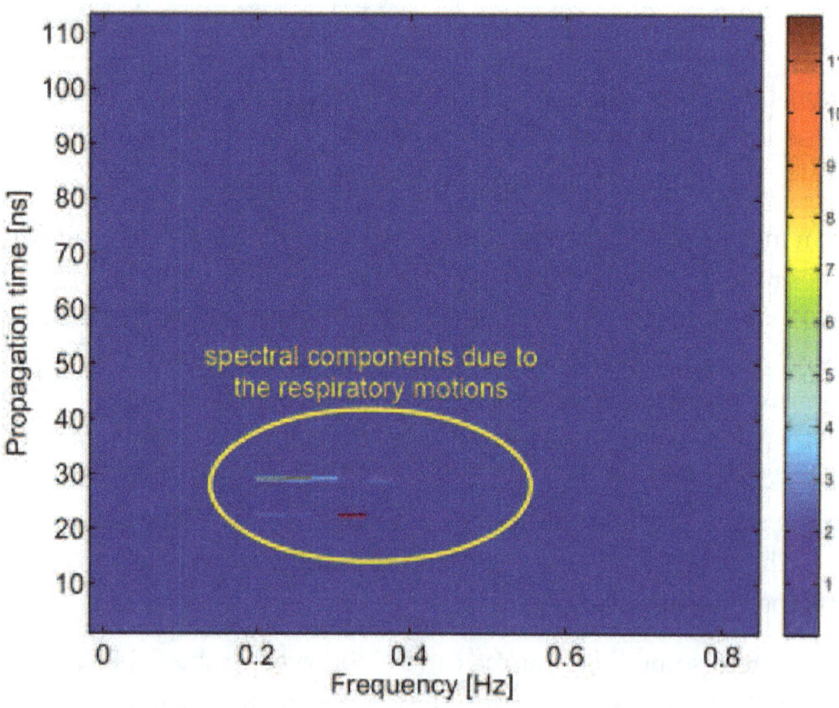

Figure 11. Power spectrum of slow-time filter output. Receiving channel Rx2.

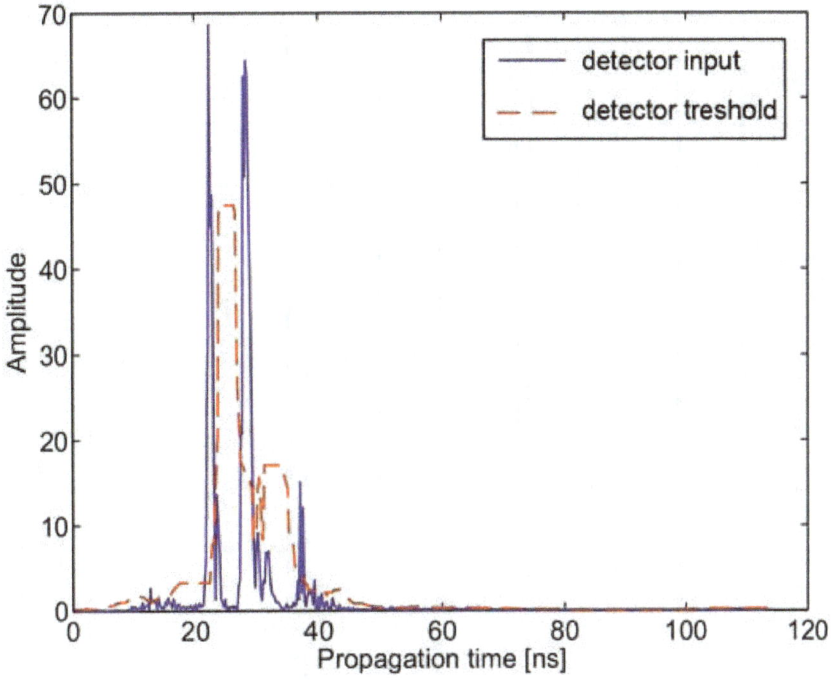

Figure 12. Total power of the signal $h_2(t_k, \tau)$ allocated in the frequency band B ($X_1(t)$, blue color) and OS-CFAR detector threshold ($\gamma_1(t)$, red color). Receiving channel Rx1.

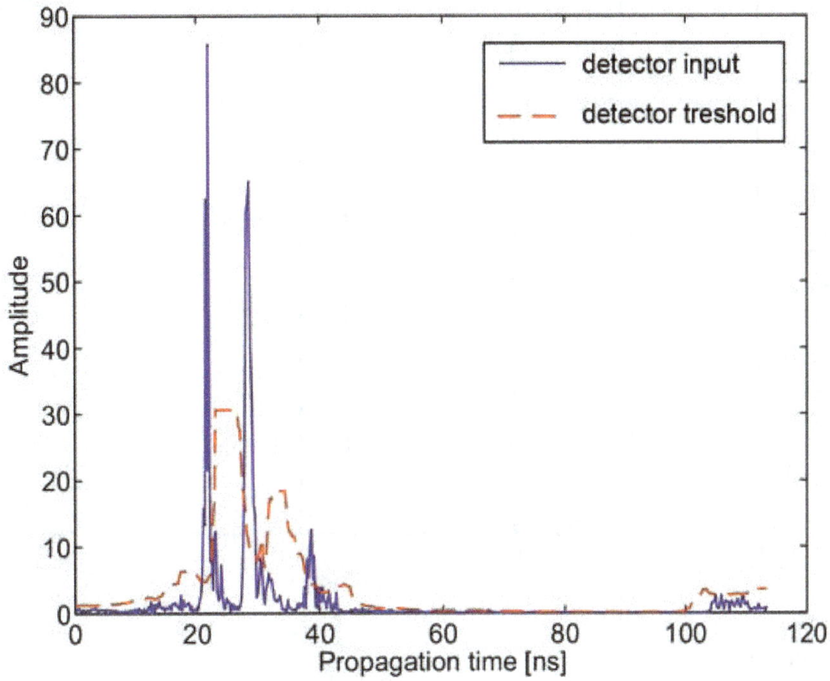

Figure 13. Total power of the signal $h_2(t_k, \tau)$ allocated in the frequency band B ($X_1(t)$, blue color) and OS-CFAR detector threshold ($\gamma_1(t)$, red color). Receiving channel Rx2.

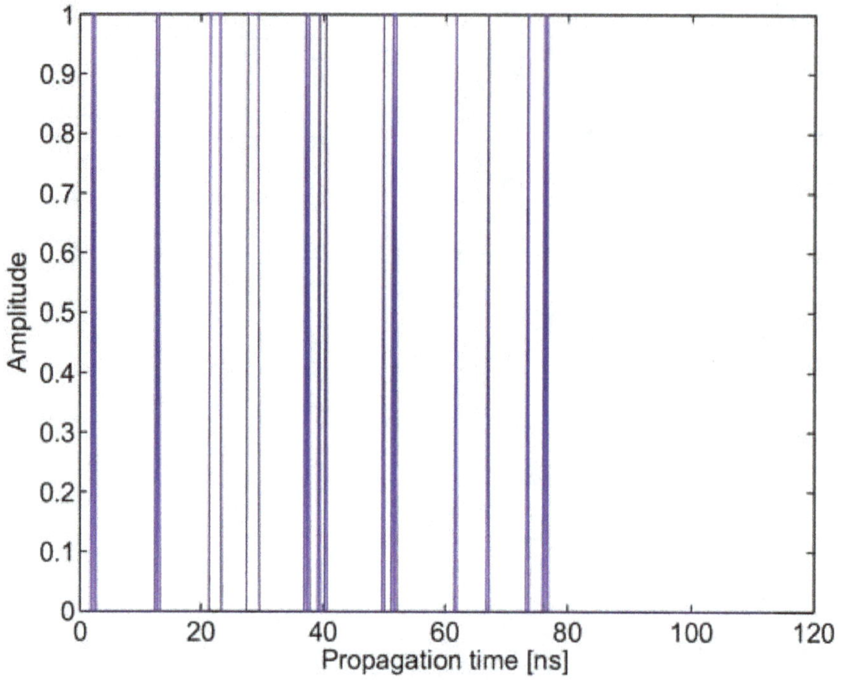

Figure 14. OS-CFAR detector output. Receiving channel Rx1.

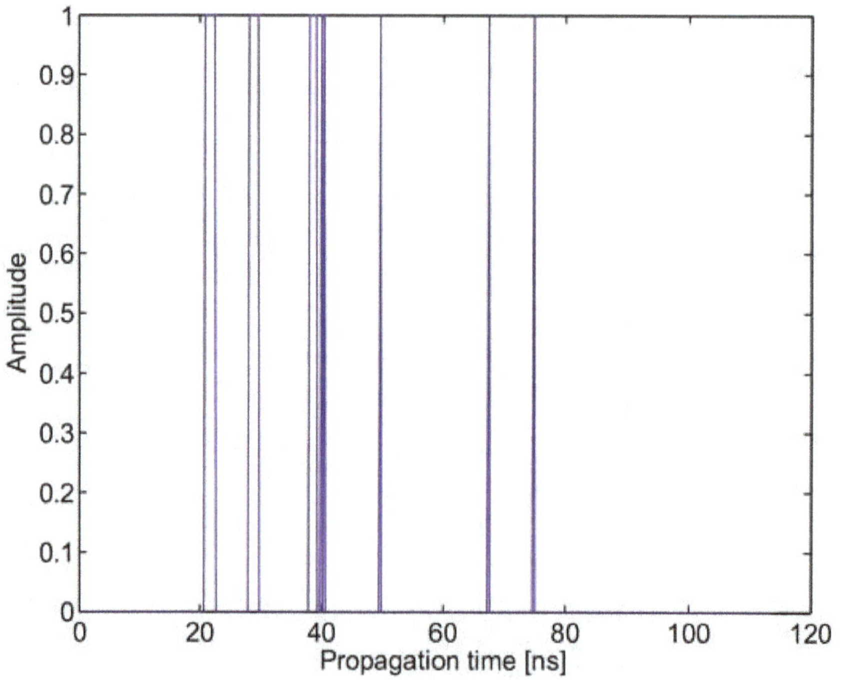

Figure 15. OS-CFAR detector output. Receiving channel Rx2.

4.2.4. Phase 4: TOA estimation and TOA association

The very efficient algorithm for TOA estimation can be found, for example, in Ref. [24]. This algorithm referred to as the trace connection method provides not only TOA estimation but also the association of the data received from two receiving channels and deghosting operation essential for multiple target detection. The trace connection method is quite complex, and hence it is beyond this paper. The details concerning this method can be found, for example, in Ref. [24]. The output of this phase consists in the pairs of TOA associated with the same target obtained for the first (Rx1) and second (Rx2) receiving channel of the radar.

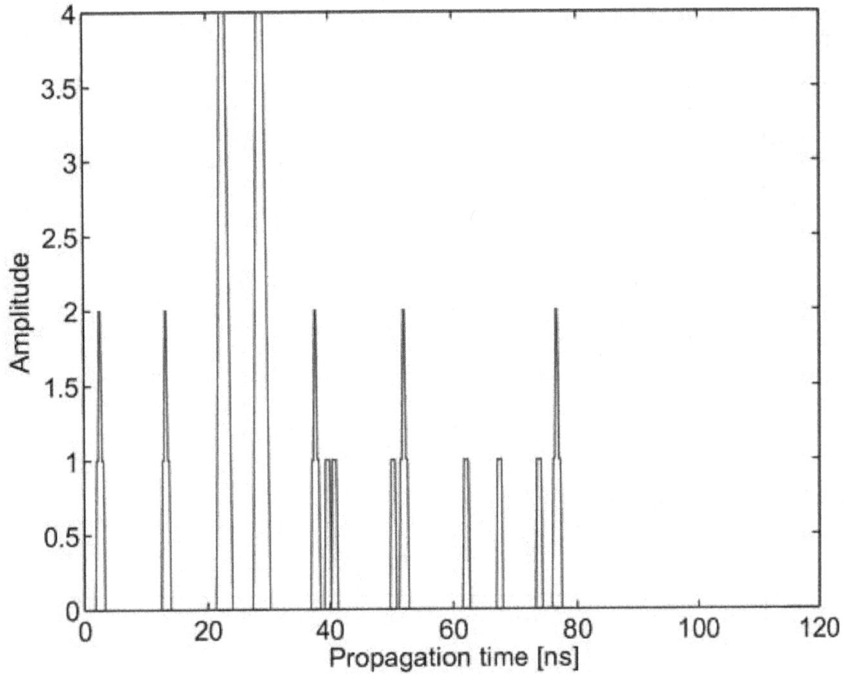

Figure 16. TD input. Receiving channel Rx1.

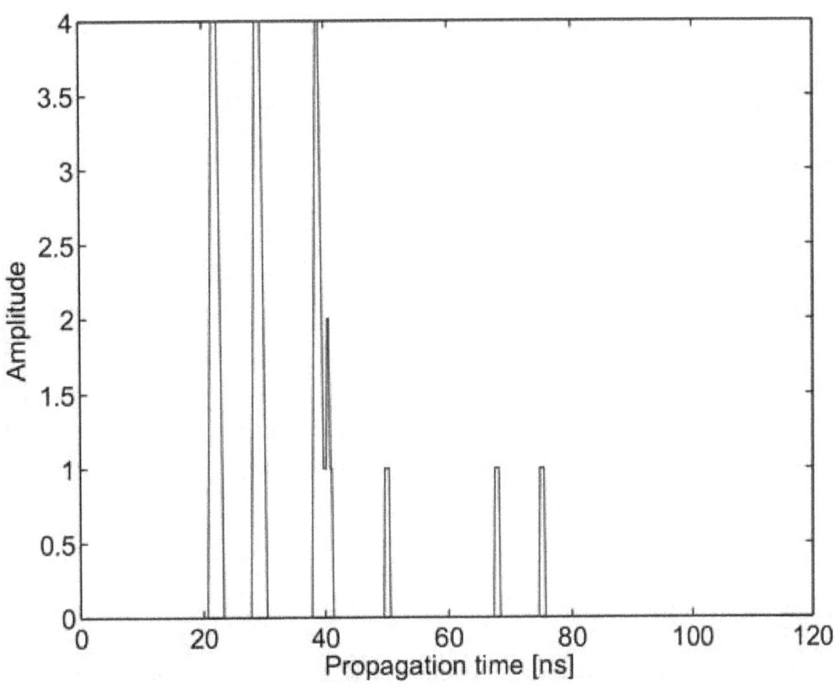

Figure 17. TD input. Receiving channel Rx2.

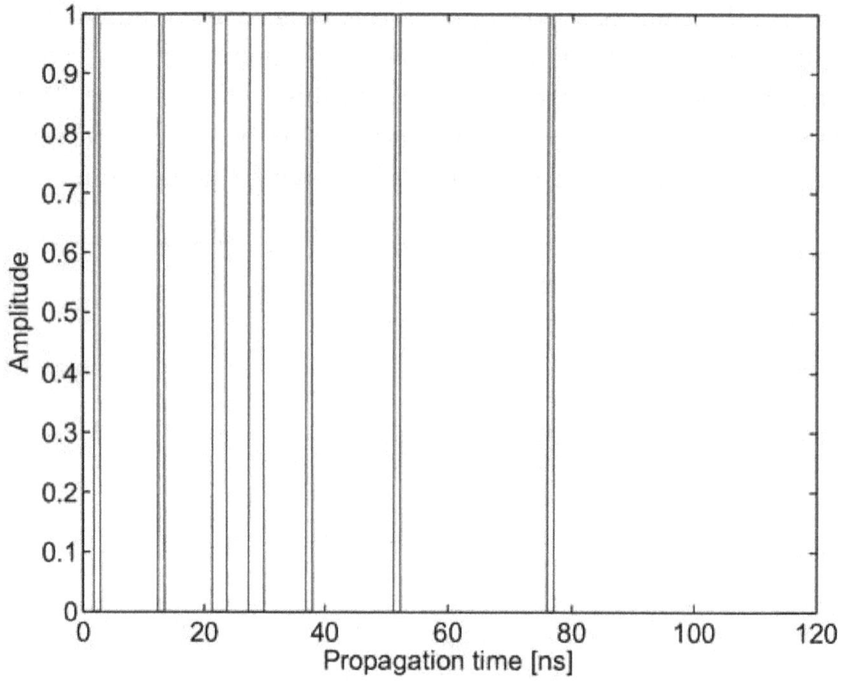

Figure 18. TD output. Receiving channel Rx1.

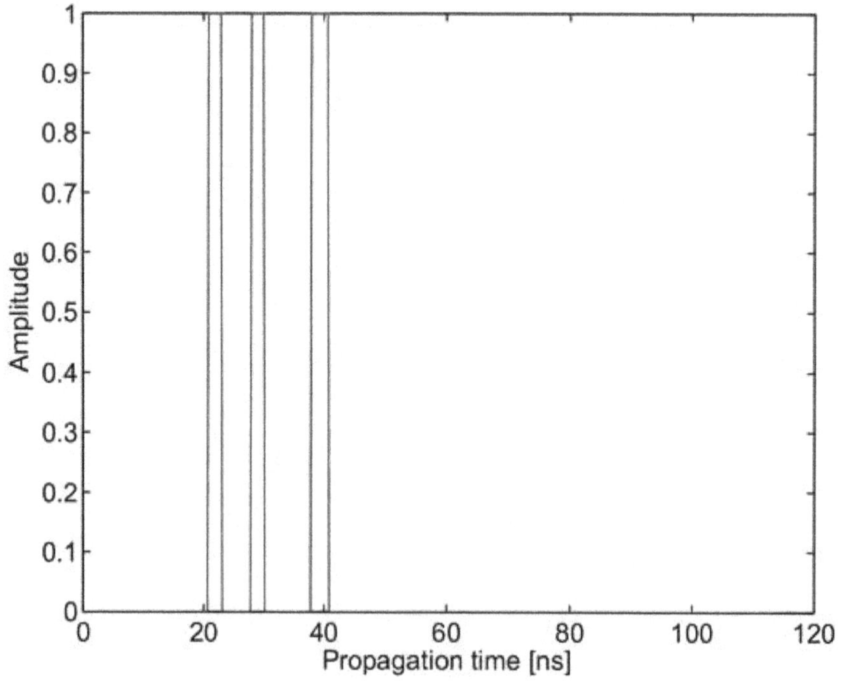

Figure 19. TD output. Receiving channel Rx2.

4.2.5. *Phase 5: target localization*

The estimations of the target coordinates represent the output of the target localization phase and at the same time the final result of WP-STAPELOC method. The results provided by the trace connection method (Phase 4) and by direct computation method (Phase 5) for the illustrative scenario discussed throughout this paper are given in **Figure 20**. Here, the true and estimated positions of the static persons are expressed by green and red marks, respectively. Moreover, tolerance areas represented by the circles with the center in the true position of the target and with the diameter 30 cm are also sketched in **Figure 20**. The tolerance areas are used to demonstrate that the person is not a pointed but distributed target with a certain non-zero size. The results depictured in **Figure 20** confirm that WP-STAPELOC method performs very well for the illustrative scenario. In spite of the complex scenario (three-person localization through the thick concrete wall), all static persons were localized with the good accuracy (all persons were localized in corresponding tolerance area).

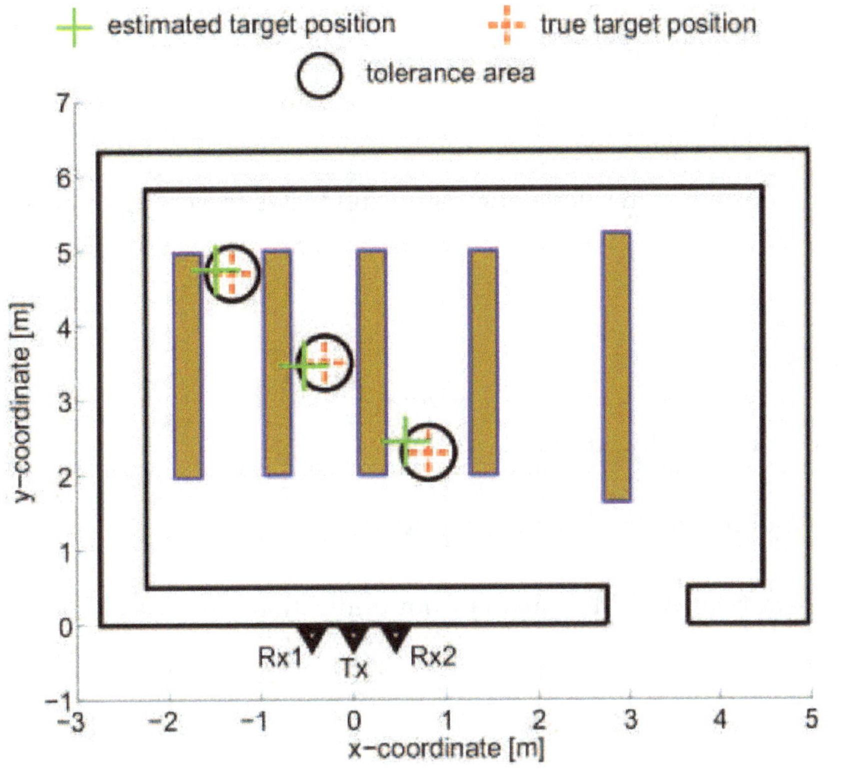

Figure 20. True (green mark) and estimated (red mark) positions of the static persons.

5. Conclusion

In this chapter, we have dealt with a problem of multiple static person localization using a single M-sequence UWB radar. For that purpose, we have introduced the novel radar signal processing procedure focused on the static person detection and localization. The key novelty of this method consists of the application of a two-stage detector based on the combination of power-spectrum estimator, OS-CFAR detector, and a simple TD. The performance of the proposed procedure has been evaluated for through-the-wall scenario of localization of three static persons. The obtained results have confirmed that the procedure described in this paper can provide a good performance also for challenging scenarios.

Unfortunately, the illustrated good performance of the WP-STAPELOC method does not mean that this method is able to provide a perfect and robust performance almost for all possible scenarios. Nevertheless, the presented approach addresses the core problems of multiple static person localization and suggests their fundamental solutions. This fact is considered of the highest importance and benefits of the proposed methods. In order to obtain a really robust system for the localization of static persons by UWB sensors, a modified version of WP-STAPELOC method has to be developed. The new modified version based on the presented principles should take into account shadowing effect, wall effect and its compensation, multipath electromagnetic wave propagation, a presence of strong reflectors in monitored area, narrowband interference, and so on. Moreover, the static person localization method should be optimized with regard to its computational complexity. We believe that the solution of this challenging and complex task will be available in the near future.

Acknowledgements

This work was supported by the Slovak Research and Development Agency under the contract No. APVV-0404-12.

Author details

Dušan Kocur*, Daniel Novák and Mária Švecová

*Address all correspondence to: dusan.kocur@tuke.sk

Department of Electronics and Multimedia Communications, Technical University of Košice, Košice, Slovakia

References

[1] Huffmann C., Ericson L. Through-the-wall sensors for law enforcement. market Survey. [Internet]. 10/2012 [Updated: 04/2014]. Available from: https://www.ncjrs.gov/pdffiles1/nij/grants/245944.pdf [Accessed: 25.8.2016]

[2] Sachs J., Helbig M., Herrmann R., Kmec M., Schilling K., Zaikov E. Remote vital sign detection for rescue, security, and medical care by ultra-wideband pseudo-noise radar. Ad Hoc Networks. 2014;13:42–53. DOI: 10.1016/j.adhoc.2012.07.002

[3] Lazaro A., Girbau D., Villarino R. Techniques for clutter suppression in the presence of body movements during the detection of respiratory activity through UWB radars. Sensors. 2014;14(2):2595–2618. DOI: 10.3390/s140202595

[4] Singh S., Liang Q., Chen D., Sheng L. Sense through wall human detection using UWB radar. EURASIP Journal on Wireless Communications and Networking. 2011;20:1–11. DOI: 10.1186/1687-1499-2011-20

[5] Zaikov E., Sachs J., Aftanas M., Rovňáková J. Detection of trapped people by UWB radar. In: Microwave Conference (GeMIC), 2008 German; 10–12 March 2008; VDE VERLAG GmbH; 2008. pp. 1–4.

[6] Yarovoy A. G., Ligthart L. P., Matuzas J., Levitas B. UWB radar for human being detection. IEEE Aerospace and Electronic Systems Magazine. 2006;21(3):10–14. DOI: 10.1109/MAES.2006.1624185

[7] Rivera N. V., Venkatesh S., Anderson Ch., Buehrer R. M. Multi-target estimation of heart and respiration rates using ultra wideband sensors. In: Signal Processing Conference, 2006 14th European; 4–8 September 2006; Florence, Italy; IEEE; 2006. pp. 1–6.

[8] Nezirović A. Trapped-Victim Detection in Post-Disaster Scenarios using Ultra-Wideband Radar [dissertation]. Delft: Delft University of Technology; 210. 165 p. Available from: http://repository.tudelft.nl/islandora/object/uuid:4416dd48-9829-4af0-b678-8fcd8e87788a/?collection=research

[9] Nezirović A., Yarovoy A. G., Ligthart L. P. Signal processing for improved detection of trapped victims using UWB radar. IEEE Transactions on Geoscience and Remote Sensing. 2010;48(4):2005–2014. DOI: 10.1109/TGRS.2009.2036840

[10] Li X., Liang Q., Lau F. Sense-through-wall human detection using the UWB radar with sparse SVD. Physical Communication. 2014;13:260–266. DOI: 10.1016/j.phycom.2013.12.002

[11] Li J., Liu L., Zeng Z., Liu F. Simulation and signal processing of UWB radar for human detection in complex environment. In: Ground Penetrating Radar (GPR), 2012 14th International Conference; 4–8 June 2014; Shanghai,China; IEEE; 2012. pp. 209–213. DOI: 10.1109/ICGPR.2012. 6254862

[12] Liu Z., Liu L., Barrowes B. The application of the Hilbert-Huang transform in through-wall life detection with UWB impulse radar. PIERS Online. 2010;6(7):695–699. DOI: 10. 2529/PIERS100217122115

[13] Li J., Zeng Z., Sun J., Liu F. Through-wall detection of human being's movement by UWB radar. IEEE Geoscience and Remote Sensing Letters. 2012;9(6):1079–1083. DOI: 10.1109/LGRS.2012.2190707

[14] Wang Y., Yu X., Zhang Y., Lv H., Jiao T., Lu G., et al. Using wavelet entropy to distinguish between humans and dogs detected by UWB radar. Progress in Electromagnetics Research. 2013;139:335–352. DOI: 10.2528/PIER13032508

[15] Huang N., Shen Z., Long S., Wu M., Shih H., Zheng Q. The empirical mode decomposition and the Hilbert spectrum for nonlinear and non-stationary time series analysis. Proceedings of the Royal Society of London A: Mathematical, Physical and Engineering Sciences. 1998;454(1971):903–995. DOI: 10.1098/rspa.1998.0193

[16] Stockwell R. G., Mansinha L., Lowe R. P. Localization of the complex spectrum: The S-transform. IEEE Transactions on Signal Processing. 1996;44(4):998–1001. DOI: 10.1109/78.492555

[17] Proakis J. G., Manolakis D. G. Digital Signal Processing. 4th ed. Michigan University: Pearson Prentice Hall; 2007. 1084 p.

[18] Kim Ch., Lee J. ToA-based multi-target localization and respiration detection using UWB radars. EURASIP Journal on Wireless Communications and Networking. 2014;2014(1):1–15. DOI: 10.1186/1687-1499-2014-145

[19] Kim C., Lee J., Cho T., Ki D., Cho B. H., Yoon J. Multi-target localization of breathing humans. In: Proceedings of the 2013 IEEE International Conference on Ultra-Wideband (ICUWB 2013); Sydney, Australia; 2013. pp. 49–54.

[20] Rovňáková J., Švecová M., Kocur D., Nguyen T. T., Sachs J. Signal processing for through wall moving target tracking by M-sequence UWB radar. In: Radioelektronika, 2008 18th International Conference; 24–25 April 2008; Prague, Czech Republic.; IEEE; 2008. p. 1–4. DOI: 10.1109 RADIOELEK.2008.4542694

[21] Kocur D., Švecová M., Rovňáková J. Through-the-wall localization of a moving target by two independent ultra wideband (UWB) radar systems. Sensors. 2013;13(9):11969–11997. DOI: 10.3390/s130911969

[22] Rohling H. Some radar topics: waveform design, range CFAR and target recognition. In: Byrnes J., Ostheimer G., editors. Advances in Sensing with Security Applications. 2nd ed. Springer, Netherlands; 2006. pp. 293–322. DOI: 10.1007/1-4020-4295-7_13

[23] Novák D., Kocur D. Multiple static person localization based on respiratory motion detection by UWB radar. In: 2016 26th International Conference Radioelektronika (RADIOELEKTRONIKA); 19–20 April 2016; Košice. IEEE; 2016. pp. 252–257. DOI: 10.1109/RADIOELEK.2016.7477386

[24] Rovňáková J., Kocur D. TOA estimation and data association for through-wall tracking of moving targets. EURASIP Journal on Wireless Communications and Networking. 2010;2010(1):1–11. DOI: 10.1155/2010/420767

Permissions

All chapters in this book were first published in MSA, by InTech Open; hereby published with permission under the Creative Commons Attribution License or equivalent. Every chapter published in this book has been scrutinized by our experts. Their significance has been extensively debated. The topics covered herein carry significant findings which will fuel the growth of the discipline. They may even be implemented as practical applications or may be referred to as a beginning point for another development.

The contributors of this book come from diverse backgrounds, making this book a truly international effort. This book will bring forth new frontiers with its revolutionizing research information and detailed analysis of the nascent developments around the world.

We would like to thank all the contributing authors for lending their expertise to make the book truly unique. They have played a crucial role in the development of this book. Without their invaluable contributions this book wouldn't have been possible. They have made vital efforts to compile up to date information on the varied aspects of this subject to make this book a valuable addition to the collection of many professionals and students.

This book was conceptualized with the vision of imparting up-to-date information and advanced data in this field. To ensure the same, a matchless editorial board was set up. Every individual on the board went through rigorous rounds of assessment to prove their worth. After which they invested a large part of their time researching and compiling the most relevant data for our readers.

The editorial board has been involved in producing this book since its inception. They have spent rigorous hours researching and exploring the diverse topics which have resulted in the successful publishing of this book. They have passed on their knowledge of decades through this book. To expedite this challenging task, the publisher supported the team at every step. A small team of assistant editors was also appointed to further simplify the editing procedure and attain best results for the readers.

Apart from the editorial board, the designing team has also invested a significant amount of their time in understanding the subject and creating the most relevant covers. They scrutinized every image to scout for the most suitable representation of the subject and create an appropriate cover for the book.

The publishing team has been an ardent support to the editorial, designing and production team. Their endless efforts to recruit the best for this project, has resulted in the accomplishment of this book. They are a veteran in the field of academics and their pool of knowledge is as vast as their experience in printing. Their expertise and guidance has proved useful at every step. Their uncompromising quality standards have made this book an exceptional effort. Their encouragement from time to time has been an inspiration for everyone.

The publisher and the editorial board hope that this book will prove to be a valuable piece of knowledge for researchers, students, practitioners and scholars across the globe.

List of Contributors

Jian Wang and Yun Long
Wuhan National Laboratory for Optoelectronics, School of Optical and Electronic Information, Huazhong University of Science and Technology, Wuhan, Hubei, China

Xi Sung Loo and Kok Wai Johnny Chew
GLOBALFOUNDRIES Singapore Pte Ltd, Singapore

Kiat Seng Yeo
Singapore University of Technology & Design, Singapore

Young Chul Lee
Department of electronic engineering, Mokpo National Maritime University (MMU), Mokpo, South Korea

Kok Yeow You
Department of Communication Engineering, Faculty of Electrical Engineering, Universiti Teknologi Malaysia, Johor, Malaysia

Hatem El Matbouly
LCIS, Grenoble-INP, Grenoble Alpes University, Valence, France

Snežana Stefanovski Pajović, Milka Potrebić and Dejan V. Tošić
School of Electrical Engineering, University of Belgrade, Belgrade, Serbia

Modar Shbat and Francisco C. Ordaz-Salazar
Telematics Engineering Department, Polytechnic University of San Luis Potosí, San Luis Potosí, México

Javier S. González-Salas
Mathematical Department, Polytechnic University of San Luis Potosi, San Luis Potosí, México

Vincenzo Stornelli, Leonardo Pantoli and Giorgio Leuzzi
Department of Industrial and Information Engineering and Economics, University of L'Aquila, L'Aquila, Italy

Sayed Alireza Sadrossadat and Farnaz Mohammadi
Department of Electrical and Computer Engineering, University of Tehran, Tehran, Iran

Imran Mohsin
Shenzhen Institutes of Advanced Technology (SIAT), Chinese Academy of Sciences, Shenzhen, China
Shenzhen College of Advanced Technology, University of Chinese Academy of Sciences, Shenzhen, China

Dušan Kocur, Daniel Novák and Mária Švecová
Department of Electronics and Multimedia Communications, Technical University of Košice, Košice, Slovakia

Index

www.ingramcontent.com/pod-product-compliance
Lightning Source LLC
LaVergne TN
LVHW080730210926
848485LV00003B/543